METHODS IN MOLECULAR BIOLOGY™

Series Editor
John M. Walker
School of Life Sciences
University of Hertfordshire
Hatfield, Hertfordshire, AL10 9AB, UK

For further volumes:
http://www.springer.com/series/7651

Cytochrome P450 Protocols

Third Edition

Edited by

Ian R. Phillips

Division of Biosciences, University College London, London, UK;
School of Biological and Chemical Sciences, Queen Mary, University of London, London, UK

Elizabeth A. Shephard

Department of Structural and Molecular Biology, University College London, London, UK

Paul R. Ortiz de Montellano

Department of Pharmaceutical Chemistry, University of California, San Francisco, CA, USA

Editors
Ian R. Phillips
Division of Biosciences
University College London
London, UK

School of Biological and Chemical Sciences
Queen Mary
University of London
London, UK

Elizabeth A. Shephard
Department of Structural and Molecular Biology
University College London
London, UK

Paul R. Ortiz de Montellano
Department of Pharmaceutical Chemistry
University of California
San Francisco, CA, USA

ISSN 1064-3745   ISSN 1940-6029 (electronic)
ISBN 978-1-62703-320-6   978-1-62703-321-3 (eBook)
DOI 10.1007/978-1-62703-321-3
Springer New York Heidelberg Dordrecht London

Library of Congress Control Number: 2013932010

© Springer Science+Business Media New York 2013
This work is subject to copyright. All rights are reserved by the Publisher, whether the whole or part of the material is concerned, specifically the rights of translation, reprinting, reuse of illustrations, recitation, broadcasting, reproduction on microfilms or in any other physical way, and transmission or information storage and retrieval, electronic adaptation, computer software, or by similar or dissimilar methodology now known or hereafter developed. Exempted from this legal reservation are brief excerpts in connection with reviews or scholarly analysis or material supplied specifically for the purpose of being entered and executed on a computer system, for exclusive use by the purchaser of the work. Duplication of this publication or parts thereof is permitted only under the provisions of the Copyright Law of the Publisher's location, in its current version, and permission for use must always be obtained from Springer. Permissions for use may be obtained through RightsLink at the Copyright Clearance Center. Violations are liable to prosecution under the respective Copyright Law.
The use of general descriptive names, registered names, trademarks, service marks, etc. in this publication does not imply, even in the absence of a specific statement, that such names are exempt from the relevant protective laws and regulations and therefore free for general use.
While the advice and information in this book are believed to be true and accurate at the date of publication, neither the authors nor the editors nor the publisher can accept any legal responsibility for any errors or omissions that may be made. The publisher makes no warranty, express or implied, with respect to the material contained herein.

Printed on acid-free paper

Humana Press is a brand of Springer
Springer is part of Springer Science+Business Media (www.springer.com)

Preface

Cytochromes P450 (CYPs) comprise a large superfamily of proteins that are of central importance in the detoxification or activation of a tremendous number of natural and synthetic hydrophobic xenobiotics, including many therapeutic drugs, chemical carcinogens, and environmental pollutants. Some CYPs catalyze the metabolism of endogenous compounds, particularly the ones involved in signaling. CYPs, therefore, are important in mediating interactions between an organism and its chemical environment and in the regulation of physiological processes. Many CYPs are inducible by the compounds they metabolize. In addition, genetic polymorphisms of *CYP* genes affect expression or activity of the enzymes, which can result in adverse drug reactions or genetic diseases. Consequently, CYPs are among the most extensively studied groups of proteins, being investigated by researchers in fields as diverse as biochemistry, molecular biology, pharmacology, toxicology, environmental biology, and genetics. The wide range of techniques that have been applied to the CYPs reflects the diverse backgrounds of the many researchers active in this field.

Previous editions of *Cytochrome P450 Protocols* contained collections of key "core" techniques, most of which are still relevant. The emphasis, however, was on methods for the investigation of individual CYPs and substrates, mostly in an in vitro context. The current edition focuses on high-throughput methods for the simultaneous analysis of multiple CYPs, substrates, or ligands. Although the emphasis is on CYPs of mammalian origin, it reflects an increasing interest in CYPs of bacterial species. However, most of the methods described are suitable for the investigation of CYPs from any source. Also included are chapters on CYP reductase (the redox partner of CYPs) and the flavin-containing monooxygenases (FMOs), another family of proteins that are important in the metabolism of foreign chemicals, and that share several substrates in common with the CYPs.

The chapters of this edition of *Cytochrome P450 Protocols*, although not formally divided into sections, are grouped loosely according to topic. Included are high-throughput methods for identification of substrates, ligands, and inhibitors of CYPs; metabolomic and lipidomic approaches for identification of endogenous substrates of CYPs ("de-orphanizing" CYP substrates); reconstitution systems for the incorporation of modified and novel metalloporphyrins into CYPs in vivo or for developing nanoparticle bioreactors for biophysical and mechanistic studies of CYPs and drug-metabolite profiling; high-throughput assays for measuring the activity of CYPs and for identification of their substrates and adducts; methods for the generation and quantification of novel CYPs and for identification of their potential substrates; techniques for phenotyping, genotyping, and identification of transcriptional regulatory sequences; a high-throughput method for the generation of libraries of redox-self-sufficient CYP biocatalysts; a guide to CYP allele nomenclature; and methods for the isolation of mouse primary hepatocytes, for the differentiation of a hepatoma cell line into cells with hepatocyte-like metabolic properties, and for transfection of such cells with DNA and siRNA constructs to investigate the function and regulation of expression of CYPs.

Each chapter is written by researchers who have been involved in the development and application of the particular technique. Protocols are presented in a step-by-step manner,

v

with extensive cross-references to notes that highlight critical steps, potential problems, and alternative methods. We hope that this format will enable researchers who have no previous knowledge of the technique to understand the basis of the method and to perform it successfully.

We are extremely grateful to all the authors who contributed so generously to this volume and to John Walker, the series editor of *Methods in Molecular Biology*, for his advice and patience.

London, UK   *Ian R. Phillips*
London, UK   *Elizabeth A. Shephard*
San Francisco, CA, USA   *Paul R. Ortiz de Montellano*

Contents

Preface.. *v*

Contributors.. *ix*

1   Bioluminescent Assays for Cytochrome P450 Enzymes........................ 1
    Douglas S. Auld, Henrike Veith, and James J. Cali

2   Simultaneous Determination of Multiple CYP Inhibition Constants
    using a Cocktail-Probe Approach.......................... 11
    Michael Zientek and Kuresh Youdim

3   High-Throughput Mass Spectrometric Cytochrome
    P450 Inhibition Screening.............................. 25
    Kheng B. Lim, Can C. Ozbal, and Daniel B. Kassel

4   The Synthesis, Characterization, and Application of ^{13}C-Methyl
    Isocyanide as an NMR Probe of Heme Protein Active Sites.................. 51
    Christopher McCullough, Phani Kumar Pullela, Sang-Choul Im,
    Lucy Waskell, and Daniel Sem

5   High-Throughput Fluorescence Assay for
    Cytochrome P450 Mechanism-Based Inactivators...................... 61
    Cesar Kenaan, Haoming Zhang, and Paul F. Hollenberg

6   Identification of Endogenous Substrates of Orphan Cytochrome
    P450 Enzymes Through the Use of Untargeted Metabolomics Approaches 71
    Qian Cheng and F. Peter Guengerich

7   Genetic and Mass Spectrometric Tools for Elucidating
    the Physiological Function(s) of Cytochrome P450 Enzymes
    from *Mycobacterium tuberculosis*........................... 79
    Hugues Ouellet, Eric D. Chow, Shenheng Guan, Jeffery S. Cox,
    Alma L. Burlingame, and Paul R. Ortiz de Montellano

8   An *Escherichia coli* Expression-Based Approach for Porphyrin
    Substitution in Heme Proteins........................... 95
    Michael B. Winter, Joshua J. Woodward, and Michael A. Marletta

9   Expression in *Escherichia coli* of a Cytochrome P450 Enzyme
    with a Cobalt Protoporphyrin IX Prosthetic Group................... 107
    Wesley E. Straub, Clinton R. Nishida, and Paul R. Ortiz de Montellano

10 Nanodiscs in the Studies of Membrane-Bound Cytochrome
    P450 Enzymes.................................... 115
    A. Luthra, M. Gregory, Y.V. Grinkova, I.G. Denisov, and S.G. Sligar

11 Rapid LC-MS Drug Metabolite Profiling Using Bioreactor Particles........... 129
    Linlin Zhao, Besnik Bajrami, and James F. Rusling

12 Fluorescence-Based Screening of Cytochrome P450 Activities
    in Intact Cells................................... 135
    M. Teresa Donato and M. José Gómez-Lechón

viii   Contents

13 Screening for Cytochrome P450 Reactivity with a Reporter Enzyme............ 149
    Kersten S. Rabe and Christof M. Niemeyer

14 High-Throughput Fluorescence Assay of Cytochrome P450 3A4 157
    Qian Cheng and F. Peter Guengerich

15 Targeted Protein Capture for Analysis of Electrophile-Protein Adducts 163
    *Rebecca E. Connor, Simona G. Codreanu, Lawrence J. Marnett,
    and Daniel C. Liebler*

16 DNA Shuffling of Cytochrome P450 Enzymes 177
    *James B.Y.H. Behrendorff, Wayne A. Johnston,
    and Elizabeth M.J. Gillam*

17 Measurement of P450 Difference Spectra Using Intact Cells................. 189
    Wayne A. Johnston and Elizabeth M.J. Gillam

18 DNA Shuffling of Cytochromes P450 for Indigoid Pigment Production 205
    Nedeljka N. Rosic

19 P450 Oxidoreductase: Genotyping, Expression, Purification of Recombinant
    Protein, and Activity Assessments of Wild-Type and Mutant Protein 225
    Vishal Agrawal and Walter L. Miller

20 LICRED: A Versatile Drop-In Vector for Rapid Generation
    of Redox-Self-Sufficient Cytochromes P450 239
    Federico Sabbadin, Gideon Grogan, and Neil C. Bruce

21 Update on Allele Nomenclature for Human Cytochromes P450 and the
    Human Cytochrome P450 Allele (CYP-Allele) Nomenclature Database 251
    Sarah C. Sim and Magnus Ingelman-Sundberg

22 Simultaneous In Vivo Phenotyping of CYP Enzymes 261
    Sussan Ghassabian and Michael Murray

23 Detection of Regulatory Polymorphisms: High-Throughput Capillary
    DNase I Footprinting... 269
    Matthew Hancock and Elizabeth A. Shephard

24 Isolation of Mouse Hepatocytes 283
    *Mina Edwards, Lyndsey Houseman, Ian R. Phillips,
    and Elizabeth A. Shephard*

25 Highly Efficient SiRNA and Gene Transfer into Hepatocyte-Like
    HepaRG Cells and Primary Human Hepatocytes: New Means
    for Drug Metabolism and Toxicity Studies 295
    *Véronique Laurent, Denise Glaise, Tobias Nübel, David Gilot,
    Anne Corlu, and Pascal Loyer*

Index.. *315*

Contributors

VISHAL AGRAWAL • *Department of Pediatrics, University of California San Francisco, San Francisco, CA, USA*

DOUGLAS S. AULD • *Novartis Institutes for Biomedical Research, Cambridge, MA, USA*

BESNIK BAJRAMI • *Department of Chemistry, University of Connecticut, Storrs, CT, USA*

JAMES B.Y.H. BEHRENDORFF • *Australian Institute for Bioengineering and Nanotechnology, University of Queensland, Brisbane, Australia*

NEIL C. BRUCE • *Department of Biology, York University, York, UK*

ALMA L. BURLINGAME • *Department of Pharmaceutical Chemistry, University of California, San Francisco, CA, USA*

JAMES J. CALI • *Promega Corp., Madison, WI, USA*

QIAN CHENG • *Department of Biochemistry and Center in Molecular Toxicology, Vanderbilt University School of Medicine, Nashville, TN, USA*

ERIC D. CHOW • *Department of Microbiology and Immunology, University of California, San Francisco, CA, USA*

SIMONA G. CODREANU • *Department of Biochemistry, Vanderbilt University School of Medicine, Nashville, TN, USA; Jim Ayers Institute for Precancer Detection and Diagnosis, Vanderbilt University School of Medicine, Nashville, TN, USA*

REBECCA E. CONNOR • *Department of Biochemistry, Vanderbilt University School of Medicine, Nashville, TN, USA*

ANNE CORLU • *Inserm UMR991 Foie, Métabolismes et Cancer, Hôpital Pontchaillou and Université de Rennes 1, Rennes, France*

JEFFERY S. COX • *Department of Microbiology and Immunology, University of California, San Francisco, CA, USA*

PAUL R. ORTIZ DE MONTELLANO • *Department of Pharmaceutical Chemistry, University of California, San Francisco, CA, USA*

I.G. DENISOV • *Department of Biochemistry, University of Illinois, Urbana, IL, USA*

M. TERESA DONATO • *Departamento de Bioquímica y Biología Molecular, Universidad de Valencia, Valencia, Spain; CIBERehd, Fondo de Investigaciones Sanitarias, Barcelona, Spain*

MINA EDWARDS • *Institute of Structural and Molecular Biology, University College London, London, UK*

SUSSAN GHASSABIAN • *Faculty of Pharmacy, University of Sydney, Sydney, NSW, Australia; Centre for Integrated Preclinical Drug Development, University of Queensland, Brisbane, Australia*

ELIZABETH M.J. GILLAM • *School of Chemistry and Molecular Biosciences, University of Queensland, Brisbane, Australia*

DAVID GILOT • *CNRS UMR6061, Institut de Génétique et Développement de Rennes, Université de Rennes 1, Rennes, France*

DENISE GLAISE • *Inserm UMR991, Foie, Métabolismes et Cancer, Hôpital Pontchaillou and Université de Rennes 1, Rennes, France*

M. JOSÉ GÓMEZ-LECHÓN • *IIS Hospital La Fe, Unidad de Hepatología Experimental, Valencia, Spain; CIBERehd, Fondo de Investigaciones Sanitarias, Barcelona, Spain*

M. GREGORY • *Department of Biochemistry, University of Illinois, Urbana, IL, USA*

Y.V. GRINKOVA • *Department of Biochemistry, University of Illinois, Urbana, IL, USA*

GIDEON GROGAN • *Department of Biology, York University, York, UK*

SHENHENG GUAN • *Department of Pharmaceutical Chemistry, University of California, San Francisco, CA, USA*

F. PETER GUENGERICH • *Department of Biochemistry and Center in Molecular Toxicology, Vanderbilt University School of Medicine, Nashville, TN, USA*

MATTHEW HANCOCK • *Institute of Structural and Molecular Biology, University College London, London, UK*

PAUL F. HOLLENBERG • *Medical Sciences Research Building III, Department of Pharmacology, University of Michigan, Ann Arbor, MI, USA*

LYNDSEY HOUSEMAN • *Institute of Structural and Molecular Biology, University College London, London, UK*

SANG-CHOUL IM • *Department of Anesthesiology, University of Michigan and VA Medical Center, Ann Arbor, MI, USA*

MAGNUS INGELMAN-SUNDBERG • *Section for Pharmacogenetics, Department of Physiology and Pharmacology, Karolinska Institute, Stockholm, Sweden*

WAYNE A. JOHNSTON • *Institute of Molecular Bioscience, University of Queensland, Brisbane, Australia*

DANIEL B. KASSEL • *Takeda San Diego Inc., San Diego, CA, USA*

CESAR KENAAN • *Medical Sciences Research Building III, Department of Pharmacology, University of Michigan, Ann Arbor, MI, USA*

VÉRONIQUE LAURENT • *Inserm U 1078, Génétique Moléculaire et Génétique Epidémiologique, Hôpital Morvan - CHU de Brest, Brest, France; SynNanoVect Platform, Brest, France*

DANIEL C. LIEBLER • *Department of Biochemistry, Vanderbilt University School of Medicine, Nashville, TN, USA; Jim Ayers Institute for Precancer Detection and Diagnosis, Vanderbilt University School of Medicine, Nashville, TN, USA*

KHENG B. LIM • *Takeda San Diego Inc., San Diego, CA, USA*

PASCAL LOYER • *Inserm UMR991 Foie, Métabolismes et Cancer, Hôpital Pontchaillou and Université de Rennes 1, Rennes, France; SynNanoVect Platform, Rennes, France*

A. LUTHRA • *Department of Biochemistry, University of Illinois, Urbana, IL, USA*

MICHAEL A. MARLETTA • *Department of Chemistry, The Scripps Research Institute, La Jolla, USA; Department of Chemistry, Department of Molecular and Cell Biology, California Institute for Quantitative Biosciences, Division of Physical Biosciences, Lawrence Berkeley National Laboratory, University of California, Berkeley, CA, USA*

LAWRENCE J. MARNETT • *Department of Biochemistry, Vanderbilt University School of Medicine, Nashville, TN, USA; Jim Ayers Institute for Precancer Detection and Diagnosis, Vanderbilt University School of Medicine, Nashville, TN, USA*

CHRISTOPHER MCCULLOUGH • *Department of Chemistry, Marquette University, Milwaukee, WI, USA*

WALTER L. MILLER • *Department of Pediatrics, University of California San Francisco, San Francisco, CA, USA*

MICHAEL MURRAY • *Faculty of Pharmacy, University of Sydney, Sydney, NSW, Australia*

CHRISTOF M. NIEMEYER • *Karlsruhe Institute of Technologie (KIT), Institute for Biological Interfaces (IBG-1), Biomolecular Micr- and Nanostructures, Eggenstein-Leopoldshafen, Germany*

CLINTON R. NISHIDA • *Department of Pharmaceutical Chemistry, University of California, San Francisco, CA, USA*

TOBIAS NÜBEL • *Lonza Cologne GmbH, Nattermannallee, Koeln, Germany*

HUGUES OUELLET • *Department of Microbiology, University of Texas, El Paso, TX, USA*

CAN C. OZBAL • *Agilent Technologies Inc., Wakefield, MA, USA*

IAN R. PHILLIPS • *School of Biological and Chemical Sciences, Queen Mary, University of London, London, UK; Division of Biosciences, University College London, London, UK*

PHANI KUMAR PULLELA • *Department of Chemistry, Marquette University, Milwaukee, WI, USA*

KERSTEN S. RABE • *Fakultät Chemie, Biologisch-Chemische Mikrostrukturtechnik, Technische Universität Dortmund, Dortmund, Germany; Division of Chemistry and Chemical Engineering, California Institute of Technology, Pasadena, CA, USA*

NEDELJKA N. ROSIC • *School of Biological Sciences, University of Queensland, St. Lucia, QLD, Australia*

JAMES F. RUSLING • *Department of Chemistry, University of Connecticut, Storrs, CT, USA; Department of Cell Biology, University of Connecticut Health Center, Farmington, CT, USA*

FEDERICO SABBADIN • *Department of Biology, York University, York, UK*

DANIEL SEM • *Department of Chemistry, Marquette University, Milwaukee, WI, USA; Department of Pharmaceutical Sciences, Concordia University of Wisconsin–School of Pharmacy, Mequon, WI, USA*

ELIZABETH A. SHEPHARD • *Department of Structural and Molecular Biology, University College London, London, UK*

SARAH C. SIM • *Section for Pharmacogenetics, Department of Physiology and Pharmacology, Karolinska Institute, Stockholm, Sweden*

S.G. SLIGAR • *Department of Biochemistry, Beckman Institute and School of Molecular and Cellular Biology, University of Illinois, Urbana, IL, USA*

WESLEY E. STRAUB • *Life Technologies, San Francisco, CA, USA*

HENRIKE VEITH • *National Center for Advancing Translational Sciences, National Institutes of Health, Bethesda, MD, USA*

LUCY WASKELL • *Department of Anesthesiology, University of Michigan and VA Medical Center, Ann Arbor, MI, USA*

MICHAEL B. WINTER • *Department of Chemistry, California Institute for Quantitative Biosciences, University of California, Berkeley, CA, USA*

Joshua J. Woodward • *Department of Chemistry, California Institute for Quantitative Biosciences, University of California, Berkeley, CA, USA*

Kuresh Youdim • *F. Hoffmann-La Roche AG, Basel, Switzerland*

Haoming Zhang • *Medical Sciences Research Building III, Department of Pharmacology, University of Michigan, Ann Arbor, MI, USA*

Linlin Zhao • *Department of Chemistry, University of Connecticut, Storrs, CT, USA*

Michael Zientek • *Pfizer Inc., La Jolla, CA, USA*

Chapter 1

Bioluminescent Assays for Cytochrome P450 Enzymes

Douglas S. Auld, Henrike Veith, and James J. Cali

Abstract

The cytochrome P450 (CYP) family contains 57 enzymes in humans. The activity of CYPs against xenobiotics is a primary consideration in drug optimization efforts. Here we describe a series of bioluminescent assays that enable the rapid profiling of CYP activity against compound collections. The assays employ a coupled-enzyme format where firefly luciferase is used to measure CYP enzyme activity through metabolism of pro-luciferase substrates.

Key words Cytochrome P450, Bioluminescent assays, Enzyme assays

1 Introduction

The cytochromes P450 (CYPs) are the main enzymes that metabolize therapeutic drugs and numerous other xenobiotics (1). Certain CYP enzymes also play a prominent role in adverse xenobiotic interactions and are often the underlying reasons for contraindications found in certain drug combinations. For example, a drug or food component that inhibits a CYP enzyme can slow the clearance of a co-ingested drug that is metabolized by the same CYP, leading to drug accumulation beyond the desired dose and eventual toxicity. Owing to the prominent role CYP enzymes have in metabolism and potential adverse outcomes, various CYP probe substrates have been developed wherein metabolism by a specific CYP enzyme leads to a fluorescent or bioluminescent signal. These probe substrates have been used to develop high-throughput assays to measure CYP activity in the presence and absence of test compounds to identify CYP-enzyme inhibitors (2–5). In such assays inhibition can arise through a number of modes. Any reduction of free enzyme in the assay, through a compound acting either as a competitive substrate or an inhibitor of the probe substrate, will slow the rate of conversion of the labeled substrate and result in inhibition. In addition, inhibition can arise in a time-dependent fashion, which can arise from either slow-binding inhibitors or

Ian R. Phillips et al. (eds.), *Cytochrome P450 Protocols*, Methods in Molecular Biology, vol. 987,
DOI 10.1007/978-1-62703-321-3_1, © Springer Science+Business Media New York 2013

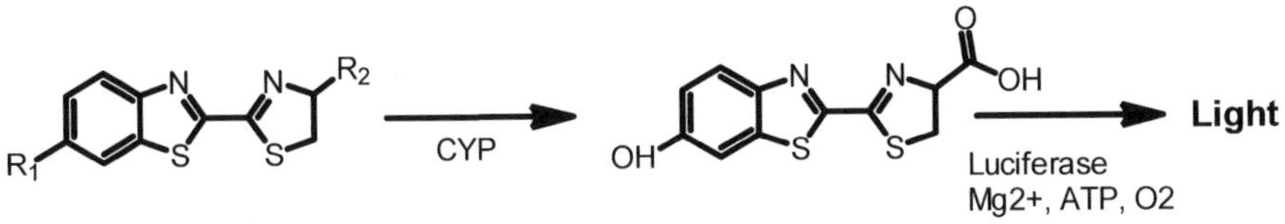

Fig. 1 Representative luminogenic CYP enzyme assay scheme. The *left* structure represents D-luciferin modified to a pro-luciferin form either by the substitution of the 6′ hydroxyl group with a group designated R₁ or through modification of the carboxylate with a group designated R₂ to form esters or acetals. CYP enzymes modify the R-group modifications to produce D-luciferin (*right* structure), which is detected in a luciferase reaction by the amount of light that is produced, which is directly proportional to the amount of D-luciferin formed

mechanism-based inhibitors where an intermediate of the CYP reaction irreversibly inhibits the enzyme.

For construction of assays specific for CYP enzymes, probe substrates have been developed which are selectivity oxidized by one or more CYP enzymes leading to production of a detectable product. Various detection methods have been used including mass spectrometry, absorbance, radioactivity, fluorescence, and biolumi-nescence (5). Validation data have been shown demonstrating that these methods can predict in vivo CYP inhibition (6–8). However, there are substantial differences among these methods in sensitiv-ity, scalability, ease of use, and vulnerability to test compound interference (3, 4). Advantages of bioluminescence assays over fluorescence assays include increased sensitivity (up to three orders of magnitude), and obviating the problem of fluorescent-compound interference (8). As well, the use of optimized luciferases and detec-tion reagents, a necessary coupling enzyme used to read out the bioluminescent signal, can reduce interference of compounds acting as direct luciferase inhibitors (3, 9). These advantages posi-tion bioluminescent CYP assays as a good choice for miniaturized high-throughput applications of CYP enzymes (3).

Bioluminescent CYP assays rely on a series of CYP substrates that are pro-substrates for a light-generating reaction with firefly luciferase. For bioluminescent assays, a luciferase pro-substrate is converted by a CYP enzyme in an initial reaction to a form that leads to light production using a second reaction mixture contain-ing firefly luciferase (Fig. 1). Light intensity is proportional to CYP activity (8). Selectivity of a pro-luciferin substrate for various CYP enzymes is determined by the nature of the modifications to a core luciferin structure and through variations of the core structure itself. The most frequently used luciferin is 2-(6-hydroxybenzo(d) thiazol-2-yl)-4,5-dihydrothiazole-4-carboxylic acid (D-luciferin), which is the natural substrate of firefly luciferases. Modifications to the luciferin that create a pro-luciferin block activity with luciferase and the modifications are reversed by reaction with a CYP enzyme. Whereas some derivatives are highly selective for a single CYP enzyme others have a broader selectivity profile (Table 1). All these

Table 1

Summary of the steps which are used for bioluminescent CYP assays in a miniaturized assay 1,536-well microtiter-plate format

Step	Parameter	Value	Description
1	Reagent	2 μL	CYP enzyme and pro-luciferin substrates
2	Library compounds	23 nl	40 μM to 0.24 nM dilution series
3	Controls	23 nl	CYP enzyme-specific inhibitors
4	Reagent	2 μL	NADPH-regeneration solution
5	Incubation time	60 min	CYP enzyme-specific temperature
6	Reagent	4 μL	Detection reagent
7	Incubation time	20 min	Room temperature
8	Detection	1 s	ViewLux luminescent read
			See steps for each step (assay steps can be scaled to larger assay volumes as necessary)

substrates are useful for assays containing purified or recombinant CYPs where only a single species of CYP enzyme is present and, therefore, cross-reactivity with multiple CYPs is not an issue. The selective substrates have additional utility when employing liver microsomes and cultured hepatocytes where multiple species of CYPs are present and a selective substrate is required to target a single enzyme (10).

2 Materials

2.1 Luciferin Substrates

Most pro-luciferin CYP substrates have single modifications to D-luciferin (Fig. 2, compound 1) at the 6′ carbon (Fig. 2, compounds 2–7). A variant of the luciferin core structure that substitutes the benzothiazole system with a quinoline unit provides a unique and selective CYP reactivity profile compared with the D-luciferin with the same CYP reactive group (Fig. 2, compound 2 compared with compound 9). Additional selectivity profiles are achieved when a second modification is used to create a carboxyl ester in place of the carboxylic acid group found in unmodified D-luciferin (Fig. 2, compounds 10–13). Activation of the pro-luciferins relies on standard CYP-catalyzed reactions including aromatic hydroxylations and *O*-dealkylations typically to generate a 6′ hydroxyl required for luciferase activity. The carboxyl esters require further de-esterification and this is achieved by including a carboxyl esterase in the second reaction with luciferase. In a unique

Luciferase & CYP Substrates	Enzyme Activities	CYP Substrates	Enzyme Activities
1. D-Luciferin 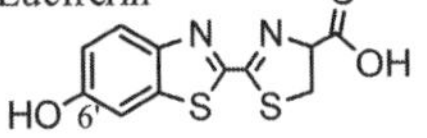	Firefly Luciferase	9. Luciferin-4A 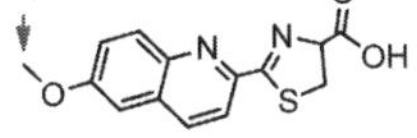	CYP4A selective
2. Luciferin-ME 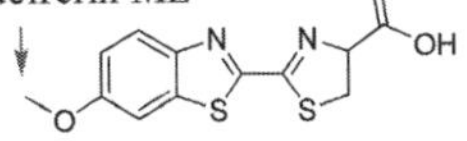	CYP1A2, -2C8, -2C9, -2J2, -4A11, -4F3B, -19	10. Luciferin-H-EGE 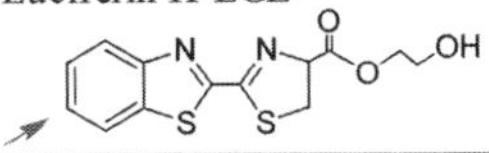	CYP1A1, -1A2, -2C19
3. Luciferin-CEE 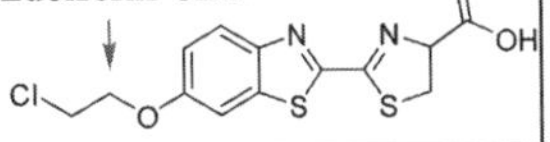	CYP1A1, -1B1, -3A7	11. Luciferin-ME-EGE 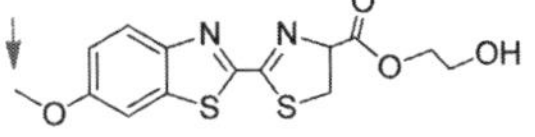	CYP1A1, -1A2, -2D6
4. Luciferin-H 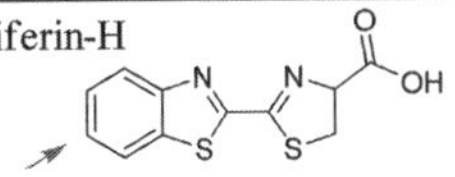	CYP2C9 selective	12. Luciferin-2J2/4F12 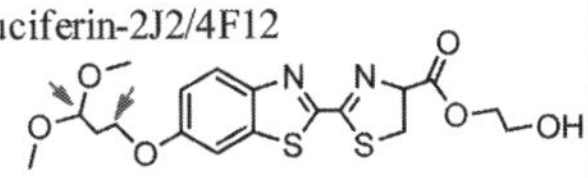	CYP2J2, -4F12, -3A4, -1A1, -1B1
5. Luciferin-PPXE 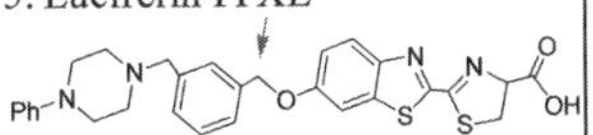	CYP3A4, -3A4, -3A7	13. Luciferin-MultiCYP 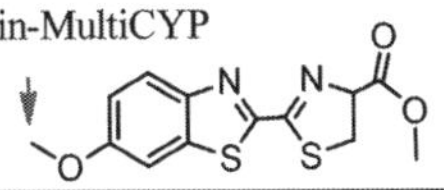	Pan-CYP substrate
6. Luciferin-4F12 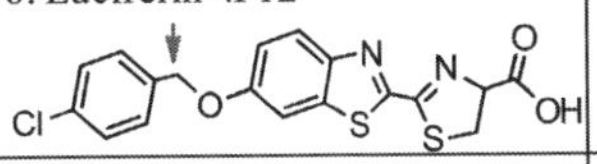	CYP4F12 selective	14. Luciferin-3A7 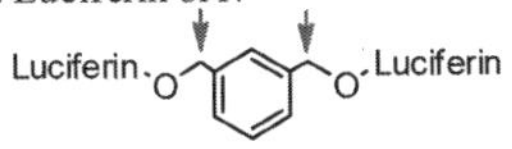	CYP3A7 selective
7. Luciferin-4F2/3 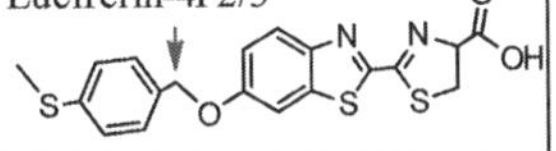	CYP4F2, -4F3B	15. Luciferin-1A1 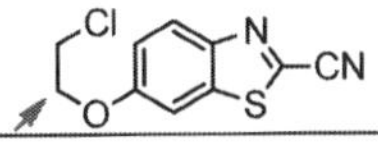	CYP1A1 selective
8. Luciferin-IPA 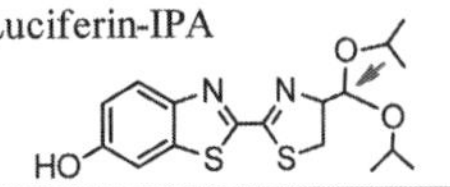	CYP3A4 selective	16. Luciferin-1A2 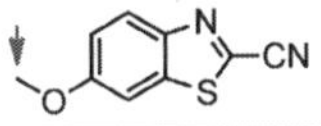	CYP1A2 selective

Fig. 2 Luminogenic CYP substrates. Compounds 2–7 are 6′ derivatives of D-luciferin, compound 8 is the diisopropyl acetal of D-luciferin, compound 9 is the methyl ether derivative of quinolyl-luciferin, compounds 10–13 are esters of 6′ derivatives of D-luciferin, compound 14 is a bis-luciferin, and compounds 15 and 16 are pre-pro-luciferins. *Arrows* indicate CYP oxidation sites that produce a luciferin, luciferin ester, or pre-luciferin. Enzyme activities are for human CYP enzymes and firefly luciferase

pro-luciferin CYP substrate that is highly selective for CYP3A4 the 6′ hydroxyl is left intact while the carboxylic acid group of D-luciferin is replaced by a di-isopropyl acetal group (Fig. 2, compound 8). In this case, CYP3A4 oxidizes the carbon between the two acetal oxygens to form an unstable hemi-orthoester, which rapidly decomposes to an ester. A bis-luciferin with a linker that joins two D-luciferin moieties at their respective 6′ hydroxyls is highly selective for CYP3A7 (Fig. 2, compound 14). Additional CYP selectivity profiles are achieved with a pre-pro-luciferin chemistry. In this case, the CYP substrates are derivatives of 6-hydroxybenzo(d)thiazole-2-carbonitrile. Although this compound is not reactive with luciferase it is rapidly converted to D-luciferin in a nonenzymatic reaction with D-cysteine at the carbonitrile. In this case, CYP converts the pre-pro-luciferin to the pre-luciferin 6-hydroxybenzo(d)thiazole-2-carbonitrile, which rapidly converts to luciferin in a luciferase reaction mixture that contains D-cysteine. Un-reacted substrate also reacts with D-cysteine, but is not detected by luciferase because it remains in the pro-luciferin form (Fig. 2, compounds 15 and 16).

2.2 Reagents for High-Throughput Screening Applications

The luciferase-based P450-Glo™ Screening Systems (Promega, Madison, WI) for CYP1A2 (V9770), CYP2C9 (V9790), CYP2C19 (V9880), CYP2D6 (V9890), and CYP3A4 using luciferin-PPXE (V9910). Control compounds which can be employed for assay validation include furafylline for CYP1A2, sulfaphenazole for CYP2C9, ketoconazole for CYP2C19, quinidine for CYP2D6, and ketoconazole for CYP3A4. These can be purchased from Sigma-Aldrich (St. Louis, MO).

3 Methods

The steps which are used for bioluminescent CYP assays in a miniaturized assay 1,536-well microtiter-plate format are summarized in Table 1.

1. Keep CYP enzyme and pro-luciferin substrate mixture on ice while dispensing (see Note 1). The working concentration is 2× final concentration. Pipette the reaction mixture into white 1,536-well solid-bottom plates using a bottle-valve dispenser (11) and incubate at the appropriate temperature. The final concentrations are CYP1A2 10 nM, 100 µM luciferin-ME as substrate, in 100 mM KPO_4 buffer; CYP2C9 10 nM 2C9, 100 µM Luciferin-H as substrate, in 25 mM KPO_4 buffer; CYP2C19 5 nM, 10 µM Luciferin-H EGE as substrate, in 50 mM PO_4 buffer; CYP2D6 5 nM final, 30 µM Luciferin-ME EGE as substrate, in 100 mM KPO_4 in buffer; and CYP3A4 10 nM final, 25 µM Luciferin-PPXE as substrate.

2. Following addition of enzyme/substrate mix, prepare a control plate. Controls are furafylline (CYP1A2), sulfaphenazole (CYP2C9), ketoconazole (CYP2C19 and CYP3A4), and quinidine (CYP2D6) (see Note 2). The first four columns of the assay plate are reserved for controls and these are added from a separate 1,536-well compound plate. For constructing the control plate, samples in only the first four columns are prepared as follows: columns 1 and 2, 16-point titrations in duplicate of the appropriate inhibitor (both beginning at 10 mM in DMSO for all isozymes, except CYP2D6, for which the starting concentration is 250 µM); column 3, a neutral control (DMSO vehicle only); and column 4, the appropriate control inhibitor for the CYP (present at a final concentration of either 57 µM or 1.4 µM for CYP2D6).

3. Compound plates containing test samples (in columns 5–48; columns 1–4 are left empty) and the control plate are then transferred to the 1,536-well assay plate, using a pin tool (12). For pin-tool-based compound transfer clean the pins using a wash sequence: DMSO, iPA, and MeOH, followed by a 3-s vacuum-drying step.

4. Incubate the enzyme/substrate/compound mixture for 10 min at the respective incubation temperature.

5. To start the reaction, add a NADPH-regeneration solution containing 1.3 mM NADP$^+$, 3.3 mM glucose-6-phosphate, 3.3 mM MgCl$_2$, and glucose-6-phosphate dehydrogenase (0.4 U/mL) for all isozymes; for CYP3A4, 200 mM KPO$_4$ was also added. Prepare this reagent at room temperature, but keep on ice while dispensing into large batches of plates.

6. Next, incubate the plates for 1 h at either room temperature, for CYP1A2, CYP2D6, and CYP3A4, or at 37°C, for CYP2C9 and CYP2C19.

7. Develop the luminescent signal by adding P450 Glo-Buffer, for CYP1A2, CYP2C9, and CYP3A4 (Promega), or Luciferin Detection Buffer, for CYP2C19 and CYP2D6 (Promega) (see Note 3). The detection reagent contains luciferase and ATP to generate a luminescent signal. Keep this reagent on ice and shield from light during dispensing. Addition of this reagent also stops the P450 reaction.

8. Incubate the plates for another 20 min before reading on a ViewLux (Perkin Elmer) with 2× binning and an exposure of 60 s.

9. Data can be reported as %activity relative to controls. If compounds were titrated at multiple concentrations then the concentration for 50% inhibition (IC_{50}) can be calculated following curve fitting. Typically, inhibition is observed but activation can also occur (see Note 4). IC_{50}s derived from the bioluminescent CYP assays described here correlate well with orthogonal CYP assay methods, as described in the Introduction, provided the methods being compared are properly implemented (see Note 5).

4 Notes

1. For preparing the pro-luciferin substrate stocks, luciferin-H EGE and luciferin-ME EGE are reconstituted in acetonitrile; other substrates are provided as DMSO stocks. Substrates stocks can be stored at –20°C or –70°C, protected from light. This caution about protection from light relates to storage of the luminogenic substrates only. The time elapsed

during a typical assay work flow is sufficiently short as to obviate the need for shielding these samples from light. Luciferin-PPXE substrate stock must be stored at −70°C. CYP enzyme stocks should be stored at −70°C. Recombinant CYP enzymes prepared for step 1 and liver microsomes should be stored at ≤−70°C. Storage at >−70°C (e.g., in a typical −20°C freezer) results in gradual loss of CYP enzyme activity. The enzymes should be thawed rapidly then placed on wet ice until ready to initiate reactions. The stability of CYP enzymes at room temperature or 37°C varies among enzymes. Among the enzymes discussed here, CYP1A2 appears the least stable and CYP2C9 the most stable.

Recombinant CYP enzymes and liver microsomes are membrane suspensions rather than soluble enzyme solutions. Therefore, care must be taken to ensure thorough mixing of reaction mixtures to avoid settling of the membranes in reaction vessels, which can lead to assay variability.

2. Test compounds should be diluted in DMSO; however, CYP3A enzymes can be highly sensitive to DMSO, and the degree of sensitivity can be substrate dependent. None of the non-CYP3A enzyme assays are affected by DMSO concentrations ≤1%. The substrate chosen here for CYP3A4 (Luc-PPXE) has only modest sensitivity to DMSO and the final concentration of DMSO in the assays described above is 0.5%, which is acceptable for the CYP3A4/Luc-PPXE 3A system. However, luciferin-BE, the original luminogenic CYP3A substrate, is inhibited substantially by DMSO, and DMSO concentrations should be ≤0.1% if this substrate is used. None of the CYP assays are affected by acetonitrile, methanol, or ethanol at concentrations ≤1.0%.

3. It is important to match the appropriate luciferase detection reagent with the specific CYP being assayed, as indicated in step 7 of the method. The CYP2D6 and CYP2C19 reactions produce luciferin esters that are not active with luciferase. Therefore, the reagent called "Luciferin Detection Buffer" supplied for CYP2C19 and CYP2D6 contains an esterase that de-esterifies the CYP reaction products. This step is essential for light generation with luciferase. In contrast, the CYP1A2, CYP2C9, and CYP3A4 substrates are not esters and the corresponding "P450 Glo-Buffer" supplied with these assays does not contain or require esterase. Promega suggests that the reconstituted Luciferase Detection Reagent can be stored at −20°C for up to 3 months and, for convenience, this reagent can be stored at room temperature (e.g., ~23°C) for 24 h or at 4°C for 1 week without loss of activity. Always avoid multiple freeze-thaw cycles of reagents.

4. When screening for CYP inhibitors certain test compounds are occasionally encountered that stimulate rather than inhibit CYP activity (e.g., α-naphthoflavone stimulates CYP3A4 activity). This phenomenon, known as heterotropic cooperativity, is most frequently observed with CYP3A4, though it also occurs with other CYPs and is consistent with the known biochemistry of these enzymes (13). Stimulation indicates that the compound interacts with the CYP and can be explained as an allosteric effect on the probe reaction.

5. A common cause of discrepancies which are occasionally observed between two CYP assay methods that use different probe substrates is the phenomenon of substrate-dependent inhibition. This occurs with CYPs that bind substrates in two or more nonoverlapping configurations that are not mutually competitive with all competitive inhibitors of the particular CYP. Substrate-dependent-inhibition profiles are most commonly observed with CYP3A4 (14). It is important to recognize that such IC_{50} differences do not reflect superiority of one assay over the other, but point to the need to use different substrates, in some cases, to explore different inhibition modes.

Acknowledgements

This research was supported by the Molecular Libraries Initiative of the NIH Roadmap for Medical Research and the Intramural Research Program of the National Institutes of Health.

References

1. Wienkers LC, Heath TG (2005) Predicting in vivo drug interactions from in vitro drug discovery data. Nat Rev Drug Discov 4:825–833

2. Sobel M, Ma D, Cali JJ (2007) P450-Glo CYP3A4 biochemical and cell-based assays. Promega Notes 98:15–18

3. Veith H, Southall N, Huang R, James T, Fayne D, Artemenko N, Shen M, Inglese J, Austin CP, Lloyd DG, Auld DS (2009) Comprehensive characterization of cytochrome P450 isozyme selectivity across chemical libraries. Nat Biotechnol 27:1050–1055

4. Foti RS, Wahlstrom JL (2008) CYP2C19 inhibition: the impact of substrate probe selection on in vitro inhibition profiles. Drug Metab Dispos 36:523–528

5. Zlokarnik G, Grootenhuis PD, Watson JB (2005) High throughput P450 inhibition screens in early drug discovery. Drug Discov Today 10:1443–1450

6. Meisenheimer PL, Uyeda HT, Ma D, Sobol M, McDougall MG, Corona C, Simpson D, Klaubert DH, Cali JJ (2011) Proluciferin acetals as bioluminogenic substrates for cytochrome P450 activity and probes for CYP3A inhibition. Drug Metab Dispos 39:2403–2410

7. Cohen LH, Remley MJ, Raunig D, Vaz AD (2003) In vitro drug interactions of cytochrome p450: an evaluation of fluorogenic to conventional substrates. Drug Metab Dispos 31:1005–1015

8. Cali JJ, Ma D, Sobol M, Simpson DJ, Frackman S, Good TD, Daily WJ, Liu D (2006) Luminogenic cytochrome P450 assays. Expert Opin Drug Metab Toxicol 2:629–645

9. Auld DS, Zhang YQ, Southall NT, Rai G, Landsman M, Maclure J, Langevin D, Thomas CJ, Austin CP, Inglese J (2009) A basis for reduced chemical library inhibition of firefly luciferase obtained from directed evolution. J Med Chem 52:1450–1458

10. Li AP (2009) Evaluation of luciferin-isopropyl acetal as a CYP3A4 substrate for human hepatocytes: effects of organic solvents, cytochrome

P450 (P450) inhibitors, and P450 inducers. Drug Metab Dispos 37:1598–1603

11. Niles WD, Coassin PJ (2005) Piezo- and solenoid valve-based liquid dispensing for miniaturized assays. Assay Drug Dev Technol 3:189–202

12. Cleveland PH, Koutz PJ (2005) Nanoliter dispensing for uHTS using pin tools. Assay Drug Dev Technol 3:213–225

13. Niwa T, Murayama N, Yamazaki H (2008) Heterotropic cooperativity in oxidation mediated by cytochrome P450. Curr Drug Metab 9:453–462

14. Wang RW, Newton DJ, Liu N, Atkins WM, Lu AYH (2000) Human cytochrome P-450 3A4: in vitro drug-drug interactions are substrate-dependent. Drug Metab Dispos 28:360–366

Chapter 2

Simultaneous Determination of Multiple CYP Inhibition Constants using a Cocktail-Probe Approach

Michael Zientek and Kuresh Youdim

Abstract

To identify cytochrome P450 (CYP) drug–drug interaction (DDI) potential of a new chemical entity, the use of a specific clinically relevant probe substrate in the presence of a test compound is common place. In early discovery of new chemical entities, a balance of rigor, the ability to predict clinical DDI, and throughput is desired in an in vitro assay. This chapter describes a high-throughput CYP-mediated DDI assay method that balances these characteristics. The method utilizes a cassette approach using a cocktail of five selective probe substrates for the major clinically relevant CYPs involved in drug interactions. CYP1A2, 2C9, 2C19, 2D6, and 3A activities are assessed with liquid chromatography/tandem mass spectrometry (LC-MS/MS) quantification of metabolite formation. The method also outlines specific inhibitors to evaluate dynamic range and as a positive control. The benefits and needs for caution of this method are noted and discussed.

Key words Drug–drug interaction, Cytochrome P450, CYP1A2, CYP2C9, CYP2C19, CYP2D6, CYP3A, Cocktail of substrates, Cassette of substrates, LC-MS/MS, Liquid chromatography/tandem mass spectrometry

1 Introduction

Drug–drug interactions (DDI) are a major liability for any new drug entering the marketplace. Adverse drug reactions, of which DDI are a significant component, are a leading cause of hospital administrations (1) and drug withdrawals (2). Coadministration of multiple drugs to patients for one disease (e.g., HIV infection or cancer) or treatment for several diseases concurrently (geriatrics) is common place. Therefore, the potential is great for DDI in a polypharmacy environment. The FDA guidance related to CYP DDIs recommends both the accepted probe substrates and measured metabolites for clinical DDI assessment (3–5). In practice, in the early stages, five major enzymes are investigated, CYP1A2, CYP2C9, CYP2C19, CYP2D6, and CYP3A4, accounting for greater than 90% of total hepatic CYPs (6, 7) and 70% of metabolism of all marketed drugs (8). While this guidance sets the effect

Ian R. Phillips et al. (eds.), *Cytochrome P450 Protocols*, Methods in Molecular Biology, vol. 987, DOI 10.1007/978-1-62703-321-3_2, © Springer Science+Business Media New York 2013

thresholds that trigger clinical studies, the agency will also accept appropriate in vitro data indicating no significant CYP inhibition as justification that subsequent in vivo studies are not necessary (4). Patient safety concerns and regulatory requirements have led the pharmaceutical industry to adopt the general practice of only progressing development candidates with acceptably low predicted CYP DDI liabilities which have been accessed in vitro (e.g., $IC_{50} \geq 1{,}000$-fold efficacious concentration). When this goal is not met, such early knowledge enables the planning of clinical studies at the appropriate time in the development program.

A competitive- and/or reversible-inhibition screen is often the first step in understanding the DDI potential of a new chemical entity. The definitive assessment of inhibition is the inhibition constant (K_i), which provides not only the inhibition potency but also information on the mechanism of inhibition (competitive, noncompetitive). However, in the lead optimization profiling environment this approach is overcomplex for the question being asked, and generates far too many samples to enable rapid screening of compound series. The substrate-cocktail approach has been developed to provide a balance of sufficient throughput without compromising data relevance. This method has been made possible due to advances in chromatographic methods and mass-spectrometry sensitivity, in addition to further understanding of specific probes for each clinically relevant CYP isoform (9–19). Immediate benefits seen from a cocktail or cassette method is a complete evaluation of the major metabolizing enzyme in a single reaction under the same conditions and the number of compounds that can be assessed by such a reaction. Chemical series modifications can be assessed, providing information on minor changes in structure-activity relationship (SAR) revealing shifts in potency toward or away from one or many isoforms. The assay consists of human liver microsomes and a cocktail of probe substrates metabolized by the five major CYP isoforms (tacrine for CYP1A2, diclofenac for CYP2C9, (S)-mephenytoin for CYP2C19, dextromethorphan for CYP2D6, and midazolam for CYP3A). The assay has been fully automated in both a 96- and a 384-well format (19, 20).

2 Materials

2.1 Assay

1. Substrates: Dextromethorphan hydrobromide monohydrate (mol wt 370.33), diclofenac sodium salt (mol wt 304.11), midazolam (mol wt 325.77) (see Note 1), (S)-mephenytoin (mol wt 218.26), tacrine hydrochloride (mol wt 234.72) (all from Sigma-Aldrich (St. Louis, MO)) (see Note 2).

2. Metabolite standards: Dextrorphan (mol wt 257.2), 4'-hydroxy-diclofenac (mol wt 311.0), 1'-hydroxymidazolam (mol wt 341.1), 4'-hydroxymephenytoin (mol wt 234.1) (all from BD Biosciences,

Discovery Labware, Woburn, MA), 1′-hydroxytacrine (mol wt 214.3) (Pfizer internal compound library, but may be purchased from TLC PharmaChem) (see Note 3).

3. Assay buffer: 100 mM potassium phosphate buffer pH 7.4 with 1 mM magnesium chloride (see Note 4).

4. Cofactor: NADPH-regeneration system (β-nicotinamide adenine dinucleotide phosphate sodium salt hydrate (NADP+), isocitric dehydrogenase from porcine heart, dl-isocitric acid trisodium salt (all from Sigma-Aldrich (St. Louis, MO))). For convenience, 1 mM NADPH can be substituted for the regeneration system (see Note 5).

5. Enzyme source: Human liver microsomes (HLM), pooled from ≥50 male and female donors (BD Gentest, Bedford, MA) (see Note 6).

6. Control inhibitor: Miconazole (mol wt 479.15), a pan-CYP inhibitor, or individual-specific inhibitors for each CYP isoform (4) (see Note 7).

7. Labware: Dilutions and reactions were prepared in deep-well, polypropylene, 1 mL/well-capacity plates (Beckman Coulter, Fullerton, CA).

8. Incubations should be conducted at 37°C in a water-heated block or heated 96/384-well shaker or for higher throughput can be adapted to a robotic system with the capability to control temperature (i.e., Peltier system, recirculating water bath, or incubator).

9. A deproteinizing agent combined with an internal standard, such as chilled acetonitrile with triazolam (mol wt = 342) Sigma RBI (Natick, MA), acts as a reaction quenching reagent and a LC-MS/MS standard.

2.2 Preexperiment Preparations

1. Initial stock solutions of CYP isoform-specific substrates:

    (a) Diclofenac (7.5 mM) in water.

    (b) Dextromethorphan (10 mM) in 90% acetonitrile/10% methanol.

    (c) (S)-mephenytoin (50 mM) in acetonitrile.

    (d) Midazolam (10 mM) in 66.7% methanol/33.3% acetonitrile.

    (e) Tacrine (10 mM) in DMSO.

2. Working stock of the cocktail of substrates: For a 10-reaction experiment, combine stocks of 5 µL of 7.5 mM diclofenac, 3.75 µL of 10 mM dextromethorphan, 6 µL of 50 mM S-mephenytoin, 1.5 µL of 10 mM midazolam, and 1.5 µL of 10 mM tacrine with 5,982 µL of 100 mM potassium phosphate buffer (pH 7.4) containing 1 mM $MgCl_2$. This may be scaled up appropriately to the number of reactions needed.

14   Michael Zientek and Kuresh Youdim

3. Test compound or control inhibitor: Can be prepared as either a single concentration assessment or as an IC_{50} determination. If a single assessment is being conducted a stock of 300 μM in 10% DMSO, 90% acetonitrile is prepared to achieve a final concentration of 3 μM in the assay. If a full IC_{50} determination is desired, a starting conc. of 3 mM in 10% DMSO, 90% acetonitrile is appropriate to achieve a 30 μM starting concentration. Half to one-third log dilutions in 10% DMSO, 90% acetonitrile of the 3 mM starting concentration will provide a wide concentration range at ≥6 concentration points. In the case of a potent control pan-inhibitor, such as miconazole, a 1 μM final concentration will inhibit at least 80% of CYP activity for all isoforms (see Note 8).

4. Individual stock solutions of 500 mM isocitric acid, 100 mM NADP+, and 100 U/mL isocitrate dehydrogenase in 100 mM potassium phosphate buffer (pH 7.4) containing 1 mM $MgCl_2$ should be prepared, and are stable in the refrigerator for 1 week. At the time of the experiment, the three reagents should be combined with 100 mM potassium phosphate buffer, pH 7.4, with 1 mM magnesium chloride in a ratio of 1:1:1:2 and pre-incubated for 15 min at 37°C.

5. Thaw stock human liver microsomes (HLM) on ice and then prepare a stock of HLM to a concentration of 0.71 mg/mL in 100 mM potassium phosphate buffer (pH 7.4) containing 1 mM $MgCl_2$ and keep on ice until use.

2.3 Bioanalysis

1. Column: Phenomenex Onyx Monolithic C18, 50×4.6 mm (product number = CH0-7644).

2. Mobile phases: Aqueous solvent (A & C) with 0.1% formic acid and acetonitrile (B) with 0.1% formic acid were used as mobile phases.

3 Methods

3.1 Experimental Methodology

1. Perform reactions in a final volume of 500 μL/well (see Note 9). To each well add:

    (a) 400 μL of the working stock of the cocktail of substrates warmed to 37°C. The final concentration of the substrates in a 500 μL reaction is 40 μM S-mephenytoin, 5 μM dextromethorphan, 5 μM diclofenac, 2 μM midazolam, and 2 μM tacrine (see Note 10).

    (b) 70 μL of 0.71 mg/mL human liver microsomes stock (0.1 mg/mL final concentration) (see Note 11).

    (c) 5 μL of test or control compound in 10% DMSO and 90% acetonitrile (see Note 12).

2. Samples (compound, microsomes, and substrate cocktail) are pre-incubated for 5 min at 37°C.

(d) After the 5-min preincubation, add 25 µL of pre-warmed NADPH regeneration system (~1 mM NADPH generated) to initiate the reaction.

3. Following an 8-min incubation (see Note 13) at 37°C, terminate reactions by the addition of 500 µL/well chilled internal standard (acetonitrile containing 0.2 µg/mL triazolam) (see Subheading 2.1, item 9).

4. CYP inhibition is quantitated by simultaneously analyzing 4-hydroxymephenytoin, dextrorphan, 4-hydroxydiclofenac, 1-hydroxymidazolam, and 1-hydroxytacrine using LC-MS (Fig. 1).

3.2 Bioanalysis

1. Standard curve preparation

(a) Prepare small volumes (500 µL) of 200 µg/mL (free base or free acid equivalent concentration) metabolite stock solutions of individual metabolites in 10% DMSO/90% acetonitrile. These can be stored in a refrigerator for up to 1 month.

(b) Add 100 µL of each solution to 500 µL acetonitrile to make a total volume of 1 mL of 20,000 ng/mL (20 µg/mL) metabolite cocktail stock solution.

(c) Prepare ½ serial dilutions from the metabolite cocktail stock solution by adding 0.5 mL to 0.5 mL acetonitrile to give standard concentration ranging from 10,000 to 2.44 ng/mL (Table 1).

(d) Prepare denatured human liver microsome solution at 0.1 mg/mL in buffer. This solution can be stored in a refrigerator for 3 months (see Note 14).

(e) Add 5 µL of each standard into 95 µL of denatured microsomes to give final concentration range for the standard curve of 500 to 0.12 ng/mL (Table 1).

(f) Centrifuge for 15 min at $2400 \times g$, 4°C prior to injecting into mass spectrometer (MS).

2. Analytical conditions (see Note 15)

(a) Multiple-reaction monitoring (MRM) LC-MS/MS analysis should be conducted on the most sensitive quadrupole mass spectrometer available, and measured in positive ion mode (see Note 16).

(b) For metabolites of interest, their MRM transitions, collision energies (CE), declustering potentials (DP), and collision cell exit potential (CXP) are shown in Table 2 (Fig. 1) (based upon Sciex API4000). Optimization of values will be dependent on specific MS. Values found in Table 2 serve as a guide.

Table 1
Analytical standard curve preparation via ½ dilutions from the metabolite cocktail stock solution

Metabolite cocktail stock solution concentration (ng/mL)	Volume of stock to be diluted (mL)	Volume of acetonitrile diluent (mL)	Final metabolite stock concentration (ng/mL)	Final metabolite standard concentration in assay matrix (ng/mL)
20,000	0.5	0.5	10,000	500
10,000	0.5	0.5	5,000	250
5,000	0.5	0.5	2,500	125
2,500	0.5	0.5	1,250	62.5
1,250	0.5	0.5	625	31.3
625	0.5	0.5	312.5	15.6
312.5	0.5	0.5	156.3	7.81
156.3	0.5	0.5	78.1	3.91
78.1	0.5	0.5	39	1.95
39	0.5	0.5	19.5	0.977
19.5	0.5	0.5	9.8	0.488
9.77	0.5	0.5	4.9	0.244
4.9	0.5	0.5	2.4	0.122
0	0	1.0	0	0

Table 2
Multiple-reaction monitoring (MRM) LC-MS/MS analysis transitions, collision energies (CE), declustering potentials (DP), and collision cell exit potential (CXP) for specific metabolites formed by each cytochrome P450 isoform

Metabolite	P450	MRM	CE	DP	CXP
1-OH-tacrine	CYP1A2	M/Z 152>110	30	50	15
4-OH-diclofenac	CYP2C9	M/Z 312.3>230.1	45	45	15
4-OH-S-mephenytoin	CYP2C19	M/Z 235.2>150.1	27	45	15
Dextrorphan	CYP2D6	M/Z 256.2>157.1	53	90	15
1-OH-midazolam	CYP3A4	M/Z 342.3>168.1	53	65	15
Internal standard/triazolam		M/Z 343.2>308.1	40	70	15

Gradient conditions

Initial conditions: Total flow 0.200 mL/min.

Pump B Pct 1.0%; Pump C flow 3.000 mL/min. (Teed into liquid path immediately before the column.)

Remaining conditions can be found in Table 3.

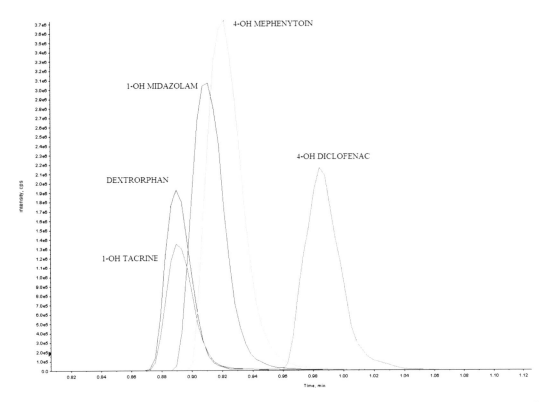

Fig. 1 Chromatogram detecting the simultaneous formation of the metabolites formed in the cocktail of substrates drug-drug interaction reaction (peak area intensity versus time eluted off the chromatography column)

Table 3
Liquid chromatography gradient conditions to separate the specific metabolites formed by each of the cytochrome P450 isoforms

Time	Events	Parameter
0.01	Pump B conc.	10
0.20	Total flow	0.20
0.21	Pump C flow	3.5
0.22	Total flow	3
0.23	Pump C flow	0.01
0.42	Pump B conc.	10
0.60	Pump B conc.	25
1.45	Pump B conc.	65
1.57	Pump B conc.	65
1.58	Pump B conc.	10
1.90	System controller	Stop

18 Michael Zientek and Kuresh Youdim

3.3 Data Analysis

3.3.1 Analytical

1. Acceptance criteria

Calibration line

Calibration standards should be within 20% of nominal conc. with at least six remaining on the line.
Linearity and model used (R2>0.95).
Same regression model throughout runs.
Calibration line the used to calculated.

Metabolite concentrations
Metabolite concentrations (nM) should be calculated using appropriate software on the MS from the standard curve using linear regression with $1/x2$ weighting. The metabolite concentrations can be converted to pmol/min/mg and should be used to ensure acceptable substrate turnover as well as assess experiment to experiment variability (see Note 17).

3.3.2 Percent Inhibition or IC$_{50}$ Determination

1. If a single concentration percent inhibition is desired, the metabolic rate of formation of the metabolite for each substrate can then be expressed as a percentage of the control (uninhibited activity) and a percent inhibition value determined by subtraction from 100%.

2. If IC$_{50}$ are being determined, estimations can be made from the data by fitting a standard four-parameter logistic using nonlinear regression Eq. 1. In this equation *Range* is the fitted uninhibited value (y_{max}) minus the *Background*, and "*s*" is a slope factor or Hill coefficient. The equation assumes that "*y*" (*enzyme velocity*) decreases with increasing "*x*" (*concentration*) (see Note 18).

$$y = \text{Range} / 1 + (x / \text{IC}_{50})^s + \text{Background} \qquad (1)$$

4 Notes

1. In the USA and the UK, midazolam and the internal standard triazolam are controlled substances requiring the facility to carry certification in order to obtain these compounds.

2. The user may want to add other specific CYP isoform substrates to the cocktail of substrates, but a similar validation should occur to that performed by Zientek et al. before use (20). Also, potential competition of individual isoforms for cytochrome b$_5$ and NADPH cytochrome P450 reductase may limit the electron transport and thus metabolic rate (21, 22). It has been shown that such issues can be overcome and in an acceptable range to provide an appropriate level of quality data (20, 23).

3. The addition of the magnesium ion from magnesium chloride is needed to stabilize the negative charge formed on the hydroxyl oxygen during dehydrogenation and then a consecutive transfer of a hydride to produce NADPH from NADP$^+$ (24).

4. Temperature may need to be adjusted to control the temperature loss between the heat source and the labware; reaction rate will suffer if not optimized.

5. If NADPH is used instead of the NADPH regeneration system, $MgCl_2$ can be removed from the buffer.

6. Human liver microsomes are from human-derived materials and require special handling and blood-borne-pathogen training.

7. In early discovery, a pan-CYP inhibitor is appropriate as a positive control inhibitor but should be left to the discretion of the scientist. There are many specific inhibitors for the five major CYP isoforms to choose from, and the FDA has made recommendations as to which ones are most appropriate (4). Some of these potent but specific inhibitors are furafylline (1A2), sulfaphenazole (2C9), ticlopidine (2C19), quinidine (2D6), and ketoconazole (CYP3A4/5). The concentration or concentration range should be appropriate to observe significant inhibition of the specified isoform while not affecting the activity of the other CYP isoforms.

8. Final concentrations of DMSO and non-DMSO organic solvents of 0.1% (v/v) and 1.0% (v/v), respectively. CYPs are quite sensitive to DMSO; therefore caution should be used when utilizing this solvent as an experimental reagent (20, 25–27).

9. This method can be miniaturized to accommodate 50 μL reactions, appropriate for a 384-well plate (19, 20).

10. All substrates utilized in the inhibition experiment are prepared at stock concentration to have final concentrations at the K_M, to assure appropriate sensitivity to the inhibitor (18, 20, 28–35). This also allows estimation of K_i values, using the Cheng-Prusoff equation, from the IC_{50} concentration (36).

11. Low microsomal protein concentrations of 0.1 mg/mL or below reduces the effect the inhibitor binding to microsomal matrix (5, 37, 38).

12. Typically, concentrations in the 1–5 μM range are used for single concentration DDI assessment (15, 39).

13. Strict adherence to initial rate and compliance with Michaelis–Menten kinetics (<10% substrate depletion) is not possible with very different reaction rates, although with an 8-min reaction time no more than 25% substrate depletion is observed for the most extensively metabolized substrate while maintaining sensitivity of the least metabolized substrate (20).

14. Denatured microsomes can be prepared using a number of approaches: (a) boiling for 3 min, (b) leaving at room temperature for 48 h, or (c) direct denaturing into acetonitrile. In each case the microsomal mix then needs to be prepared such that the organic:aqueous ratio matches that in the samples. This ensures all samples are matrix matched, thus mitigating the effects of any ion suppression on metabolite intensities (40).

15. Given the difference in analytical setup in different laboratories, users have flexibility to adapt analytical conditions to ensure maximum sensitivity achieved with their particular instrument. The conditions described below serve only as a guide (17, 19, 41).

16. Single-stage quadrupole mass spectrometers (SSQMS) as well as triple-stage quadrupole mass spectrometers (TSQMS) are commonly used for this type of analysis (41). Although quadrupole mass analyzers have the ability to operate in both negative and positive ion modes, specific advantages of SSQMS instruments include low cost and their relatively small size, whereas TSQMS instruments have greater discrimination against chemical background, resulting in real gains in selectivity and sensitivity. In TSQMS, the Q1 mass analyzer filters the desired ions such that they are fragmented by argon or nitrogen within Q2, and their fragment ions are subsequently scanned by Q3 before reaching the mass detector. Consequently, given that TSQMS acquires much richer, higher-value datasets than SSQMS in selected reaction monitoring (SRM) or multiple-reaction monitoring (MRM) modes would suggest it to be the instrument of choice in routine and high-throughput quantitative bioanalysis. However, detection sensitivity decreases dramatically when wide mass range is analyzed in a scanning mode, which can be a limitation in its application for screening of "*unknown*" drug metabolites.

17. It is suggested that a control activity chart be kept to monitor inter-experiment rate variability, and those experiments that lie outside 50% CV should be repeated.

18. When using Eq. 1, the maximal percent control (y_{max}) incorporated into the *Range* term refers to the lack of inhibition. Therefore, y_{max} or 100% control activity must be observed in the span of concentrations tested or the equation should be modified. Specifically, the y_{max} should be fixed at 100% of control to assure a proper IC_{50} determination. Another point to consider is when only a few points define the dose–response curve, the assumption is that the inhibitor binds to the enzyme via the law of mass action; therefore the Hill coefficient may be set to a standard slope of -1.0, although in most cases the data should dictate the appropriate fit.

References

1. Smith CC, Bennett PM, Pearce HM, Harrison PI, Reynolds DJ, Aronson JK, Grahame-Smith DG (1996) Adverse drug reactions in a hospital general medical unit meriting notification to the Committee on Safety of Medicines. Br J Clin Pharmacol 42:423–9
2. Issa AM, Phillips KA, Van Bebber S, Nidmarthy HG, Lasser KE, Haas JS, Alldredge BK, Wachter RM, Bates DW (2007) Drug withdrawals in the United States: a systematic review of the evidence and analysis of trends. Curr Drug Saf 2:177–185
3. Bjornsson TD, Callaghan JT, Einolf HJ, Fischer V, Gan L, Grimm S, Kao J, King SP, Miwa G, Ni L, Kumar G, McLeod J, Obach RS, Roberts S, Roe A, Shah A, Snikeris F, Sullivan JT, Tweedie D, Vega JM, Walsh J, Wrighton SA (2003) The conduct of in vitro and in vivo drug-drug interaction studies: a Pharmaceutical Research and Manufacturers of America (PhRMA) perspective. Drug Metabol Dispos 31:815–32
4. Huang S, Stifano T (2006) C.f.D.E.a.R. (CDER), C.f.B.E.a.R. (CBER) (ed) Draft guidance for industry: drug interaction studies—study design, data analysis, and implications for dosing and labeling. U.S. Department of Health and Human Services Food and Drug Administration.
5. Tucker G, Houston B, Huang S (2001) Optimizing drug development: strategies to assess drug metabolism/transporter interaction potential—toward a consensus. Pharmaceut Res V18:1071–1080
6. Shimada T, Yamazaki H, Mimura M, Inui Y, Guengerich FP (1994) Interindividual variations in human liver cytochrome P-450 enzymes involved in the oxidation of drugs, carcinogens and toxic chemicals: studies with liver microsomes of 30 Japanese and 30 Caucasians. J Pharmacol Exp Ther 270:414–23
7. Yuan R, Madani S, Wei XX, Reynolds K, Huang SM (2002) Evaluation of cytochrome P450 probe substrates commonly used by the pharmaceutical industry to study in vitro drug interactions. Drug Metab Dispos 30:1311–9
8. Williams JA, Hyland R, Jones BC, Smith DA, Hurst S, Goosen TC, Peterkin V, Koup JR, Ball SE (2004) Drug-drug interactions for UDP-glucuronosyltransferase substrates: a pharmacokinetic explanation for typically observed low exposure (AUCi/AUC) ratios. Drug Metab Dispos 32:1201–8
9. Bu HZ, Knuth K, Magis L, Teitelbaum P (2000) High-throughput cytochrome P450 inhibition screening via cassette probe-dosing strategy. IV. Validation of a direct injection on-line guard cartridge extraction/tandem mass spectrometry method for simultaneous CYP3A4, 2D6 and 2E1 inhibition assessment. Rapid Commun Mass Spectrom 14:1943–8
10. Bu HZ, Knuth K, Magis L, Teitelbaum P (2001) High-throughput cytochrome P450 (CYP) inhibition screening via cassette probe-dosing strategy: III. Validation of a direct injection/on-line guard cartridge extraction-tandem mass spectrometry method for CYP2C19 inhibition evaluation. J Pharm Biomed Anal 25:437–42
11. Bu HZ, Knuth K, Magis L, Teitelbaum P (2001) High-throughput cytochrome P450 (CYP) inhibition screening via a cassette probe-dosing strategy. V. Validation of a direct injection/on-line guard cartridge extraction–tandem mass spectrometry method for CYP1A2 inhibition assessment. Eur J Pharm Sci 12:447–52
12. Bu HZ, Magis L, Knuth K, Teitelbaum P (2000) High-throughput cytochrome P450 (CYP) inhibition screening via cassette probe-dosing strategy. I. Development of direct injection/on-line guard cartridge extraction/tandem mass spectrometry for the simultaneous detection of CYP probe substrates and their metabolites. Rapid Commun Mass Spectrom 14:1619–24
13. Bu HZ, Magis L, Knuth K, Teitelbaum P (2001) High-throughput cytochrome P450 (CYP) inhibition screening via a cassette probe-dosing strategy. VI. Simultaneous evaluation of inhibition potential of drugs on human hepatic isozymes CYP2A6, 3A4, 2C9, 2D6 and 2E1. Rapid Commun Mass Spectrom 15:741–8
14. Bu HZ, Magis L, Knuth K, Teitelbaum P (2001) High-throughput cytochrome P450 (CYP) inhibition screening via cassette probe-dosing strategy. II. Validation of a direct injection/on-line guard cartridge extraction-tandem mass spectrometry method for CYP2D6 inhibition assessment. J Chromatogr B Biomed Sci Appl 753:321–6
15. Gao F, Johnson DL, Ekins S, Janiszewski J, Kelly KG, Meyer RD, West M (2002) Optimizing higher throughput methods to assess drug-drug interactions for CYP1A2, CYP2C9, CYP2C19, CYP2D6, rCYP2D6, and CYP3A4 in vitro using a single point IC(50). J Biomol Screen 7:373–82
16. Obach RS, Walsky RL, Venkatakrishnan K, Houston JB, Tremaine LM (2005) In vitro cytochrome P450 inhibition data and the prediction of drug-drug interactions: qualitative

relationships, quantitative predictions, and the rank-order approach. Clin Pharmacol Ther 78:582–92

17. Smith D, Sadagopan N, Zientek M, Reddy A, Cohen L (2007) Analytical approaches to determine cytochrome P450 inhibitory potential of new chemical entities in drug discovery. J Chromatogr B Analyt Technol Biomed Life Sci 850:455–63

18. Walsky RL, Obach RS (2004) Validated assays for human cytochrome P450 activities. Drug Metab Dispos 32:647–60

19. Youdim KA, Lyons R, Payne L, Jones BC, Saunders K (2008) An automated, high-throughput, 384 well Cytochrome P450 cocktail IC50 assay using a rapid resolution LC-MS/MS end-point. J Pharm Biomed Anal 48:92–9

20. Zientek M, Miller H, Smith D, Dunklee M, Heinle L, Thurston A, Lee C, Hyland R, Fahmi O, Burdette D (2008) Development of an in vitro drug-drug interaction assay to simultaneously monitor five cytochrome P450 isoforms and performance assessment using drug library compounds. J Pharmacol Toxicol Methods 58:206–214

21. Cawley GF, Batie CJ, Backes WL (1995) Substrate-dependent competition of different P450 isozymes for limiting NADPH-cytochrome P450 reductase. Biochemistry 34:1244–7

22. West SB, Lu AY (1972) Reconstituted liver microsomal enzyme system that hydroxylates drugs, other foreign compounds and endogenous substrates. V. Competition between cytochromes P-450 and P-448 for reductase in 3,4-benzpyrene hydroxylation. Arch Biochem Biophys 153:298–303

23. Weaver R, Graham KS, Beattie IG, Riley RJ (2003) Cytochrome P450 inhibition using recombinant proteins and mass spectrometry/multiple reaction monitoring technology in a cassette incubation. Drug Metab Dispos 31:955–66

24. Hurley JH, Dean AM, Koshland DE Jr, Stroud RM (1991) Catalytic mechanism of NADP(+)-dependent isocitrate dehydrogenase: implications from the structures of magnesium-isocitrate and NADP+ complexes. Biochemistry 30:8671–8

25. Busby WF Jr, Ackermann JM, Crespi CL (1999) Effect of methanol, ethanol, dimethyl sulfoxide, and acetonitrile on in vitro activities of cDNA-expressed human cytochromes P-450. Drug Metab Dispos 27:246–9

26. Easterbrook J, Lu C, Sakai Y, Li AP (2001) Effects of organic solvents on the activities of cytochrome P450 isoforms, UDP-dependent glucuronyl transferase, and phenol sulfotransferase in human hepatocytes. Drug Metab Dispos 29:141–4

27. Chauret N, Gauthier A, Nicoll-Griffith DA (1998) Effect of common organic solvents on in vitro cytochrome P450-mediated metabolic activities in human liver microsomes. Drug Metab Dispos 26:1–4

28. Leemann T, Transon C, Dayer P (1993) Cytochrome P450TB (CYP2C): a major monooxygenase catalyzing diclofenac 4′-hydroxylation in human liver. Life Sci 52:29–34

29. Wrighton SA, Stevens JC, Becker GW, VandenBranden M (1993) Isolation and characterization of human liver cytochrome P450 2C19: correlation between 2C19 and S-mephenytoin 4′-hydroxylation. Arch Biochem Biophys 306:240–5

30. Goldstein JA, Faletto MB, Romkes-Sparks M, Sullivan T, Kitareewan S, Raucy JL, Lasker JM, Ghanayem BI (1994) Evidence that CYP2C19 is the MAJOR (S)-mephenytoin 4′-hydroxylase in humans. Biochemistry 33:1743–1752

31. Rodrigues AD (1996) Measurement of human liver microsomal cytochrome P450 2D6 activity using (O-methyl-14C)dextromethorphan as substrate. Methods Enzymol 272:186–95

32. Kronbach T, Mathys D, Umeno M, Gonzalez FJ, Meyer UA (1989) Oxidation of midazolam and triazolam by human liver cytochrome P450IIIA4. Mol Pharmacol 36:89–96

33. Dierks EA, Stams KR, Lim HK, Cornelius G, Zhang H, Ball SE (2001) A method for the simultaneous evaluation of the activities of seven major human drug-metabolizing cytochrome P450s using an in vitro cocktail of probe substrates and fast gradient liquid chromatography tandem mass spectrometry. Drug Metab Dispos 29:23–9

34. Obach RS, Reed-Hagen AE (2002) Measurement of Michaelis constants for cytochrome P450-mediated biotransformation reactions using a substrate depletion approach. Drug Metab Dispos 30:831–7

35. Becquemont L, Le Bot MA, Riche C, Funck-Brentano C, Jaillon P, Beaune P (1998) Use of heterologously expressed human cytochrome P450 1A2 to predict tacrine-fluvoxamine drug interaction in man. Pharmacogenetics 8:101–8

36. Cheng Y, Prusoff WH (1973) Relationship between the inhibition constant (K1) and the concentration of inhibitor which causes 50 per cent inhibition (I50) of an enzymatic reaction. Biochem Pharmacol 22:3099–108

37. Walsky RL, Obach RS, Gaman EA, Gleeson JP, Proctor WR (2005) Selective inhibition of human cytochrome P4502C8 by montelukast. Drug Metab Dispos 33:413–8

38. Margolis JM, Obach RS (2003) Impact of nonspecific binding to microsomes and phospholipid on the inhibition of cytochrome

P4502D6: implications for relating in vitro inhibition data to in vivo drug interactions. Drug Metab Dispos 31:606–11

39. Jenkins KM, Angeles R, Quintos MT, Xu R, Kassel DB, Rourick RA (2004) Automated high throughput ADME assays for metabolic stability and cytochrome P450 inhibition profiling of combinatorial libraries. J Pharm Biomed Anal 34:989–1004

40. Hewavitharana AK (2011) Matrix matching in liquid chromatography-mass spectrometry with stable isotope labelled internal standards—is it necessary? J Chromatogr A 1218:359–61

41. Youdim KA, Saunders KC (2010) A review of LC-MS techniques and high-throughput approaches used to investigate drug metabolism by cytochrome P450s. J Chromatogr B Analyt Technol Biomed Life Sci 878:1326–36

Chapter 3

High-Throughput Mass Spectrometric Cytochrome P450 Inhibition Screening

Kheng B. Lim, Can C. Ozbal, and Daniel B. Kassel

Abstract

We describe here a high-throughput assay to support rapid evaluation of drug discovery compounds for possible drug–drug interaction (DDI). Each compound is evaluated for its DDI potential by incubating over a range of eight concentrations and against a panel of six cytochrome P450 (CYP) enzymes: 1A2, 2C8, 2C9, 2C19, 2D6, and 3A4. The method utilizes automated liquid handling for sample preparation, and online solid-phase extraction/tandem mass spectrometry (SPE/MS/MS) for sample analyses. The system is capable of generating two 96-well assay plates in 30 min, and completes the data acquisition and analysis of both plates in about 30 min. Many laboratories that perform the CYP inhibition screening automate only part of the processes leaving a throughput bottleneck within the workflow. The protocols described in this chapter are aimed to streamline the entire process from assay to data acquisition and processing by incorporating automation and utilizing high-precision instrument to maximize throughput and minimize bottleneck.

Key words CYP inhibition, Liquid handling, Automation, Probe substrate, Post-assay pooling, RapidFire, SPE/MS/MS, MRM, High throughput

1 Introduction

In vitro cytochrome P450 (CYP) enzyme inhibition assays have been used routinely to assess potential drug–drug interactions in drug discovery and development. Potent CYP inhibitors can alter the pharmacokinetics, efficacy, and safety profile of co-administered drugs and may cause undesired or even fatal adverse events. Many pharmaceutical companies have therefore implemented high-throughput in vitro CYP inhibition screening at very early stages of drug discovery, along with other absorption, distribution, metabolism, excretion, and toxicity (ADME/TOX) assays, to reduce the likelihood of drug–drug interaction and to assess if further in vivo studies are necessary.

Although more than 40 human CYP enzymes have been identified, five enzymes are responsible for about 95% of human drug metabolism: CYP1A2, CYP2C9, CYP2C19, CYP2D6, and

Ian R. Phillips et al. (eds.), *Cytochrome P450 Protocols*, Methods in Molecular Biology, vol. 987,
DOI 10.1007/978-1-62703-321-3_3, © Springer Science+Business Media New York 2013

CYP3A4 (1). A large number of xenobiotics have been studied and classified as substrates or inhibitors for the major human CYP isozymes (2, 3). The Food and Drug Administration (FDA) (4) and Pharmaceutical Research and Manufacturers of America (PhRMA) (5) have established general guidelines regarding studies for a selected set of CYP enzymes with preferred probe substrates and control inhibitors for use in CYP inhibition screening. Both the human liver microsomes and recombinant heterologously expressed CYP enzymes have been used for the CYP assays in various formats (6–9). The number of compounds in a typical early-stage discovery project is quite large, and, with multiple concentrations required for testing, the number of test samples can quickly become overwhelming. In order to improve the screening capacity of CYP inhibition assays for early-stage compounds, various high-throughput liquid chromatography-mass spectrometry (LC/MS) methods (10–14) and non-chromatographic-based MS methods (15, 16) have been developed for analyzing samples generated from the CYP assays.

Despite the many "flavors" of CYP assays being practiced in the field, our approach for CYP inhibition is to evaluate each test compound using a panel of six recombinantly expressed CYP enzymes (1A2, 2C8, 2C9, 2C19, 2D6, and 3A4) separately and over a range of eight test concentrations. Assay samples from different isozymes are then pooled for final analysis. We have implemented robotic liquid handling to increase throughput and improve precision of the CYP inhibition assay. We also utilize a commercially available rapid solid-phase extraction (SPE) instrument coupled to a triple-quadrupole MS system to analyze these samples quickly and accurately. Details for the workflow will be explained below.

The general workflow of the CYP inhibition assay discussed in this chapter is shown in Fig. 1. The assay utilizes probe substrates to measure the extent of CYP enzyme inhibition by test compounds. A compound with a weaker inhibitory property, for example, will allow a more robust conversion of probe substrates to products by the CYP enzymes. By monitoring the formation of products using a mass spectrometer, the IC_{50} value of a test compound against each CYP isozyme can be measured. The probe substrates and products selected for their respective CYP isozyme are listed in Table 1. Extensive efforts have been invested to measure the K_m values of these isozymes against their respective probe substrates in order to optimize the isozyme concentrations, probe substrate concentrations, reaction durations, and other assay parameters. Table 1 also lists the reference inhibitors and their respective IC_{50} values used as controls. Concurrent experiments utilizing reference inhibitors are critical to serve as references for the entire workflow. These reference inhibitors can be used for assay troubleshooting and ensuring consistency in the assay quality.

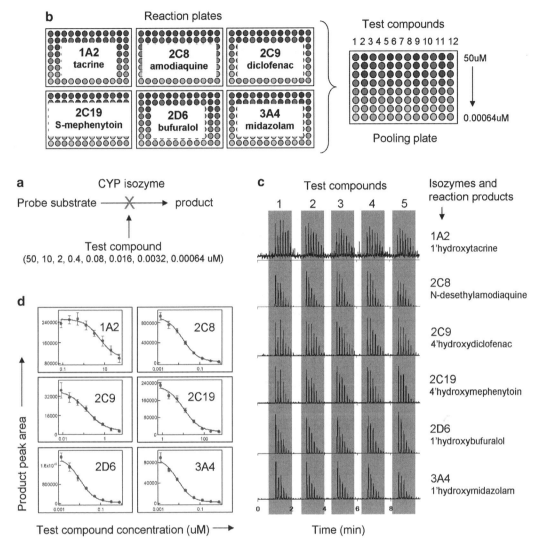

Fig. 1 An overview of the CYP inhibition assay. (**a**) Experimental design. (**b**) Implementation of the automated assay using a programmed liquid handler. Each test compound was evaluated using a panel of six recombinantly expressed CYP enzymes (1A2, 2C8, 2C9, 2C19, 2D6, and 3A4) and over a range of eight test concentrations separately. Assay samples from different isozymes are then pooled for analysis. (**c**) The analysis of the pooled mixture by a rapid SPE instrument coupled to a triple-quadrupole MS acquiring in MRM mode targeting only the six products. The raw data from five test compounds are shown. (**d**) Eight-concentration dose–response curves allowing IC$_{50}$ for each of the six CYP isozymes to be determined. The dose–response curves for one test compound are shown

We have implemented the CYP inhibition assay in the past at medium throughput using robotic liquid handling and 2-min LC/MS (17). Recently, we have made substantial improvement to the assay throughput and MS analysis by implementing more efficient liquid handling and a streamlined MS data acquisition workflow (18). The protocols described here are based on these improvements.

Table 1
Lists of isozyme-specific probe substrates and reference inhibitors for use in CYP inhibition assay

CYP	Probe substrate	Product	Precursor m/z	Fragment m/z	Reference inhibitor	IC_{50} (µM)
1A2	Tacrine	1'-Hydroxy-tacrine	215	182	Furafylline	5.1 ± 1.0
2C8	Amodiaquine	N-Desethyl-amodiaquine	328	283	Montelukast	0.015 ± 0.002
2C9	Diclofenac	4'-Hydroxy-diclofenac	312	231	Sulfaphenazole	0.2 ± 0.04
2C19	S-Mephenytoin	4'-Hydroxy-mephenytoin	235	150	Tranylcypromine	12 ± 2
2D6	Bufuralol	1'-Hydroxy-bufuralol	278	186	Quinidine	0.01 ± 0.002
3A4	Midazolam	1'-Hydroxy-midazolam	342	203	Ketoconazole	0.013 ± 0.004

The respective products monitored by MS, along with their precursor and fragment m/z, are also listed. The precursor and fragment m/z information is useful to create MS methods based on multiple-reactant monitoring (MRM) scans (also see Table 5 for more MRM parameters). The listed IC_{50} values from the reference inhibitors are typical and can be used as a guide during assay development. The user should keep track of these IC_{50} data, obtained from their daily or weekly assays, for the purpose of monitoring quality and consistency

The use of a robotic liquid handler enables two 96-well plates (24 test compounds at 8 concentrations) to be assayed in about 30 min. The use of online SPE coupled to a triple-quadrupole MS enables an analysis rate of about 8 s per sample (or 30-min total analysis time for two 96-well plates). The protocols for assays, data acquisition, and analysis will be described in detail below and the methods described can also be adapted to any hardware, software system, or manual approach.

2 Materials

2.1 CYP Inhibition Assay

Always wear personal protective gear, such as gloves and safety glasses, while working in the lab. Prepare and store all reagents at room temperature unless indicated otherwise. Use Milli-Q water (Millipore, Billerica, MA) for all solution preparations. The stock concentrations of enzymes/reagents for use in each CYP isozyme assay are summarized in Table 2.

1. Recombinant CYP enzymes: 1A2 (BD Biosciences, 456203), 2C8-b5 (BD Biosciences, 456252), 2C9 (BD Biosciences, 456218), 2C19 (BD Biosciences, 456219), 2D6 (BD Biosciences, 456217), and 3A4 (BD Biosciences, 456207). Store at −20°C (see Note 1).

Table 2
List of stock reagents and their concentrations

CYP isozyme	1A2	2C8	2C9	2C19	2D6	3A4
CYP450 enzyme (µM)	1	1	2	1	1	2
Probe substrate (µM)	30 (tacrine)	4 (amodiaquine)	25 (diclofenac)	500 (S-mephenytoin)	50 (bufuralol)	20 (midazolam)
Reference inhibitor (mM)	10 (furafylline)	1 (montelukast)	1 (sulfaphenazole)	100 (tranylcypromine)	1 (quinidine)	1 (ketoconazole)
Phosphate buffer	1 M stock					
NADPH-A (25 mM NADP+ and 66 mM Glc-6-PO$_4$)	20×					
NADPH-B (40 U/mL Glc-6-PO$_4$ dehydrogenase)	100×					

These stocks will be further diluted (see Table 3) for use in CYP inhibition assays

Table 3
Final reagent concentrations in CYP inhibition assays

CYP isozyme	1A2	2C8	2C9	2C19	2D6	3A4
CYP450 enzyme (nM)	12.5	2.5	12.5	12.5	0.6	25
Probe substrate (µM)	30 (tacrine)	0.4 (amodiaquine)	2.5 (diclofenac)	50 (S-mephenytoin)	5 (bufuralol)	2 (midazolam)
Reference inhibitor (µM) (highest concentration)	50 (furafylline)	5 (montelukast)	5 (sulfaphenazole)	500 (tranylcypromine)	5 (quinidine)	5 (ketoconazole)
Phosphate buffer (M)	0.1	0.1	0.1	0.1	0.1	0.1
NADPH-A (25 mM $NADP^+$ and 66 mM Glc-6-PO_4)	0.5×	0.5×	0.5×	0.5×	0.5×	0.5×
NADPH-B (40 U/mL Glc-6-PO_4 dehydrogenase)	0.5×	0.5×	0.5×	0.5×	0.5×	0.5×

2. NADPH-A (BD Biosciences, 451220) and NADPH-B (BD Biosciences, 451200). Store at –20°C.

3. Probe substrate tacrine, 30 µM: weigh 7.0 mg of tacrine hydrochloride (Sigma, A3773, FW 234.7) and dissolve in 1 mL of methanol. Add 9 mL of water to make 10 mL of 3-mM tacrine in 10% methanol and 90% water. Further dilute 0.1 mL of the 3-mM tacrine into 9.9 mL of water to make 10 mL of 30-µM stock solution. Store at 4°C.

4. Probe substrate amodiaquine, 4 µM: weigh 3.7 mg of amodiaquine (Sigma, A2799, FW 464.8) and dissolve in 20 mL of water to make a 400 µM solution. Further dilute 0.1 mL of the 400 µM amodiaquine into 9.9 mL of water to make 10 mL of 4 µM stock solution. Store at 4°C.

5. Probe substrate diclofenac, 25 µM: weigh 5.2 mg of diclofenac (Sigma, D6899, FW 318.1) and dissolve in 16 mL of water to make a 1 mM solution. Further dilute 0.25 mL of the 1 mM diclofenac into 9.75 mL of water to make a 10-mL 25 µM stock solution. Store at 4°C.

6. Probe substrate S-mephenytoin, 500 µM: weigh 10 mg of S-mephenytoin (Biomol, Plymouth Meeting, PA, J120, FW 218.2) and dissolve in 1.8 mL of acetonitrile. Add 7.4 mL of water to make 9.2 mL of 5 mM S-mephenytoin in 20% acetonitrile and 80% water. Further dilute 1 mL of the 5 mM S-mephenytoin into 9 mL of water to make a 10-mL 500 µM stock solution. Store at 4°C (see Note 2).

7. Probe substrate bufuralol, 50 µM: weigh 5 mg of bufuralol (BD Bioscience, Woburn, MA, UC168, FW 298) and dissolve in 16 mL of water to make a 1 mM solution. Further dilute 0.5 mL of the 1 mM bufuralol into 9.5 mL of water to make a 10-mL 50 µM stock solution. Store at 4°C.

8. Probe substrate midazolam, 20 µM: break the 1-mL glass ampule containing 1 mg/mL midazolam (AllTech, Deerfield, IL, #015713, FW 326) and mix with 14 mL of water to make a 15-mL 200 µM solution. Further dilute 1 mL of the 200 µM midazolam into 9 mL of water to make a 10-mL 20 µM stock solution. Store at 4°C.

9. Reference inhibitor furafylline, 60 mM: weigh 156.0 mg of furafylline (Sigma F124, FW 260.3) and dissolve in 10 mL of dimethylsulfoxide (DMSO).

10. Reference inhibitor montelukast, 6 mM: weigh 36.6 mg of montelukast (Sequoia Research Product, Pangbourne, UK, FW 608.2) and dissolve in 10 mL of DMSO.

11. Reference inhibitor sulfaphenazole, 6 mM: weigh 18.6 mg of sulfaphenazole (Sigma S0758, FW 314.4) and dissolve in 10 mL of DMSO.

12. Reference inhibitor tranylcypromine, 600 mM: weigh 1018.2 mg of tranylcypromine (Sigma P8511, FW 169.7) and dissolve in 10 mL of DMSO.

13. Reference inhibitor quinidine, 6 mM: weigh 22.8 mg of quinidine (Sigma Q0750, FW 378.9) and dissolve in 10 mL of DMSO.

14. Reference inhibitor ketoconazole, 6 mM: weigh 31.8 mg of ketoconazole (Sigma K1003, FW 531.4) and dissolve in 10 mL of DMSO.

15. Stock mixture of reference inhibitors: mix 1 mL from each of the six reference inhibitor stocks to prepare a 6-mL mixture of 10 mM furafylline, 1 mM montelukast, 1 mM sulfaphenazole, 100-mM tranylcypromine, 1 mM quinidine, and 1 mM ketoconazole (see Note 3).

16. Potassium phosphate dibasic, 1 M: weigh 174.2 g of K_2HPO_4 and transfer to a 1-L cylinder. Add 500 mL of water and mix to dissolve. Make up to 1 L with water.

17. Potassium phosphate monobasic, 1 M: weigh 136.1 g of KH_2PO_4 and transfer to a 1-L cylinder, add 500 mL of water, and mix to dissolve. Make up to 1 L with water.

18. Phosphate buffer 1 M, pH 7.4: Mix at 25°C 80.2 mL of dibasic and 19.8 mL of monobasic potassium phosphate.

19. Trichloroacetic acid (TCA), 1 M: weigh 163.4 g of TCA and transfer to a 1-L cylinder. Add 500 mL of water and mix to dissolve. Make up to 1 L with water.

20. Tecan Evo 200 robotic liquid handler (Tecan Group Ltd., Mannedorf, Switzerland) with the following modules and options installed: liquid-handling arm (LiHA), robotic manipulator arm (ROMA), multichannel arm 96 (MCA-96), monitored-incubator option (MIO), and 6-slot module with shaking capability.

21. MCA-96 disposable tips (DITI): 200 μL.

22. Assay plate: 96-well "V" bottom 500 μL each well (Thomson Instrument Company, 9356045).

23. Enzyme-reservoir plate: low-profile reagent reservoir with dimples (Thomson Instrument Company, 981915).

24. Reagent reservoir plate: deep-profile reagent reservoir with dimples (Thomson Instrument Company, 981900).

25. Plate seal: 96-well solid cap mat (Thomson Instrument Company, 359747B).

26. 96-well filter plate: 0.45-μm multiscreen 96-well filter plate (Millipore, MAHVN4550).

High-Throughput Mass Spectrometric Cytochrome P450 Inhibition Screening 33

2.2 Online SPE/Mass Spectrometry Analysis

1. Mass spectrometer: API4000 triple-quadrupole mass spectrometer (AB Sciex) with ESI source.

2. RapidFire system: Original system (equivalent to current model RapidFire 200) (Agilent).

3. Aqueous solvent for pump 1 = water + 0.1% trifluoroacetic acid: pipette 1.0 mL of trifluoroacetic acid (Thermo Scientific, 28904) into 1 L of water (Optima LC/MS grade, Fisher Scientific).

4. Organic solvent for pumps 2 and 3 = acetonitrile + 0.1% trifluoroacetic acid: pipette 1.0 mL of trifluoroacetic acid (Thermo Scientific, 28904) into 1 L of acetonitrile (Optima grade, Fisher Scientific).

5. Wash solvent for peristaltic pump = 50% acetonitrile in water: measure 500 mL of acetonitrile and mix thoroughly with 500 mL of water.

2.3 Data Analysis

1. Analyst 1.5.1 (AB Sciex).

2. RapidFire controlling software: HTMS 1.11 (Agilent).

3. RapidFire data-integration software: version 2.0.1 (Agilent).

4. Microsoft Office Excel 2003 SP3 (Microsoft).

5. XLFit: curve-fitting software for Excel version 5.2.0.0 (IDBS, Guildford, UK).

3 Methods

The incubation conditions are based on established in-house assay protocols that were validated with optimized kinetic parameters; the K_m of each of the six CYP isozymes was determined and probe substrate concentrations at or below K_m were used for the assays. The assay protocols also ensured a linear product formation relative to enzyme concentration, with 20% or less turnover of probe substrates to products. The assay was automated using a Tecan robotic liquid handler equipped with a multichannel arm and was programmed to provide IC_{50} data for six CYP isozymes for up to 12 compounds at 8 concentrations per assay on 96-well plates.

To simplify Tecan automation and facilitate bulk pipetting, the reagents were premixed into two separate mixtures before the start of assay:

- Reagent Mix-1 = CYP isozyme + probe substrate + buffer + water

- Reagent Mix-2 = NADPH-A + NADPH-B + water

Reagent Mix-1 and 2 are prepared manually ahead of the assay. During the assay, 66 μL of Reagent Mix-1 is dispensed into each well of a 96-well plate, followed by 10 μL of test compound. Finally 24 μL of Reagent Mix-2 is added to start the reaction (see Note 4).

34 Kheng B. Lim et al.

The recipes listed in Table 4 are designed to prepare enough reagents to perform assays for 6 isozymes × 24 compounds × 8 concentrations. Reference inhibitors are placed in each assay plate and co-assayed with the test compounds for quality control.

Before the assay, all 12 assay plates (AP-xxx-#), 6 enzyme/substrate plates (ESP-xxx), 2 compound plates (CP-#), 2 pooling plates (PP-#), and 2 deep-well reservoir plates for TCA (TCA) and NADPH (NADPH) must be labeled and laid out on the deck of Tecan as illustrated in Fig. 2.

3.1 Preparation of Test Compounds and Standard Inhibitors

1. Starting with a concentration of 5 μM, prepare a set of seven consecutive fivefold serial dilutions of inhibitors, in acetonitrile/DMSO (1:1, v/v), to produce a total of eight different concentrations of each inhibitor (Fig. 3).

2. On a 96-well dilution plate ("DP-1"), pipette 80 μL of 50/50 acetonitrile/DMSO into all wells between and including rows B through H.

3. Pipette 20 μL of 10 mM test compounds in DMSO into wells A1 through A11.

4. Pipette 20 μL of the reference inhibitor mixture, prepared in Subheading 2.1, item 15, into well A12.

5. Using a 12-channel pipette, add 20 μL of acetonitrile to wells A1 through A12 and mix thoroughly. Then, with the same 12-channel pipette, transfer 20 μL from row A and place into row B and mix thoroughly.

6. Repeat the serial dilution process (taking 20 μL from row B and mix into C, then C into D and so on) until row H.

7. On a separate 96-well dilution plate ("DP-2") fill in all wells with 180 μL of water.

8. Perform a plate-stamping operation, transferring 20 μL from each well in DP-1 into DP-2, and mix thoroughly (see Note 5).

9. Save the DP-2 for use in the CYP inhibition assay described below. (This DP-2 will be called CP-1 starting in Subheading 3.5).

10. Repeat the dilution procedure for the next set of 11 test compounds and one reference inhibitor mixture so that two plates (24 compounds × 8 concentrations) can be assayed together.

3.2 Preparation of Reagent Mix-1

1. Label six 50-mL Falcon tubes as "1A2," "2C8," "2C9," "2C19," "2D6," and "3A4."

2. Prepare the reagents in the 50-mL Falcon tubes according to Table 4 for each isozyme (see Note 6).

3. Mix the reagents by capping the Falcon tubes and inverting a few times (see Note 7).

4. Keep all Falcon tubes on ice until immediately before the assay.

Table 4

Reagent volumes required to perform CYP inhibition assays for 6 isozymes × 24 compounds × 8 concentrations

Reagent Mix-1						
66 µL used for each reaction	1A2	2C8	2C9	2C19	2D6	3A4
Water	10,024	10,248	10,164	10,024	10,290	10,024
Buffer	2,240	2,240	2,240	2,240	2,240	2,240
Substrate	2,240 (tacrine)	2,240 (amodiaquine)	2,240 (diclofenac)	2,240 (S-mephenytoin)	2,240 (bufuralol)	2,240 (midazolam)
Enzyme	280	56	140	280	14	280
Reagent Mix-2						
24 µL used for each reaction						For all isozymes
Water						31,424
NADPH-A (25 mM $NADP^+$ and 66 mM Glc-6-PO_4)						3,360
NADPH-B (40 U/mL Glc-6-PO_4 dehydrogenase)						672

All volumes are measured in microliters

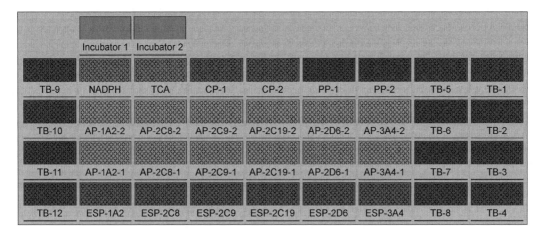

Fig. 2 Layout of reaction plates, reagent reservoirs, and pipette tips on the deck of Tecan. *ESP* enzyme/substrate plates, *AP* assay plates, *CP* compound plates, *PP* pooling plates, *TB* tip boxes

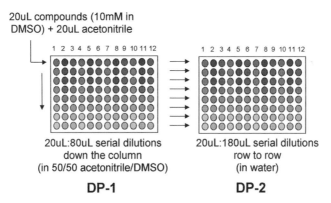

Fig. 3 Dilution scheme of test compounds. The stocks at 10 mM in DMSO were mixed with an equal volume of acetonitrile in row A1-A12 in plate DP-1. A 5× serial dilution was performed using a 12-channel pipette, diluting the test compound using 50/50 acetonitrile/DMSO from rows A through H. Finally, the test compounds were diluted 10× in water by stamping plate DP-1 to plate DP-2. Test compounds in DP-2 were used for the CYP inhibition assay

5. Immediately before the assay mix the reagents again (see Note 8), by gently inverting the tubes, and place the reagents in their respective enzyme-/substrate-reservoir plates labeled as "ESP-xxx" (see Note 9), and as shown in Fig. 2.

3.3 Preparation of Reagent Mix-2

1. Label one 50-mL Falcon tube as "NADPH."
2. Prepare the reagents in the 50-mL Falcon tube according to Table 4 (see Note 6).
3. Mix the reagents by capping the Falcon tube and inverting a few times (see Note 7).
4. Keep the Falcon tube on ice until immediately before the assay.

High-Throughput Mass Spectrometric Cytochrome P450 Inhibition Screening   37

5. Immediately before the assay mix the reagent again (see Note 8), by gently inverting the tube, and place the reagent in the NADPH-reservoir plate labeled as "NADPH," and as shown in Fig. 2.

3.4 Preparation of TCA Quench Solution

1. Mix 5 mL of 1 M TCA (prepared in Subheading 2.1, item 19) and 20 mL of water in a deep-well reservoir plate to make 25 mL of 200 mM TCA quench solution.

2. Place the reservoir in the location labeled as "TCA," as shown in Fig. 2.

3.5 Automated CYP Inhibition Assay

1. Place the two CP-2 plates prepared in Subheading 3.1, step 10 (now called compound plates CP-#), on the Tecan deck as shown in Fig. 2.

2. Place the assay plates (AP-xxx-#) (see Note 10) on the Tecan deck as shown in Fig. 2.

3. Place the filter plates (as described in Subheading 2.1, item 26) in the locations for pooling plates (PP-#) (see Note 11) on the Tecan deck as shown in Fig. 2.

4. Make sure the tip boxes containing DITI (TB-#) are loaded (see Note 12) on deck as shown in Fig. 2.

5. Check the Tecan calibration (see Note 13).

6. Start the Tecan program (see Note 14) to run the assay. It will take about 30 min for the entire run. The sequence of the assay as performed by the Tecan is outlined in Subheading 3.6.

3.6 Sequence of Automated Assay

1. MCA arm move to pick up tips from TB-1.

2. Aspirate reagents from ESP-1A2 and sequentially dispense 66 μL into AP-1A2-1 and AP-1A2-2.

3. Discard the used tips.

4. Repeat steps 2–3 for the enzymes/substrates of 5 other isozymes, using TB-2, TB-3, TB-4, TB-5, and TB-6 (see Note 15).

5. MCA arm move to pick up tips from TB-7.

6. Aspirate the test compounds from CP-1 and sequentially dispense 10 μL into AP-###-1 (see Note 16).

7. Discard the used tips.

8. MCA arm move to pick up tips from TB-8.

9. Aspirate the test compounds from CP-2 and sequentially dispense 10 μL into AP-###-2 (see Note 17).

10. Discard the used tips (see Note 18).

11. MCA arm move to pick up tips from TB-9.

12. Aspirate the NADPH from "NADPH" and sequentially dispense 24 μL into AP-###-1.

13. Discard the used tips.

14. Using ROMA move the group-1 assay plates (AP-xxx-1) into the incubator (MIO), which is set at 37°C. Start Tecan timer #1.

15. MCA arm move to pick up tips from TB-10.

16. Aspirate the NADPH reagent from "NADPH" and sequentially dispense 24 µL into AP-###-2.

17. Discard the used tips.

18. Using ROMA move the group-2 assay plates (AP-xxx-2) into the incubator (MIO), which is set at 37°C. Start Tecan timer #2 (see Note 19).

19. After 10 min has elapsed on timer #1 use ROMA to pull assay plates AP-1A2-1, AP-2C8-1, AP-2C9-1, AP-2D6-1, and AP-3A4-1 from the incubator and return to their original locations, as shown in Fig. 2.

20. MCA arm move to pick up tips from TB-11.

21. Aspirate the TCA reagent from "TCA" and sequentially dispense 100 µL into assay plates AP-1A2-1, AP-2C8-1, AP-2C9-1, AP-2D6-1, and AP-3A4-1.

22. Place the tips back into TB-11 (see Note 20).

23. After 10 min has elapsed on timer #2, use ROMA to pull assay plates AP-1A2-2, AP-2C8-2, AP-2C9-2, AP-2D6-2, and AP-3A4-2 from the incubator and return to their original locations, as shown in Fig. 2.

24. MCA arm move to pick up tips from TB-11.

25. Aspirate the TCA reagent from "TCA" and sequentially dispense 100 µL into assay plates AP-1A2-2, AP-2C8-2, AP-2C9-2, AP-2D6-2, and AP-3A4-2.

26. Place the tips back into TB-11 (see Note 20).

27. After 20 min has elapsed on timer #1, use ROMA to pull assay plate AP-2C19-1 from the incubator and return to its original location, as shown in Fig. 2.

28. MCA arm move to pick up tips from TB-11.

29. Aspirate the TCA reagent from "TCA" and dispense 100 µL into assay plate AP-2C19-1.

30. Place the tips back into TB-11 (see Note 20).

31. After 20 min has elapsed on timer #2, use ROMA to pull assay plate AP-2C19-2 from the incubator and return to its original location, as shown in Fig. 2.

32. MCA arm move to pick up tips from TB-11.

33. Aspirate the TCA reagent from "TCA" and dispense 100 µL into assay plate AP-2C19-2.

34. Discard the used tips.

High-Throughput Mass Spectrometric Cytochrome P450 Inhibition Screening 39

35. MCA arm move to pick up tips from TB-12.

36. Sequentially aspirate 60 µL from AP-1A2-1, 40 µL from AP-2C8-1, and 40 µL from AP-2C9-1, with air gaps in between aspirations, and then dispense all content into pooling plate 1 (PP-1) (see Note 21).

37. Using the same tips from TB-12, sequentially aspirate 40 µL from AP-2C19-1, 40 µL from AP-2D6-1, and 40 µL from AP-3A4-1, with air gaps in between aspirations, and then dispense all content into pooling plate 1 (PP-1).

38. Using the same tips from TB-12, sequentially aspirate 60 µL from AP-1A2-2, 40 µL from AP-2C8-2, and 40 µL from AP-2C9-2, with air gaps in between aspirations, and then dispense all content into pooling plate 2 (PP-2) (see Note 21).

39. Using the same tips from TB-12, sequentially aspirate 40 µL from AP-2C19-2, 40 µL from AP-2D6-2, and 40 µL from AP-3A4-2, with air gaps in between aspirations, and then dispense all content into pooling plate 2 (PP-2).

40. Centrifuge PP-1 and PP-2 at $1,000 \times g$ for 4 min to filter the contents into two separate 96-well receiving plates (MS-1 and MS-2).

41. Place seals (described in Subheading 2.1, item 25) to cover MS-1 and MS-2. These two final plates will be analyzed starting on Subheading 3.7, step 12, described below.

3.7 Data Acquisition

1. Create a MS acquisition method (see Note 22) in Analyst under the manual tune page using the following parameters:
   - Acquisition duration = 120 min (see Note 23)
   - Scan type = MRM
   - Polarity = positive
   - CUR = 35
   - IS = 4,500
   - TEM = 600
   - GS1 = 50
   - GS2 = 60
   - Ihe = ON
   - CAD = 8
   - DP = 70
   - EP = 10
   - CXP = 10
   - MRM transitions = see Table 5.

2. Save the method with file name "CYP Assay.dam."

Table 5
MRM parameters for the simultaneous detection of products from the CYP inhibition assay

CYP	Q1 mass (amu)	Q3 mass (amu)	Dwell (ms)	DP	CE
1A2	215	182	50	60	40
2C8	328	283	50	70	40
2C9	312	231	50	70	30
2C19	235	150	50	70	30
2D6	278	159	50	70	30
3A4	342	203	50	70	45

These parameters are specific to the API4000QT mass spectrometer

3. From the RapidFire, install a Phenyl cartridge and a 10-μL sample loop. Connect the output tubing to the ESI source of the mass spectrometer.

4. From the RapidFire instrument control software, under the "HTMS" tab, use the following parameters:
   - Current Row = a
   - Current Column = 1
   - Sip Height = 2 mm
   - Wash Time = 500 ms
   - Pre Wash = 2
   - Post Wash = 2
   - Wash Between Sips = checked
   - Current Plate Name = Plate1
   - Start New Set of Plates = checked
   - 96 Well Plate = selected

5. Press the update button (orange upward pointing arrow) to save the settings.

6. From the RapidFire instrument control software, under the "MassSpec" tab, use the following parameters:
   - State 1 (sample aspirate time) = 200 ms
   - State 2 (sample wash time) = 2,000 ms
   - State 3 (sample elute time) = 5,000 ms
   - State 4 (column re-equilibrate time) = 5,000 ms

7. Turn on pump 1 and set flow to 1.5 mL/min.

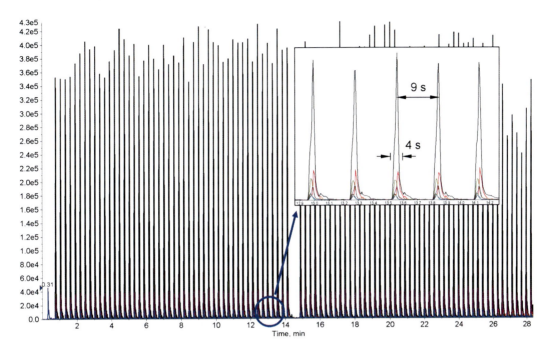

Fig. 4 Representative MRM data from two 96-well plates of pooled samples from CYP inhibition assay. The data acquisition was accomplished using a RapidFire system coupled to an API4000QT mass spectrometer. The sampling cycle is 9 s per well and the peaks detected by the mass spectrometer are approximately 4 s wide at baseline. It took a little over 14 min to finish analyzing samples from one 96-well plate with the simultaneous quantitation of 6 CYP products

8. Turn on pump 2 and set flow to 1.5 mL/min.
9. Turn on pump 3 and set flow to 0.5 mL/min.
10. Turn on wash pump and set the flow dial to "10."
11. Turn on the vacuum line to the sipper.
12. Place the analysis plate (MS-1), from Subheading 3.6, step 41 above, onto the RapidFire sample stage.
13. Click on the "file acquisition" button in Analyst manual tune page and type in MS data file name "CYP MS Data.wiff."
14. Click "OK" to start the data acquisition on the mass spectrometer.
15. Click "start" (the button with green triangle symbol) on the RapidFire instrument control software to start sample analysis on the RapidFire (see Note 24).
16. When the analysis for the first plate is finished, swap in the second analysis plate (MS-2) and click "continue" to start the data acquisition for the second plate (see Note 25).
17. Repeat step 16 for additional plates.
18. When all the plates have been analyzed, stop the MS data acquisition by clicking the "stop" button in Analyst manual

42   Kheng B. Lim et al.

tune page. Also click the "cancel" button on the RapidFire instrument control software to stop the sample analysis on the RapidFire.

19. Turn off pumps 1, 2, and 3 as well as the peristaltic pump that provides solvent to the wash troughs. Turn off the vacuum line and place the mass spectrometer in standby mode.

3.8 Data Processing

Please refer to the user manual of the RapidFire software package for more detailed information on data processing. Data extraction from individual MRM channels that contain quantitative inhibition information for each CYP isozyme is accomplished using Analyst data-processing software. Peak integrations for these MRM data are performed using the RapidFire proprietary integration software. Details for data processing are outlined below:

1. Open the MS data file "CYP MS Data.wiff" using Analyst.

2. Click on the "XIC" button on Analyst and select all six MRM channels to display.

3. Right click on each of the MRM trace and select "Save to Text File...." For the 1A2 MRM channel, name the text file as "1A2.txt."

4. Repeat step 3 for the remaining five MRM channels, export the data, and name the text file accordingly.

5. Locate the RapidFire data files from the RapidFire computer (see Note 26).

6. Place all the RapidFire data files located in step 5 and the text files exported from Analyst in step 4 in a folder named "RapidFire Data."

7. Start the RapidFire integrator software. Click on "main" and select "Load Data." Select the folder "RapidFire Data" and click "OK" (see Note 27).

8. To begin data analysis, select the MRM trace from 1A2.txt. Zoom onto one of the known landmark peaks by click-dragging the selected region (see Note 28).

9. Click "Assay" from the menu bar and select "Set Peak Width." Click and drag a box around the landmark peak to define the peak width (see Note 29).

10. Click "Assay" from the menu bar and select "Set Peak Number." Type in the dialog window the location of the sample (plate #, row #, and column #) that corresponds to the selected landmark peak (see Note 30).

11. Click "Assay" and select "Process." The integrated peaks will be outlined in blue. Manually inspect the results to make sure all peaks are integrated (see Note 31).

High-Throughput Mass Spectrometric Cytochrome P450 Inhibition Screening 43

12. Click "Assay" and select "Generate Peak File." Click "No" to the next dialog window so that the default names will be used for the comma-delineated result file. The result file can be viewed using Microsoft Excel (see Note 32).

3.9 Computing IC$_{50}$

There are many approaches to calculating IC$_{50}$ values. It is recommended to incorporate programming into the data-processing workflow so that the automation aspect not only increases throughput but also reduces errors typically introduced during manual data handling. The steps outlined below illustrate how the IC$_{50}$ value can be computed from data obtained from Subheading 3.8, step 12, above, using Excel and XLfit software. The same steps can be incorporated into simple Excel macros so that a large set of data can be processed automatically. The IC$_{50}$ results from assays using reference inhibitors should be similar to those listed in Table 1; the absolute IC$_{50}$ values will vary slightly from different laboratories, but should be consistent within the same laboratory. The users are encouraged to keep a record of the IC$_{50}$ data from these control assays as a mean to track the consistency of their daily or weekly CYP inhibition assays.

1. Open one of the peak-result files, generated in Subheading 3.8, step 12 above, using Microsoft Excel.

2. In cells O3 to O10, type in the compound concentrations 50, 10, 2, 0.4, 0.08, 0.016, 0.0032, and 0.00064, as shown in Fig. 5 insert (see Note 33).

3. In cells P3 to P10, calculate the % activity, using data from cells M3 to M10. Type the formula "=M3/M$10*100" in cell P3. Copy P3 and paste the formula into cells P4 to P10, as shown in Fig. 5 insert (see Note 34).

4. Highlight cells O3 to P10 by clicking on cell O3 and dragging the mouse cursor to cell P10.

5. Open the XLfit "Fit Designer" by clicking "tools" and select "XLfit" and then "Fit Designer."

6. Select "data" tab on the Fit Designer. The X and Y values should already be populated with values from steps 2 and 3 (see Fig. 6a).

7. For the "Return Value Details" select "X value" and type in "50" in the box labeled "Where Y equals" (see Fig. 6a).

8. Check the box label "Log X" (see Fig. 6a).

9. Select cell "Q3" for the "Fit Cell", and cell "Q4" for the "Chart Cell" (see Fig. 6a).

10. Select "Model" tab on the Fit Designer, and click on fit model 205, under "Dose Response One Site." Click the "Apply" button (see Fig. 6b) (see Note 35).

11. Repeat steps 2–10 for the next data set.

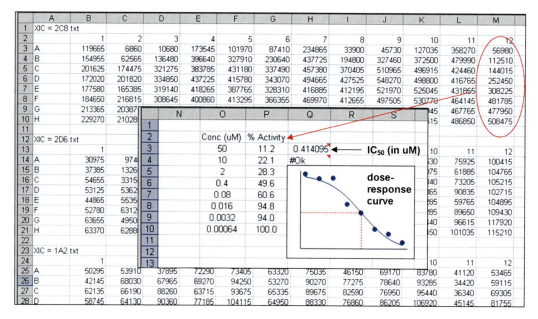

Fig. 5 Example illustrating how IC$_{50}$ value can be computed using Excel and XLfit. In this example data set obtained from the CYP inhibition assay of test compound #12 against CYP2C8 (*data circled in red*) was used. The peak-area data was first converted to % activity by normalizing all data against the data obtained at lowest test compound concentration. Using XLfit (see Fig. 6) both the IC$_{50}$ value and a fitted dose–response curve can be generated on the Excel sheet. The IC$_{50}$ value computed for this example is 0.41 µM

4 Notes

1. CYP 2C8 preparation containing cytochrome b5 is used to increase the enzyme activity and reduce the incubation time. All other CYP enzyme preparations are non-b5 containing, since those enzymes already exhibit high activity against their respective probe substrates.

2. S-mephenytoin has limited solubility in water. Dissolve it in 100% acetonitrile first then add water to make up to a final solution of 20% acetonitrile.

3. The concentration ranges and dilution factors for the reference inhibitors are customized so as to bracket their known IC$_{50}$ values. The mixture format simplifies the incorporation of controls into the assay. Each isozyme will still be assayed using its respective inhibitor. Based on extensive testing, we have concluded that the measured IC$_{50}$ value is unaffected by the presence of other reference inhibitors.

4. The concentrations for each assay component are listed in Table 3. The final volume for the individual reaction consisting of the following components is 100 µL and contains 0.5% DMSO and 0.5% acetonitrile. Each individual reaction requires 10 µL probe substrate + 10 µL buffer + 10 µL test compound/

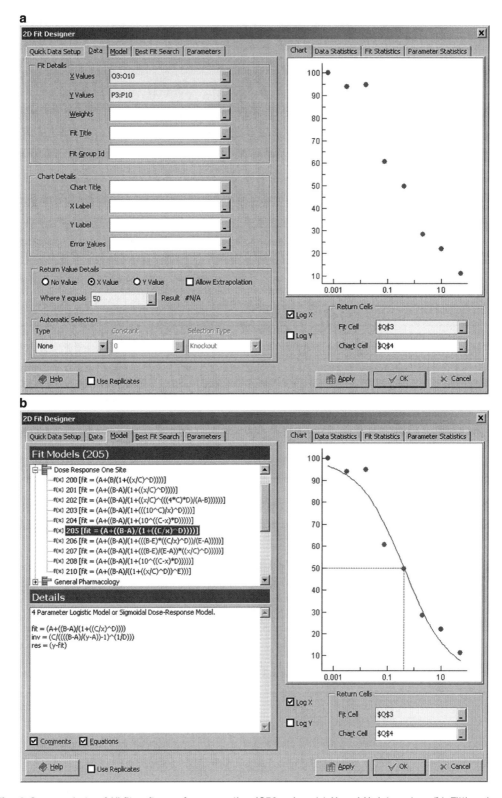

Fig. 6 Screen shots of XLfit software for computing IC50 value. (**a**) X and Y data setup. (**b**) Fitting dose–response model to the data

reference inhibitor + 2.5 μL NADPH-A + 0.5 μL NADPH-B + CYP isozyme + water. Depending on the CYP isozyme, the volume of water used is adjusted accordingly so that the final volume is 100 μL:

- 66 μL of Reagent Mix-1 (CYP isozyme + probe substrate + buffer + water)
- 10 μL of test compound or reference inhibitor
- 24 μL of Reagent Mix-2 (NADPH-A + NADPH-B + water)

5. The plate-stamping operation can be performed in several ways: (a) in an automated manner, using the Tecan equipped with MCA-96, or (b) manually, using an 8- or 12-channel pipette. In all cases, pipetting the solution up/down several times is sufficient to ensure thorough mixing.

6. After thawing the P450 enzymes and NADPH solutions according to the manufacturer's guidelines, keep the enzymes and all reagents on ice until ready for use. The reagents are prepared in slight excess to ensure sufficient volume for aspiration by the liquid handler. The leftover reagents can be combined with fresh ones for the next assay. Discard any leftover reagents after the entire CYP inhibition assay is completed.

7. Mix by gently inverting the Falcon tubes. Do not shake, as the formation of foam/bubble will affect the aspirate precision of the Tecan.

8. The protein or non-soluble components may have settled during the waiting period. Gently mix the reagents again before pouring them into their respective reservoirs.

9. There are six enzyme/substrate plates "ESP-xxx," with "xxx" indicating the respective CYP isozymes such as 1A2 and 2C8.

10. There are 12 assay plates (AP-xxx-#), with "xxx" indicating the respective CYP isozymes such as 1A2 and 2C8; "#" designation of "1" or "2" is used to track assays performed using test compounds from compound plate 1 or 2 (CP-1 or CP-2), respectively.

11. There are two pooling plates (PP-#); "#"designation of "1" or "2" is used to track assays performed using test compounds from compound plate 1 or 2 (CP-1 or CP-2), respectively.

12. There are twelve 200-μL disposable tip boxes (TB-#), with "#" ranging from 1 to 12.

13. It is a good practice to check the robotic calibration of the Tecan before starting the assay. Ensure that the calibration for the MCA-96 arm is correct and that it can pick up the DITI without crashing. Ensure that the ROMA arm can pick up a plate and insert into the MIO.

14. The Tecan program should be written by someone knowledgeable in Tecan programming language. The procedure outlined

in Subheading 3.6, steps 1–41, can be used as a guide to write the program. Alternatively, the procedure can be adapted to any liquid handler in your laboratory or even performed manually, using multichannel pipetting, if the steps outlined are followed.

15. At this point all enzymes and substrates should now be delivered to each well of the 12 assay plates (AP-xxx-#). There will be some leftover enzyme/substrate in the ESP-xxx, which can be mixed with fresh enzyme/substrate for the next run.

16. Test compounds from compound plate 1 (CP-1) will be dispensed to assay plates AP-1A2-1, AP-2C8-1, AP-2C9-1, AP-2C19-1, AP-2D6-1, and AP-3A4-1.

17. Test compounds from compound plate 2 (CP-2) will be dispensed to assay plates AP-1A2-2, AP-2C8-2, AP-2C9-2, AP-2C19-2, AP-2D6-2, and AP-3A4-2.

18. At this point the test compounds from the two compound plates have been added to their respective enzyme/substrate mix. The CYP isozymes are still unable to initiate the conversions of substrates to products, since the NADPH is still absent.

19. At this point the conversions of probe substrates to products by the respective CYP isozymes have started. The rate and extent of product conversions will be dependent on the inhibitory potencies of the test compounds and their concentrations. We want to limit the maximum conversion of products to less than 20% of the substrate to avoid nonlinearity. Based on earlier experiments, the reaction duration for 1A2, 2C8, 2C9, 2D6, and 3A4 has been set to 10 min; the 2C19 activity in our hands is much lower and we allow the reaction to run for 20 min so that more product is formed to improve the detection sensitivity by subsequent MS analysis.

20. Tips from TB-11 will be reused to dispense TCA to stop the reactions.

21. Due to size limitation of the 200-μL tips, the pooling of the samples will have to be split into two batches, with samples from three isozymes each. A larger tip size can be used to accommodate all six isozymes in one run, but at the expense of aspiration precision. The pooling volume from each isozyme plate can be adjusted to meet the MS sensitivity requirement. If, for example, the MS signal from the 1A2 sample is weak, the pooling volume for 1A2 can be increased (with an equivalent volume decrease for the other isozymes) to improve the detection sensitivity for 1A2.

22. It is assumed that the user has received proper training to operate the API4000QT mass spectrometer and is familiar with the Analyst software. The parameters provided can be adapted to other types of mass spectrometers, with minor retuning of the

instrument, using commercially available standard for the products.

23. The data acquisition duration of the mass spectrometer is purposely set to far exceed the RapidFire run time so that the MS is allowed to passively and continuously acquire data in the background while the user can focus on the operation of the RapidFire. The MS acquisition is started and stopped manually.

24. A delay between the start times of the MS and RapidFire is expected and can be corrected during data processing. It is advised to check the MS real-time spectrum display at the beginning of the acquisition to verify that signal similar to Fig. 4 is observed. The sampling cycle is 9 s per well and the peaks detected by mass spectrometer are approximately 4 s wide at the baseline. Check for leaks in the RapidFire or clogging of the sipper tube, if no signal is observed. Refer to the RapidFire user manual for additional troubleshooting recommendations. The analysis should take about 15 min per plate and this includes the time used for washing the aspirator tip between sample aspirations.

25. The MS acquires data in the passive mode, relying on the RapidFire to feed the samples. The MS data acquisition will continue until the last plate is analyzed and acquisition stopped manually. Data from multiple plates can be acquired into one MS data file.

26. The plate log files contain time-stamp information for each sample analyzed. For a two-plate experiment there will be two files named "Plate1.txt" and "Plate2.txt."

27. All six MRM traces will be loaded and displayed in the integrator software. The traces will be outlined in red.

28. You can select any MRM trace to start the integration setup. Once the peak parameters are selected from one trace, they will automatically apply to the five remaining MRM traces. Any peak within the MRM trace can be selected as the landmark peak. This may be the first or last peak within a plate or a control peak known to produce high or low signal.

29. The peak-width window will be used to select all other peaks and the selection of this window size will determine in some cases whether to include regions containing a peak shoulder or tailing, when calculating the peak area.

30. Pick the first or last peak of the MRM trace so that it is easier to identify the plate and well location of the sample from where the peak is derived.

31. If for some reason part of the MRM remains red it is likely that the peak width is set incorrectly in step 9 or an incorrect peak number was entered in step 10. Repeat steps 8–11 to correct

the problem. If the integration failure is due to split peaks or peaks containing shoulders, return to step 3 and perform peak smoothing in Analyst before exporting to text files.

32. The peak-area information will be exported to an Excel .csv file format, one file per plate. Each file will contain data from all six CYP isozyme assays arranged in 96-well-plate format.

33. The test compound concentrations constitute the "x" values for the dose–response curve.

34. The % activity constitutes the "y" values for the dose–response curve.

35. The IC_{50} value will be output to cell Q3 of the Excel sheet and a fitted dose–response curve will also be plotted starting in cell Q5 (see Fig. 5 insert).

Acknowledgements

We would like to thank Ms. Melinda Manuel, Mr. Joshua Cramlett, and Dr. Beverly Knight from Takeda San Diego for their support in the development and testing of the high-throughput P450 inhibition assay.

References

1. Guengerich FP (2003) Cytochromes P450, drugs, and diseases. Mol Interv 3:194–204
2. Omiecinski CJ, Remmel RP, Hosagrahara VP (1999) Concise review of the cytochrome P450s and their roles in toxicology. Toxicol Sci 48:151–156
3. Tredger JM, Stoll S (2002) Cytochromes P450—their impact on drug treatment. Hosp Pharmacist 9:167–173
4. Drug Metabolism/Drug Interaction Studies in the Drug Development Process: Studies In Vitro (1997) http://www.fda.gov/downloads/Drugs/GuidanceComplianceRegulatory Information/Guidances/UCM072104.pdf. Accessed 10 Oct 2011
5. Bjornsson TD et al (2003) The conduct of in vitro and in vivo drug-drug interaction studies: A Pharmaceutical Research and Manufacturers of America (PhRMA) perspective. Drug Metab Dispos 31:815–832
6. Walsky RL, Obach RS (2004) Validated assays for human cytochrome P450 activities. Drug Metab Dispos 32:647–660
7. Di L, Kerns EH, Li SQ, Carter GT (2007) Comparison of cytochrome P450 inhibition assays for drug discovery using human liver microsomes with LC-MS, rhCYP450 isozymes

with fluorescence, and double cocktail with LC-MS. Int J Pharm 335:1–11
8. Weaver R, Graham KS, Beattie IG, Riley RJ (2003) Cytochrome P450 inhibition using recombinant proteins and mass spectrometry/multiple reaction monitoring technology in a cassette incubation. Drug Metab Dispos 31:955–966
9. Smith D et al (2007) Analytical approaches to determine cytochrome P450 inhibitory potential of new chemical entities in drug discovery. J Chromatogr B 850:455–463
10. Lin T, Pan K, Mordenti J, Pan L (2007) In vitro assessment of cytochrome P450 inhibition: Strategies for increasing LC/MS-based assay throughput using a one-point IC_{50} method and multiplexing high-performance liquid chromatography. J Pharm Sci 96:2485–2493
11. Janiszewski JS et al (2001) A high-capacity LC/MS system for the bioanalysis of samples generated from plate-based metabolic screening. Anal Chem 73:1495–1501
12. Jenkins KM et al (2004) Automated high throughput ADME assays for metabolic stability and cytochrome P450 inhibition profiling of combinatorial libraries. J Pharm Biomed Anal 34:989–1004

13. Kim MJ et al (2005) High-throughput screening of inhibitory potential of nine cytochrome P450 enzymes in vitro using liquid chromatography/tandem mass spectrometry. Rapid Commun Mass Spectrom 19:2651–2658

14. Roddy TP et al (2007) Mass spectrometric techniques for label-free high-throughput screening in drug discovery. Anal Chem 79:8207–8213

15. Gobey J et al (2005) Characterization and performance of MALDI on a triple quadrupole mass spectrometer for analysis and quantification of small molecules. Anal Chem 77:5643–5654

16. Wu J et al (2007) High-throughput cytochrome P450 inhibition assays using laser diode thermal desorption-atmospheric pressure chemical ionization-tandem mass spectrometry. Anal Chem 79:4657–4665

17. Lim KB, Chien EYT, Xu R, Kassel DB (2005) MS based cytochrome P450 inhibition screening during "hit-to-lead": validation and automation. Proceedings of the 53rd American Society for Mass Spectrometry Conference MP115

18. Lim KB, Ozbal CC, Kassel DB (2010) Development of a high-throughput online solid-phase extraction/tandem mass spectrometry method for cytochrome P450 inhibition screening. J Biomol Screen 15:447–452

Chapter 4

The Synthesis, Characterization, and Application of ^{13}C-Methyl Isocyanide as an NMR Probe of Heme Protein Active Sites

Christopher McCullough, Phani Kumar Pullela, Sang-Choul Im, Lucy Waskell, and Daniel Sem

Abstract

The cytochromes P450 (CYPs) play a central role in a variety of important biological oxidations, such as steroid synthesis and the metabolism of xenobiotic compounds, including most drugs. Because CYPs are frequently assayed as drug targets or as anti-targets, tools that provide confirmation of active-site binding and information on binding orientation would be of great utility. Of greatest value are assays that are reasonably high throughput. Other heme proteins, too—such as the nitric oxide synthases (NOSs), with their importance in signaling, regulation of blood pressure, and involvement in the immune response—often display critical roles in the complex functions of many higher organisms, and also require improved assay methods. To this end, we have developed an analog of cyanide, with a ^{13}CH$_3$-reporter group attached to make methyl isocyanide. We describe the synthesis and use of ^{13}C-methyl isocyanide as a probe of both bacterial (P450cam) and membrane-bound mammalian (CYP2B4) CYPs. The ^{13}C-methyl isocyanide probe can be used in a relatively high-throughput 1-D experiment to identify binders, but it can also be used to detect structural changes in the active site based on chemical shift changes, and potentially nuclear Overhauser effects between probe and inhibitor.

Key words NMR, Heme, P450cam, CYP2B4, Screening

1 Introduction

Mammalian cytochromes P450 (CYPs) are the enzymes primarily responsible for drug and xenobiotic metabolism in humans, so have been of particular interest in the pharmaceutical industry. They are relatively large membrane-bound proteins with molecular mass of ~65 kDa (1). Accordingly, despite some recent successes, it is difficult to obtain structural information on the mammalian CYPs, either via X-ray or current multidimensional NMR techniques. And the X-ray structures that have been reported of mammalian CYPs are of proteins that lack the membrane-binding N-terminal region and have other solubilizing mutations (2–5).

Ian R. Phillips et al. (eds.), *Cytochrome P450 Protocols*, Methods in Molecular Biology, vol. 987,
DOI 10.1007/978-1-62703-321-3_4, © Springer Science+Business Media New York 2013

Studies of an additional group of heme proteins, like the above mentioned NOSs, though they may already have had their crystal structures determined, could benefit from the development of a quick and reliable assay that could enable unambiguous screening for potential binders or inhibitors of these proteins in either a primary or secondary assay. Clearly, there is a need for improved techniques that rapidly probe at least part of the active-site structure, and better define binding interactions with substrates or inhibitors for heme proteins. With this goal in mind, we have capitalized on the fact that cyanide is a promiscuous ligand for heme iron (binding through its carbon), and that alkyl isocyanides (alkyl-NC) are known to be ligands for a number of heme proteins (6, 7). The absorption spectra of the alkyl isocyanide-bound heme proteins are typical of type-II binding spectra, where the heme Soret band is shifted upon binding of the alkyl isocyanide, with maxima at 430 and/or 455 nm, depending upon the extent of back-bonding from iron into the triple bond of the isonitrile. In fact, resonance Raman experiments (6, 7) have shown that the heme-bound isocyanide binds in two conformations, either linear or bent, depending on C–Fe bond order. These two conformations are expected to be in fast exchange on the NMR timescale, but slow exchange on the resonance Raman timescale. The two orientations are shown in Fig. 1 for the methyl isocyanide ligand, which is the probe described in this chapter. The linear and bent forms are related to the resonance structures of methyl isocyanide (Fig. 2), although the imine is unfavored because of electron donation that stabilizes it via formation of a dative bond (as in Fig. 1, where electron donation is via back-bonding from iron). Alkyl isocyanides bind strongest when the heme iron is reduced ($K_d \times 10^{-5}$–10^{-6} M), although with increasing alkyl chain length, even the ferric form can bind with a micromolar range dissociation constant (8). The previously established utility of $^{13}CH_3$-methyl probes for NMR structural and dynamical studies in large proteins (9–11) is what led to our interest in developing $^{13}CH_3$-based alkyl isocyanides as probes for heme proteins, focused on the simplest member of this family, methyl isocyanide ($^{13}CH_3NC$). As the smallest member of the alkyl isocyanide family, methyl isocyanide is least likely to interfere with substrate or inhibitor binding (being only slightly larger than dioxygen, the natural ligand). The ^{13}C-label enables selective observation of the attached methyl protons, providing both ^1H and ^{13}C chemical-shift perturbation probes of structural (and potentially dynamical) changes in the binding pocket, due to substrate or inhibitor binding. And the lack of adjacent methylene protons in $^{13}CH_3NC$ avoids dipolar relaxation, which would produce line broadening. Furthermore, the rapidly rotating methyl group has a short internal correlation time, even when bound to a large heme protein, so the proton line width will be narrow, like that of a small molecule.

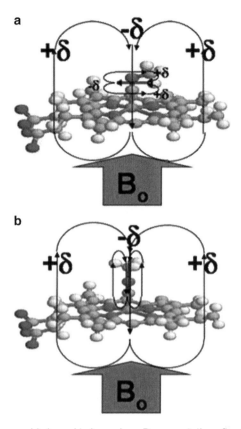

Fig. 1 Methyl isocyanide bound to heme iron. Representation of methyl isocyanide bound to the iron of the heme with (**a**) an iron–carbon bond order of two ("*bent*") and (**b**) an iron–carbon bond order of one ("*linear*"). The pyrrole rings and the CN triple (or double) bond produce a very anisotropic environment capable of producing large chemical-shift perturbations for the labeled methyl of $^{13}CH_3NC$

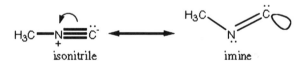

Fig. 2 Resonance structures for $^{13}CH_3NC$, methyl isocyanide

Together, these various features of the $^{13}CH_3NC$ probe allow it to serve as a sensitive structural probe of the heme active site in experiments such as 2-D ^1H-^{13}C HSQC (heteronuclear single-quantum coherence) and ^{13}C-filtered ^1H 1D (10, 12) which, given the large size of CYP enzymes, especially when complexed to micelles, would not normally be feasible. Herein, we present the synthesis and application of this ^{13}C-methyl isocyanide probe.

54   Christopher McCullough et al.

2 Materials

Prepare all solutions using deionized water and analytical grade reagents. Prepare and store all reagents at room temperature (unless indicated otherwise). Diligently follow all waste disposal regulations when disposing of waste materials.

2.1 Synthesis of ^{13}C-Methyl Isocyanide

1. ^{13}C-Methyliodide, a liquid (Cambridge Isotope Laboratories).

2. Silver cyanide salt (Sigma).

3. 8 M potassium cyanide solution, purchased as a salt (Sigma) and dissolved in water.

4. Thick-walled (approx. 1 in.) glass tube.

5. Acetylene torch to seal glass tube.

6. Mineral oil and hot plate.

7. Vacuum distillation apparatus.

8. Bulb-to-bulb distillation apparatus.

2.2 Sample Preparation of ^{13}C-Methyl Isocyanide-Bound Protein

1. Tank of argon gas.

2. Pure ^{13}C-Methyl isocyanide, as prepared above.

3. 300–500 μL of 75–100 μM solution of CYP450 or heme protein in PBS buffer comprised of 137 mM NaCl, 2.7 mM KCl, 10 mM Na_2HPO_4, and 2.0 mM KH_2PO_4 (pH 7.4). Concentration determined using a UV–Vis spectrophotometer. Protein solution stored short term at 4°C. For long-term storage (>1 week) flash freeze in liquid nitrogen and store at −80°C.

4. Fresh, solid sodium dithionite (Sigma). Stored under argon in a parafilm-sealed tube to prevent premature oxidation (see Note 1).

5. Rubber hose, hose/needle metal adapter, two needles—one long and one short—and NMR tube-sized rubber septum.

6. Optional: UV–Vis spectrophotometer such as CARY 50 BIO UV–Vis spectrophotometer.

7. 5 mm thin-walled glass NMR tube with plastic NMR tube cap (Norell).

2.3 Collection of NMR Data

1. Medium- to high-field NMR spectrometer such as a 500 or 600 MHz Varian Inova spectrometer.

2. Sample as prepared above.

3 Methods

Carry out all procedures at room temperature or 25°C unless otherwise specified.

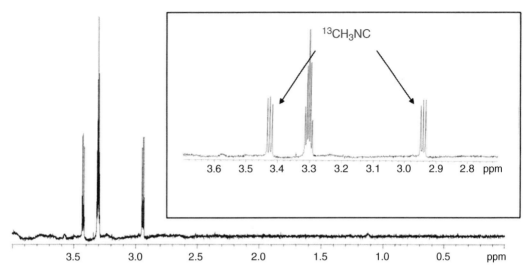

Fig. 3 NMR spectrum of $^{13}CH_3NC$ in d_4-methanol. The ^{13}C-coupled ^1H resonances of $^{13}CH_3NC$ are located on either side of the solvent resonance. The *inset* shows the additional splitting (1:1:1 triplet) of the two ^{13}C-split proton resonances. This triplet is from the ^2J-coupling between the quadrupolar ^{14}N nucleus ($I=1$) and the methyl protons. The spectrum was acquired at 300 MHz

3.1 Synthesis of ^{13}C-Methyl Isocyanide

1. Mix 1 mmol of ^{13}C-methyliodide and 2 mmol of silver cyanide, to form a solution/suspension, in a thick-walled glass tube so that the glass tube is about ¾ full (13).

2. Seal the glass tube to create a closed system by using an acetylene torch to melt and fuse the glass at open end of glass tube.

3. Heat the mixture for 1 h at 95°C in a mineral oil bath, which is being warmed on a hot plate in a closed fume hood. Be sure to exercise caution, work in hood and wear protective eye wear when working with sealed tubes.

4. Cool the tube to room temperature slowly; when cool, break tube carefully (etch glass) at one end to open it, and quench reaction for 12 h by adding 5 mL of 8 M KCN solution and mixing carefully to decompose resulting complex. Exercise extreme caution in working with cyanide solutions, and in disposal of excess KCN solution (see Note 2).

5. Distill the quenched mixture under vacuum to obtain the initial crude ^{13}C-methyl isocyanide.

6. Follow the initial vacuum distillation by a second purification step, of bulb-to-bulb distillation, using a rotary evaporator. Yield should be close to 21%.

7. Purity can be confirmed by taking a ^1H-NMR spectrum in d_4-methanol. Spectrum should look like Fig. 3, with the 1:1:1 triplet of the doublet of triplets confirming the identity of the compound to be that of ^{13}C-methyl isocyanide, as opposed to

^{13}C-methyl-labeled acetonitrile (due to the ^2J-coupling between the quadrupolar ^{14}N and the methyl protons) (see Note 3).

8. Store at $-80°C$ in sealed container when not in use, to prevent vaporization of this volatile compound.

3.2 Sample Preparation of ^{13}C-Methyl Isocyanide-Bound Protein

1. Attach rubber hose from tank of argon to needle/hose adapter and screw long needle onto end of adapter.

2. Insert NMR tube-sized rubber septum into NMR tube containing protein sample and fold down top edge to secure and seal tube.

3. Insert long needle attached to end of adapter into NMR tube through septum all the way to bottom of tube. Insert short needle through septum into NMR tube to serve as exhaust valve during subsequent bubbling of argon to remove O_2.

4. Slowly open regulator valve on argon tank to allow just enough pressure to get a steady stream of argon bubbles rising through solution, but not enough pressure to push liquid to top of tube. Allow argon to gently bubble through solution for 15 min to completely remove all oxygen (see Note 4).

5. Add a pinch of solid sodium dithionite to the NMR tube containing the sample—just enough to cover the end of a lab spatula should be more than enough. Cap the NMR tube and invert several times to ensure complete dissolution and hence reduction of the entire sample.

6. Add enough ^{13}C-methyl isocyanide to have a slight excess of it relative to the heme protein (^{13}C-methyl isocyanide: heme protein (assuming 1 heme/protein) ratio of about 1.2: 1) (see Note 5). Solution should change color from reddish brown to reddish orange. This spectral change can be observed more precisely with a UV–Vis spectrometer by observing that the A_{max} redshifts or shifts towards a value of either 430 or 455 nm.

3.3 Collection of NMR Data

1. Place the sample in the NMR spectrometer.

2. Lock, tune and match, and shim on the sample, as would be done for any NMR sample.

3. Find the 90° pulse width, using either the residual water peak or a sample peak.

4. Set up a normal 2-D ^1H-^{13}C HSQC experiment, but with the number of increments set to one ("ni" = 1) so that the HSQC becomes a ^{13}C-filtered ^1H 1-D experiment.

5. After checking all the parameters to make sure that they are reasonable, enter the value of the 90° pulse width that was just determined on the sample and ensure that the number of scans (e.g., "ns") is sufficient for about a 10-min experiment.

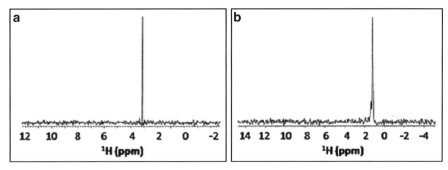

Fig. 4 ^{13}C-filtered ^{1}H 1-D spectra of ^{13}CH$_3$NC that is (**a**) unbound or (**b**) bound to Fe^{2+}-P450cam

6. Process the data as usual, with little to no line broadening, and check spectrum. It should be similar to that shown in Fig. 4, with any remaining "free" methyl isocyanide giving a peak at ~3.1 ppm and any additional peaks, probably in slow exchange with the "free" peak, representing "bound" form(s).

7. Next, take the NMR sample and, after removing the cap, quickly add an aliquot of the ligand solution, whose binding is being screened for.

8. Run the experiment again, as in steps 4–6, to see what effect the addition of ligand had on the bound ^{13}C-methyl isocyanide peak (see Note 6). One can titrate and monitor chemical-shift change as a function of ligand concentration, fitting data to obtain a dissociation constant (see Note 7).

9. Once back in the lab, the presence of inactive protein (P420) in the sample can be checked by bubbling CO through the tube, as was done with argon in preparation of the sample, and checking for the presence of an absorbance band at 420 nm, indicating inactive protein.

4 Notes

1. Sodium dithionite is readily hydrated to bisulfate. For longer-term storage, it is best to store in a desiccator.

2. All cyanide salts are very dangerous, but especially labile cyanide salts such as potassium cyanide. Do *not* mix with acid, as this would form gaseous HCN, which can be readily inhaled. While handling KCN, a solution of hydroxocobalamin and sodium thiosulfate is kept nearby (antidote) and a colleague must observe the researcher at all times for any accidental toxicity. The hydroxocobalamin and sodium thiosulfate are part of the typical cyanide antidote kits available commercially and their use is approved in the USA. The hydroxocobalamin reacts with cyanide to form a stable and nontoxic cyanocobalamin that is

eliminated from the body through urine. As stated above, KCN is highly toxic and releases hydrogen cyanide gas upon reaction with acids. Potassium cyanide solutions and glassware with which these have been in contact must be treated with an excess of potassium permanganate (4–5 g of $KMnO_4$ for 1 g of KCN) and rinsed with plenty of water. After treatment, liquid waste should be disposed of by a certified cyanide waste disposal company (e.g., one which deals with cyanide waste generated from metal plating companies).

3. The unusual ^{14}N-^{1}H scalar coupling (^{2}J-coupling, producing the triplet) was only apparent in d_4-methanol and D_2O; in other solvents that were used, including d_3-acetonitrile, d_1-chloroform, and d_6-benzene, and when complexed with protein, the spectrum only showed the large doublet splitting from the ^{13}C-^{1}H coupling. Although this unusual quadrupolar coupling—and implied long quadrupolar relaxation time—stems from the high degree of axial symmetry and short correlation time of this molecule and, as such, is only an interesting side note (i.e., would not be observed when bound to protein), this property has practical importance in that it can help confirm the identity of $^{13}CH_3NC$, allowing it to be easily distinguished from the most common side product—^{13}C-methyl-labeled acetonitrile.

4. We find that with this step it pays to be patient, as it is the easiest way to lose your sample—and weeks of hard work—throughout this procedure. We advise that the flow rate of argon is increased very slowly to ensure that the sample is not lost with excess bubbling, popping off of the septum, and spilling, or simply denaturing protein as a result of excessive bubbling.

5. We find it helpful to have a slight molar excess of ^{13}C-methyl isocyanide so that we can distinguish unambiguously the free form and the bound form, provided they are in slow exchange on the chemical-shift timescale. In addition, this slight excess of free ^{13}C-methyl isocyanide allows us to determine whether, upon addition of substrate or ligand to the mixture, the bound ^{13}C-methyl isocyanide is perturbed or simply displaced, once again yielding free ^{13}C-methyl isocyanide.

6. The chemical shift of the ^{13}C-methyl isocyanide protons will be affected by the anisotropic environment above the heme group, in addition to the relative population of linear and bent forms of the molecule itself. Both of these factors contribute to the increased shielding or deshielding these protons experience and, hence, whether the bound resonance will be upfield-shifted or downfield-shifted.

7. If ligand (inhibitor or substrate) binds while $^{13}CH_3NC$ remains bound, one can monitor changes in chemical shift of $^{13}CH_3NC$

due to the presence of simultaneously bound ligand. But if ligand and $^{13}CH_3NC$ sterically exclude each other, titration with ligand will lead to displacement of $^{13}CH_3NC$, which can also be monitored, due to its distinct chemical shift.

Acknowledgements

We thank Professors Rajendra Rathore, Milo Westler, and Bill Donaldson for helpful discussions. This research was supported, in part, by National Institutes of Health grants S10RR019012 (D.S.S.), GM085739 (D.S.S.), and GM3553 (L.W.), and also made use of the National Magnetic Resonance Facility at Madison (NIH: P41RR02301, P41GM66326). Remote connectivity was with Abilene/Internet2, funded by NSF (ANI-0333677).

References

1. Furge LL, Guengerich FP (2006) Cytochrome P450 enzymes in drug metabolism and chemical toxicology. Biochem Mol Biol Educ 34: 66–74
2. Williams PA, Cosme J, Sridar V, Johnson EF, McRee DE (2000) Mammalian microsomal cytochrome P450 monooxygenase: structural adaptations for membrane binding and functional diversity. Mol Cell 5:121–131
3. Williams PA, Cosme J, Ward A, Angove HC, Matak Vinkovic D, Jhoti H (2003) Crystal structure of human cytochrome P450 2C9 with bound warfarin. Nature 424:46–464
4. Williams PA, Cosme J, Vinkovic DM, Ward A, Angove HC, Day PJ, Vonrhein C, Tickle IJ, Jhoti H (2004) Crystal structure of human cytochrome P450 3A4 bound to metyrapone and progesterone. Science 305:683–686
5. Scott EE, He YA, Wester MR, White MA, Chin CC, Halpert JR, Johnson EF, Stout CD (2003) An open conformation for mammalian cytochrome P450 2B4 at 1.6 A° resolution. Proc Natl Acad Sci USA 100: 13196–13201
6. Tomita T, Ogo S, Egawa T, Shimada H, Okamoto N, Imai Y, Watanabe Y, Ishimura Y, Kitigawa T (2001) Elucidation of the differences between the 430- and 455-nm absorbing forms of P450-isocyanide adducts by resonance Raman spectroscopy. J Biol Chem 276: 36261–36267
7. Lee DS, Park SY, Yamane K, Obayashi E, Hori H, Shiro Y (2001) Structural characterization of n-butyl-isocyanide complexes of cytochromes P450nor and P450cam. Biochemistry 40: 2669–2677
8. Simonneaux G, Bondon A (2000) Isocyanides and phosphines as axial ligands in heme proteins and iron porphyrin models. In: Kadish KM, Smith KM, Guilard R (eds) The porphyrin handbook. Academic Press, San Diego, CA, pp 299–311
9. Pellecchia M, Meininger D, Dong Q, Chang E, Jack R, Sem DS (2002) NMR-based structural characterization of large protein- ligand interactions. J Biomol NMR 22:165–173
10. Showalter SA, Johnson E, Rance M, Bruschweiler R (2007) Toward quantitative interpretation of methyl side-chain dynamics from NMR and molecular dynamics simulations. J Am Chem Soc 129:14146–14147
11. Tugarinov V, Kanelis V, Kay LE (2006) Isotope labeling strategies for the study of high-molecular-weight proteins by solution NMR spectroscopy. Nat Protoc 1:749–754
12. Cavanagh J, Fairbrother WJ, Palmer AGIII, Rance M, Skelton NJ (2007) Protein NMR spectroscopy, 2nd edn. Academic, New York
13. Freedman TB, Nixon ER (1972) Matrix isolation studies of methyl cyanide and methyl isocyanide in solid argon. Spectrochim Acta Biochim 28A:1375–1391

Chapter 5

High-Throughput Fluorescence Assay for Cytochrome P450 Mechanism-Based Inactivators

Cesar Kenaan, Haoming Zhang, and Paul F. Hollenberg

Abstract

The mechanism-based inactivation (MBI) of the human cytochrome P450 (P450 or CYP) drug-metabolizing enzymes may lead to adverse drug–drug interactions, especially for drugs with narrow therapeutic windows. Unlike reversible inhibitors of P450, drug–drug interactions originating from MBI may persist in patients for some time after the body eliminates the offending drug because P450 enzymatic activity can be recovered only after de novo synthesis of the P450. In a pharmaceutical setting, a substantial amount of effort is often expended to understand the potential for mechanism-based inactivation and its possible contribution to the drug–drug interaction profile of drug candidates. Therefore, in vitro assays that identify and characterize which drug candidates are P450 MBIs are critically important in preclinical drug metabolism and pharmacokinetic studies. A detailed method is described for the adaptation of a 7-ethoxytrifluoromethyl coumarin O-deethylation fluorescence activity assay to a 96-well plate format to characterize the K_I and k_{inact} values for an MBI. The advantages of this microtiter format compared with the conventional method include a significant reduction in the amount of enzyme used, a reduction in assay time, and an increase in experimental throughput.

Key words Cytochrome P450, Mechanism-based inactivation, Time-dependent inhibition, High-throughput assay, Fluorescence

1 Introduction

The reactions catalyzed by the human cytochromes P450 account for the majority of biotransformations of clinically administered drugs (1). Although most P450-mediated oxidative reactions lead to the formation of more hydrophilic products that can either be eliminated directly from the body or undergo further modification by phase II enzymes, some reactions lead to the formation of reactive intermediates. These intermediates often lead to irreversible covalent binding of the chemically reactive group to components of the P450 structure, such as amino-acid residues or the heme, that are essential to P450 function. Since the generation of these reactive intermediates is dependent on P450 catalytic activity and the subsequent chemical modification of critical P450 components

Ian R. Phillips et al. (eds.), *Cytochrome P450 Protocols*, Methods in Molecular Biology, vol. 987,
DOI 10.1007/978-1-62703-321-3_5, © Springer Science+Business Media New York 2013

will often lead to loss in catalytic activity, this process is referred to as mechanism-based inactivation (MBI). Thorough reviews of the chemical and physical principles that govern mechanisms of inactivation are available in the literature, and the reader is referred to those for further information (2, 3).

MBI of human P450s has important implications in drug discovery and development. For example, drug candidates that inactivate a particular drug-metabolizing P450 may lead to significant drug–drug interactions when co-administered with one or more drugs. In the case of prodrugs, for instance, if the coadministered drug requires bioactivation by a particular P450 for the generation of the pharmacologically active species then inactivation of that particular P450 by the MBI may lead to significantly lower levels of the biologically active metabolite and a marked reduction in therapeutic effect. In addition, inactivation of a particular P450 that is primarily responsible for the metabolism of a drug leading to elimination from the body may contribute to a significant increase in the plasma concentrations and AUC of the drug and thereby lead to adverse, often lethal drug–drug interactions (4–6). For certain human P450s, however, inactivation by MBI is a desirable outcome. CYP19A and CYP17A are known to play important roles in the progression of breast and prostate cancer, respectively, and highly selective MBIs of CYP19A by exemestane (7) and CYP17A by abiraterone (8) are currently used to treat patients with cancer.

High-throughput assays that quantitatively assess whether new molecular entities are mechanism-based inactivators or time-dependent inhibitors of P450s are of considerable interest in drug discovery efforts. These methods gain much greater utility if they can be extended to a high-throughput format, which minimizes enzyme and reagent consumption and greatly decreases turnaround time and effort. The most widely used method today to determine the K_I and k_{inact} values for a P450 involves reconstituting the P450 (1 nmol) with cytochrome P450 reductase (CPR) in the presence of dilauryl phosphatidyl choline (DLPC) and varying concentrations of the MBI in a total reaction volume of 1 mL (9, 10). NADPH is then added to this "pre-incubation" or "primary reaction" step to initiate the inactivation reaction and, at various time points, an aliquot is then transferred from the "primary reaction" to a "secondary reaction" mixture containing saturating concentrations of 7-ethoxy trifluoromethyl coumarin (7-EFC), a probe substrate, and additional NADPH to support the metabolism reaction. The majority of human P450s can catalyze the O-deethylation of 7-EFC to 7-hydroxy trifluoromethyl coumarin (7-HFC), which is highly fluorescent and can be measured fluorometrically. This method, developed by Buters et al. (11) based on previous work by DeLuca et al. (12), is highly sensitive and has become the method of choice for laboratories investigating MBI of P450s.

The method outlined in this chapter represents an improvement on the previous methods of Buters et al. (11) and DeLuca et al. (12). Specifically, this method involves determining the O-deethylase activity of the P450 for 7-EFC in microtiter wells, where the minimal volume required for the fluorescence reading is significantly less than in a 1-mL cuvette. Therefore, the amount of 7-EFC converted to 7-HFC that is required to achieve a significant signal-to-noise ratio is substantially less and, thus, significantly less P450 and CPR are required for the primary and secondary reactions. In addition, the ability to perform this assay using microtiter plates allows it to be adapted to a high-throughput format. Simple modifications of this assay protocol can be introduced to determine the K_I and k_{inact} values for any recombinant (purified, Supersomes, or Baculosomes) P450 that catalyzes the O-deethylation of 7-EFC and it has been used with human CYP2B6 in the reconstituted system to characterize a number of mechanism-based inactivators in our lab including bergamottin (13), clopidogrel (14), and methadone (15).

For simplicity, our protocol outlines the use of stock purified P450 and CPR with concentrations of 6.5 μM and 12 μM, respectively, although this protocol does not require that identical stock concentrations be used. For the purpose of this demonstration, we are assuming that the MBI under investigation has a K_I of approximately 2 μM, although K_I values vary greatly from one inactivator to another and are also highly dependent on which P450 is being investigated. Additionally, the times chosen for the preincubation assume that the rate of the inactivation is moderate and this also varies greatly depending on the identities of both the MBIs and P450s.

2 Materials

1. Purified recombinant human P450 (rhCYP). For this example we will use a stock concentration of 6.5 μM.

2. Purified recombinant rat NADPH-P450 reductase (rCPR). For this example, we will use a stock concentration of 12 μM. Rat and human CPR possess virtually identical amino-acid sequences; however, rat CPR has a significantly higher expression yield than human CPR in *Escherichia coli* cells. For this reason we include rCPR in this protocol, although human CPR can be used as well.

3. Potassium phosphate, dibasic and monobasic: 1 M stocks in water (Fisher Scientific).

4. Nicotinamide adenine dinucleotide phosphate (NADPH): 20 mM (Sigma-Aldrich) (see Note 1).

5. MBI under investigation. This is typically prepared at a concentration of 10 mM in either methanol or DMSO (see Notes 2 and 3).

64 Cesar Kenaan et al.

6. 1,2-didodecanoyl-*sn*-glycero-3-phosphatidylcholine (DLPC): 1 mg/mL (Sigma-Aldrich).

7. Catalase from bovine liver: 1 mg/mL (Sigma-Aldrich).

8. 7-EFC: 25 mM in dimethyl sulfoxide (DMSO) (Invitrogen, Molecular Probes) (see Note 4).

9. Methanol (Fisher Scientific).

10. DMSO (Sigma-Aldrich).

11. Acetonitrile (Fisher Scientific).

12. Multichannel pipettors (Fisherbrand).

13. Sonicator equipped with a 10-mm flat-tip probe (Branson).

14. 96-Well black, low-volume microtiter plates (ThermoFisher).

15. Fluorescence plate reader—Victor II 1420–042 (PerkinElmer).

16. Shaking water bath—Precision (ThermoFisher).

17. THERMOstar shaking incubator (BMG Labtech).

18. Digital timer.

Information on sources and models of laboratory equipment and reagents is included to provide helpful information and similar products from other vendors may be used instead.

---

3 Methods

1. To achieve final concentrations of 1, 2, 4, 6, 10, and 15 μM MBI in 50 μL primary reaction mixtures without varying the volume of the MBI stock added, make the following corresponding MBI working stocks by dilution of a 555.6 μM (bergamottin) stock: 18.5, 37, 74.1, 111.1, 185.2, 277.8 μM.

2. The reconstituted system will contain 20.4 pmol purified rhCYP, 40.8 pmol purified rCPR, and 0.07 mg/mL DLPC per reaction well. For this assay approximately 3.14 μL of 6.5 μM rhCYP, 3.4 μL of 12 μM rCPR, and 3.66 μL of sonicated DLPC (1 mg/mL) are reconstituted for each inactivator concentration.

3. 1 M potassium phosphate buffer (pH 7.4) can either be purchased directly from commercial vendors or prepared by mixing 1 M potassium phosphate dibasic with the same concentration of potassium phosphate monobasic such that the final percentage composition by volume is 81% and 19%, respectively. Verify the pH of the buffer with a pH meter and adjust by adding either mono- or dibasic potassium phosphate as needed.

4. Prepare 245.8 μL of the primary reaction mixture by mixing 17.5 μL of 1 M potassium phosphate buffer (pH 7.4) and 0.75 μL of catalase (1 mg/mL) and add water to 245.8 μL. Since 35.1 μL of this reaction mixture will be added to a final

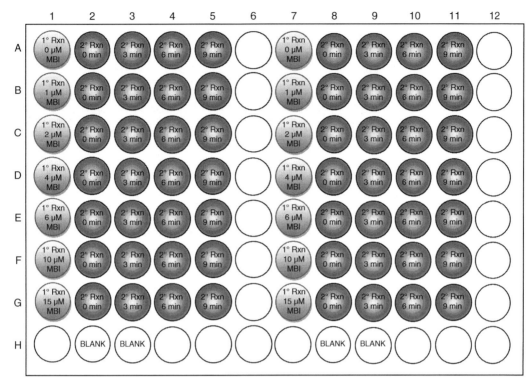

Fig. 1 Layout of 96-well plate for high-throughput characterization of the K_i and k_{inact} values for an MBI of a P450. Column 1 (rows A–G) contains the primary reaction (1° Rxn), which includes the reconstituted P450 system and increasing concentrations of MBI. At the indicated time points aliquots from the primary reaction are transferred to the secondary reactions (2° Rxn) in columns 2–5, rows A–G. After each transfer the reaction is allowed to proceed for 10 min, after which it is quenched with acetonitrile. The assay can be performed in duplicate, as depicted by columns 7–11, rows A–G. The remaining vacant wells can be used to measure background fluorescence of the samples

primary reaction volume of 50 µL, the final potassium phosphate concentration will be 50 mM.

5. For a sufficient supply of secondary reaction mixture to determine the kinetic parameters of an MBI using seven different concentrations at four different time points, mix 225 µL of 1 M potassium phosphate buffer (pH 7.4), 67.5 µL of 20 mM NADPH, and 18 µL of 25 mM 7-EFC and then add water to a final volume of 4.5 mL.

6. Sonicate the DLPC stock solution on ice for 1 min by using pulsed sonication at 40% sonicator output (see Note 5).

7. Pipette 71.4 µL of the P450, CPR, and DLPC reconstitution mixture into a 1-mL Eppendorf tube and keep at room temperature (~23°C) for 30 min (see Note 6).

8. Using a multichannel pipettor, dispense 35.1 µL of the primary reaction mixture into each of seven different inactivator concentration wells in columns 1 and 7, as illustrated in Fig. 1.

9. Using a multichannel pipettor, dispense 2.7 μL of the MBI stocks into their respective wells in columns 1 and 7 (Fig. 1). The control well receives the same volume of MBI solvent in lieu of the MBI (see Note 7).

10. When the reconstitution step is complete, dispense 10.2 μL of the reconstitution mixture to the primary reaction wells in columns 1 and 7.

11. Using a multichannel pipettor, dispense 150 μL of the secondary reaction mixtures into each secondary reaction well (columns 2–5 and 8–11) and an additional four wells (H2, 3 and H8, 9) which will serve as background fluorescence measurements (Fig. 1).

12. When not making additions, the 96-well plate should be shielded from light to prevent the photodegradation of 7-EFC.

13. Allow the primary and secondary reactions to reach thermal equilibrium by placing the 96-well plate in a shaking incubator set at 37°C for approximately 10 min.

14. When the reaction mixtures are equilibrated, add 2 μL of stock NADPH to each primary reaction well using a multichannel pipettor.

15. Using a multichannel pipettor, transfer 12.5 μL from the primary reaction wells to the secondary reaction wells of the 0-min time point (columns 2 and 8) and start the timer (see Note 8).

16. At 3 min, use a multichannel pipettor to transfer 12.5 μL from the primary reaction wells to the secondary reaction wells of the 3-min time-point columns (3 and 9) (see Note 9).

17. Similarly, at 6 and 9 min, transfer 12.5 μL from the primary reaction wells to the secondary reaction wells of the corresponding time-point columns (4 and 10, and 5 and 11, respectively).

18. Stop the secondary reactions 10 min after each transfer by adding 50 μL of ice-cold acetonitrile using a multichannel pipettor and pipetting 6–8 times to obtain thorough mixing. For example, the 0-min time point is stopped at 10 min and the 3-min time point is stopped at 13 min (see Note 10).

19. Add 50 μL of ice-cold acetonitrile to the four designated background wells and mix as above.

20. Immediately begin measuring the fluorescence intensity of the secondary reactions with a fluorescence plate reader using an excitation wavelength of 410 nm and measuring the emission at 510 nm.

21. Analyze the results by subtracting the average background reading intensity from the fluorescence intensity measurements of the secondary reaction wells. Take the control sample at the 0-min time point, to indicate the maximal activity, and calculate the log of percentage activity remaining for the remainder of

the samples. Plot the log or natural log (ln) of percentage activity remaining as a function of time.

22. Use linear regression analysis of the time-course data to estimate the initial rate constants (k_{obs}) for the inactivation of the CYP by the MBI. k_{obs} is then plotted against the BG concentration and the data are fitted to Eq. 1, from which K_I and k_{inact} are determined. If a log scale was used in step 21 then multiply inactivation rates by 2.3. However, if the data are plotted on a natural-log scale then this correction factor is not needed:

$$k_{obs} = \frac{k_{inact} \times [MBI]}{K_I + [MBI]} \tag{1}$$

4 Notes

1. Since repeated freeze-thawing will contribute to the degradation of NADPH, it is recommended that 20 mM NADPH stock solutions be stored at –20°C as several low-volume aliquots that contain only enough NADPH for a single series of assays. Alternatively, vials of pre-weighed solid NADPH can be used.

2. It is advantageous to prepare fresh stocks of the MBI under investigation as solvent evaporation over extended periods of time may result in erroneous stock concentrations.

3. Prescreening of the inactivator being tested should be performed to ensure that the compound itself does not exhibit fluorescence properties that would interfere with the fluorometric detection of 7-HFC.

4. The choice of a probe substrate in the secondary reaction is based on several criteria. In a high-throughput format such as the one presented here, EFC is an excellent candidate as it is metabolized by several human P450s, including CYP1A2, 2B6, 2C19, 2D6, 2E1, and 3A4 to produce HFC, which can be detected spectrofluorometrically using a plate reader. However, since the MBI competes with the probe substrate in the secondary reaction for binding to the P450 under investigation, it is important that the probe substrate has a relatively low K_M value for the P450 under investigation and that saturating concentrations of the probe substrate are used in the secondary reaction.

5. Thorough sonication of DLPC will ensure that preformed micelles are disrupted into monomeric units and will facilitate the proper formation of liposomes and the efficient incorporation of P450 and CPR into the liposomes.

6. The reconstitution step allows complex formation between the P450 under investigation and CPR. Although a 30-min reconstitution step at room temperature is recommended in

this method, this may vary from one P450 to another. For example, some P450s are less stable than others at room temperature and a reconstitution step on ice for longer periods of time (e.g., 1 h) may be required. The optimal reconstitution time and temperature can be evaluated by assessing P450 activity after a range of reconstitution conditions. However, if the use of Supersomes or Baculosomes is desired, the reconstitution step can be bypassed and, once the Supersomes or Baculosomes are thawed, they can be dispensed directly into the primary reaction wells in place of the reconstituted mixture.

7. The effect of methanol, as well as any other solvents that may be used to dissolve the inactivator, on the enzymatic activity must be assessed in the control sample since inactivation is calculated as a percent of the control and some organic solvents are known to inhibit the activities of some P450s. In general, it is best to keep the amount of organic solvent added to the primary reactions less than 1 or 2% of the total volume although the effect of organic solvents on P450 activity is known to vary among P450 isoforms.

8. To further minimize the competition between the probe substrate and MBI in the secondary reaction, optimization of the dilution factor is recommended. For example, an MBI which behaves as a potent competitive inhibitor of the P450 under investigation will likely require a higher dilution factor than the one outlined in this protocol. This can be achieved in several ways including decreasing the volume transferred from the primary to the secondary reaction, increasing the volume of the secondary reaction, or both.

9. MBIs that display fast inactivation kinetics, such as clopidogrel for CYP2B6 (14) or mibefradil (16) and ritonavir (17) for CYP3A4, will generally produce a rapid loss of activity within the first 3–5 min of the preincubation. In this case, it is important that shorter time points and narrower time increments, between 0 and 5 min, are assessed to capture the initial rates of inactivation.

10. The NADPH present in the secondary reaction mixture is sufficient for the reaction time in this protocol (10 min). However, if significantly longer reaction times are required, an NADPH-regenerating system is recommended (18).

11. If more than one inactivator concentration is to be tested with a particular P450, simply multiply the volumes used by the number of inactivator concentrations being tested and reconstitute P450, CPR, and DLPC in a single Eppendorf tube, to minimize variation between separately reconstituted samples, and then aliquot appropriate volumes. It is best to perform triplicate runs for each inactivator concentration and time point and to determine the average and standard deviation of the results.

References

1. Evans WE, Relling MV (1999) Pharmacogenomics: translating functional genomics into rational therapeutics. Science 286:487–491
2. Hollenberg PF, Kent UM, Bumpus NN (2008) Mechanism-based inactivation of human cytochromes p450s: experimental characterization, reactive intermediates, and clinical implications. Chem Res Toxicol 21:189–205
3. Fontana E, Dansette PM, Poli SM (2005) Cytochrome p450 enzymes mechanism based inhibitors: common sub-structures and reactivity. Curr Drug Metab 6:413–454
4. Bachmann KA, Ring BJ, Wrighton SA (2003) Drug-drug interactions and cytochrome P450. FontisMedia SA, Switzerland
5. Walsky RL, Obach RS (2007) A comparison of 2-phenyl-2-(1-piperidinyl)propane (ppp), 1,1′,1″-phosphinothioylidynetrisaziridine (thioTEPA), clopidogrel, and ticlopidine as selective inactivators of human cytochrome P450 2B6. Drug Metab Dispos 35:2053–2059
6. Grimm SW, Einolf HJ, Hall SD, He K, Lim HK, Ling KH, Lu C, Nomeir AA, Seibert E, Skordos KW, Tonn GR, Van Horn R, Wang RW, Wong YN, Yang TJ, Obach RS (2009) The conduct of in vitro studies to address time-dependent inhibition of drug-metabolizing enzymes: a perspective of the pharmaceutical research and manufacturers of America. Drug Metab Dispos 37:1355–1370
7. Hong Y, Yu B, Sherman M, Yuan YC, Zhou D, Chen S (2007) Molecular basis for the aromatization reaction and exemestane-mediated irreversible inhibition of human aromatase. Mol Endocrinol 21:401–414
8. Vasaitis TS, Bruno RD, Njar VC (2011) CYP17 inhibitors for prostate cancer therapy. J Steroid Biochem Mol Biol 125:23–31
9. Blobaum AL, Harris DL, Hollenberg PF (2005) P450 active site architecture and reversibility: inactivation of cytochromes P450 2B4 and 2B4 T302A by tert-butyl acetylenes. Biochemistry 44:3831–3844
10. Lin HL, Kent UM, Hollenberg PF (2005) The grapefruit juice effect is not limited to cytochrome P450 (P450) 3A4: evidence for bergamottin-dependent inactivation, heme destruction, and covalent binding to protein in P450s 2B6 and 3A5. J Pharmacol Exp Ther 313:154–164
11. Buters JT, Schiller CD, Chou RC (1993) A highly sensitive tool for the assay of cytochrome P450 enzyme activity in rat, dog and man. Direct fluorescence monitoring of the deethylation of 7-ethoxy-4-trifluoromethylcoumarin. Biochem Pharmacol 46:1577–1584
12. DeLuca JG, Dysart GR, Rasnick D, Bradley MO (1988) A direct, highly sensitive assay for cytochrome P-450 catalyzed O-deethylation using a novel coumarin analog. Biochem Pharmacol 37:1731–1739
13. Kenaan C, Zhang H, Hollenberg PF (2010) A quantitative high-throughput 96-well plate fluorescence assay for mechanism-based inactivators of cytochromes P450 exemplified using CYP2B6. Nat Protoc 5:1652–1658
14. Zhang H, Amunugama H, Ney S, Cooper N, Hollenberg PF (2011) Mechanism-based inactivation of human cytochrome P450 2B6 by clopidogrel: involvement of both covalent modification of cysteinyl residue 475 and loss of heme. Mol Pharmacol 80:839–847
15. Amunugama H, Zhang H, Hollenberg PF (2012) Mechanism-Based Inactivation of Cytochrome P450 2B6 by Methadone through Destruction of Prosthetic Heme. Drug Metab Dispos 40:1765–1770
16. Foti RS, Rock DA, Pearson JT, Wahlstrom JL, Wienkers LC (2011) Mechanism-based inactivation of cytochrome P450 3A4 by mibefradil through heme destruction. Drug Metab Dispos 39:1188–1195
17. Ernest CS 2nd, Hall SD, Jones DR (2005) Mechanism-based inactivation of CYP3A by HIV protease inhibitors. J Pharmacol Exp Ther 312:583–591
18. Yamazaki H, Shimada T (2006) Cytochrome P450 reconstitution systems. Methods Mol Biol 320:61–71

Chapter 6

Identification of Endogenous Substrates of Orphan Cytochrome P450 Enzymes Through the Use of Untargeted Metabolomics Approaches

Qian Cheng and F. Peter Guengerich

Abstract

Metabolomics provides an invaluable means to interrogate the function of "orphan" enzymes, i.e., those whose endogenous substrates are not known. Here we describe a high-performance liquid chromatography-coupled mass spectrometry (HPLC-MS)-based metabolomics approach to identify an endogenous substrate of an orphan cytochrome P450.

Key words Orphan enzyme, Cytochrome P450, Metabolomics, HPLC-MS

1 Introduction

With the progress of recent genome-sequencing efforts, more than 15,000 cytochromes P450 (P450s) have been registered in the database (http://drnelson.uthsc.edu/CytochromeP450.html). Most of these P450s are considered "orphans" because no substrates or functions are associated with these enzymes. A need exists to explore these orphan P450s in order to bridge the wide knowledge gap between amino-acid sequence information and biochemical function. Metabolomics provides an efficient approach to identify the substrates of these orphan P450s (1–3). Here we describe an untargeted metabolomics approach to identify substrates of orphan P450s from *Streptomyces coelicolor*. Briefly, an organic extract of the organism (or tissue) where the orphan P450 is normally found was prepared as a chemical library that may contain endogenous substrates. The extract was then incubated with the recombinant P450 in the presence of NADPH and redox partners, with subsequent metabolite profiling by LC-MS. The metabolic profile obtained after the reaction was compared with that before the reaction using specialized metabolomics software, e.g., XCMS (4, 5). Molecules that were depleted due to the enzymatic reaction were identified as

Ian R. Phillips et al. (eds.), *Cytochrome P450 Protocols*, Methods in Molecular Biology, vol. 987, DOI 10.1007/978-1-62703-321-3_6, © Springer Science+Business Media New York 2013

Qian Cheng and F. Peter Guengerich

potential substrates. A successful example of this approach is the recent study of P450 154A1 (6). This method can also be applied in the study of mammalian orphan P450s.

2 Materials

2.1 P450 Assay

1. Recombinant orphan P450 (see Note 1).

2. Redox partners of P450: 200 μM spinach ferredoxin and 20 μM spinach NADPH-ferredoxin reductase. These enzymes can be purchased from Sigma-Aldrich. Rat NADPH-P450 reductase (NPR) can be prepared as described previously (7).

3. 10 mM $NADP^+$ stock solution: 382 mg $NADP^+$ in 50 ml of Milli-Q water. Store at 4°C.

4. 100 mM glucose 6-phosphate: 3.4 g in 100 ml of Milli-Q water. Store at 4°C.

5. Yeast glucose 6-phosphate dehydrogenase (10^3 IU/ml): 1.0 mg in 1 ml of 10 mM Tris–acetate buffer (pH 7.4), containing 20% (v/v) glycerol and 1 mM EDTA. Store at 4°C.

6. Extraction solvent: ethyl acetate with 0.1% (v/v) acetic acid (see Note 2).

7. Tissue extract: Extract tissue with extraction solvent by using a ratio of 200 ml/g for dry tissue or 10/1 (vol/vol) for tissue culture. For preparing mammalian tissue extracts, homogenization before extraction is needed. The organic portion is filtered and dried in a rotary evaporator. Dissolve the extract in minimal acetonitrile so that no obvious pellet is observed. Store the extract at −20°C.

8. DLPC solution: 5 mg L-α-dilauroyl-*sn*-glycero-3-phosphocholine in 5 ml of Milli-Q water. Sonicate (with probe, 2×10 s, full power). Store at 4°C. Sonicate each day before use and hold at 23°C until use.

9. 100 mM potassium phosphate buffer, pH 7.4.

10. NADPH-generating system: combine 50 parts of 10 mM $NADP^+$, 100 parts of 100 mM glucose 6-phosphate, and 2 parts of yeast glucose 6-phosphate dehydrogenase (1 mg/ml). Prepare fresh daily and store on ice when not in use.

2.2 Metabolic Profiling by HPLC-MS

1. HPLC: Waters Acquity UPLC system with photodiode array (PDA) detector (see Note 3).

2. Column: Acquity UPLC BEH (C_{18}) column (2×100 mm, 1.7 μm, Waters) (see Note 4).

3. Mobile phase: Solvent A: 95% H_2O, 5% CH_3CN, 0.1% HCO_2H (v/v/v); solvent B: 5% H_2O, 95% CH_3CN, 0.1% HCO_2H (v/v/v).

Identification of Endogenous Substrates of Orphan Cytochrome P450 Enzymes... 73

4. Mass spectrometer: LTQ (Thermo). An ion-trap instrument is preferred in this protocol due to the broader scan range compared with quadrupole-based mass spectrometers. A q-TOF instrument (Waters) is used for high-resolution mass spectrometry (HRMS).

2.3 Processing HPLC-MS Dataset Using XCMS

Dell Dimension XPS-Gen 5 with Intel Pentium Quad Processor, 4 GB memory, 150 GB hard drive with R language, and XCMS installed (see Note 5).

3 Methods

3.1 Orphan P450 Enzyme Assay

Skip step 1 if working with bacterial P450s; mix bacterial P450 with ferredoxin and NADPH-ferredoxin reductase in potassium phosphate buffer (pH 7.4) so that the reaction mix (before addition of the NADPH-generating system) contains 1 μM P450, 2 μM NADPH-ferredoxin reductase, and 10 μM ferredoxin. All reactions should be performed at least in duplicate.

1. Reconstitute the mammalian P450 system: Sonicate the DLPC solution for 2×10-s bursts or until the solution is clear. Add reagents in following sequence: recombinant P450, NPR, and DLPC. Gently mix the reaction components for 20–30 s and then leave the mix at room temperature for 10 min. Add an appropriate amount of 100 mM potassium phosphate buffer. A typical reaction has a volume of 1.0 ml and contains 0.1 μM P450, 0.5–1 μM NPR, and 20 μg/ml DLPC. Omit the P450 from the negative controls.

2. Add an appropriate amount of tissue extract so that the total organic content is ≤1% (vol/vol). Initiate the reaction by the addition of 100 μl of the NADPH-generating system (10% of reaction volume). Incubate the reaction at 37°C for 1 h.

3. Extract the reaction mix 3× with 5 ml of ethyl acetate, combine all organic phases, and evaporate to dryness under a nitrogen stream. Dissolve the extract in 50 μl of methanol.

3.2 Metabolic Profiling by HPLC-MS

Tune and calibrate the mass spectrometer according to the manufacturer's instructions before the experiment. Positive electrospray (+ESI) mode was used in this protocol. Other ionization methods, e.g., atmospheric chemical ionization (APCI), may be employed if the orphan P450s are suspected to metabolize compounds that do not ionize well in the ESI mode, e.g., steroids.

1. HPLC gradient: Flow rate 0.35 ml/min, 0–1 min, 15% B (v/v); 1–9 min, 15–95% (v/v) B; 9–12 min, 95% (v/v) B: 12–13 min, 95–15% B (v/v); 13–15 min, 15% B (v/v). The column temperature is set at 40°C.

74   Qian Cheng and F. Peter Guengerich

2. Each sample (10 µl) is injected using the partial-loop mode. After each injection, it is recommended to run a blank sample (100% methanol) to ensure there is no carryover of the previous sample.

3.3 Processing HPLC-MS Dataset Using XCMS

1. Convert the raw LCMS data to CDF or mzXML for processing by XCMS. The manufacturer normally provides a file-convertor function in their instrument-operation software. For example, Xcalibur (Thermo) has a "File convertor" option under the menu of "Tools," which can convert .RAW files to .CDF files.

2. Build two file folders under the same directory and name these folders as "Assay" and "NegativeControl," respectively. Move the converted assay data to the Assay folder and the control data to the NegativeControl folder.

3. Start R, load an XCMS library, and run a standard XCMS analysis (see Note 6). Set the output number n to choose how many statistically significant differences should be reported. A good starting point is about 50.

4. Verify the reported metabolic difference by checking the original LCMS data under the extracted-ion mode. Make sure that the difference is significant and consistent among replicates (see Note 7). The metabolites that are consumed in the reaction are considered putative endogenous substrates and worthy of further follow-up. For example, a putative substrate was identified in our study of P450 107U1 (Fig. 1a).

3.4 Verification of the Identified Substrates

Partially purify the substrate candidates. This can be accomplished by collecting the HPLC elute or using solid-phase extraction. Repeat the assay using partially purified candidates to confirm the substrate will be depleted in an enzyme-dependent manner. With fewer interfering compounds present in the partially purified substrate, it is not rare to see the formation of product. In our studies of P450 107U1, a product peak derived from the desaturation of the putative substrate was very obvious (−2 amu, Fig. 1b), lending further support that a bona fide substrate was identified.

3.5 Structural Elucidation of Identified Substrate

After a substrate is identified, considerable effort may be required to elucidate the structure of the compound. Detailed instructions for structure elucidation for all compounds are beyond the scope of this protocol. Here we describe a simplified and basic systematic approach for structural characterization.

1. Obtain the HRMS of substrate (error <5 ppm, using a q-TOF, Orbitrap, or magnetic sector instrument) and deduce the molecular formula. Sometimes multiple formulae may be possible. The use of MS fragmentation and other information, e.g., UV absorbance, can often help with selecting the correct molecular formula.

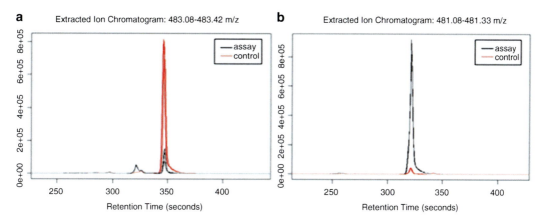

Fig. 1 Metabolic difference reported by XCMS in a study of P450 107U1. (**a**) A putative substrate ((M + H)⁺ *m/z* 483) was found to be depleted in the reaction (*black*). (**b**) Accumulation of the product derived from the desaturation of the putative substrate ((M + H^{-2})⁺ *m/z* 481)

2. Search chemical databases using the derived formula (e.g., chemspider: http://www.chemspider.com/, NCBI pubchem: http://pubchem.ncbi.nlm.nih.gov/, NIST Chemistry Webbook: http://webbook.nist.gov/chemistry/, and Scifinder: https://scifinder.cas.org/scifinder); determine whether the substrate formula matches any known compound. After the integration of all known information, if any known compound appears to be the unidentified substrate then obtain the standard compound (if available commercially) and compare its HPLC elution time and MS fragmentation pattern with those of the unidentified substrate.

3. If the substrate appears to be a novel chemical entity, collect enough material and identify the structure by crystallography or (more realistically) NMR. A minimum amount for one-dimensional ^{1}H-NMR work is ~1 μg, but an order of magnitude more is required for two-dimensional ^{1}H studies and even more is needed if natural abundance ^{13}C-NMR is required.

4 Notes

1. Recombinant P450s are required for this experiment to minimize the metabolic interference conferred by other enzymes. Expression of bacterial P450s in *Escherichia coli* is generally straightforward. On the other hand, expression of mammalian P450s in *E. coli* often requires sequence optimization and may involve other expression hosts (e.g., baculovirus-based insect cells).

2. Extraction with ethyl acetate will enrich moderately hydrophobic compounds. To enrich highly hydrophobic compound, e.g.,

lipids and steroids, solvents such as chloroform or hexane should be used. If the suspected substrate is relatively hydrophilic, a polar solvent (e.g., methanol) should be considered. Avoid using plastic containers or pipettes when preparing tissue extract because plastic polymers may leach plasticizers (especially in organic solvents) and contaminate the prepared extract.

3. UPLC is recommended for its short gradient times, high resolution, and highly reproducible retention time. If not available, other analytical HPLC systems can also be used in these experiments. PDA is not required but highly recommended because the additional UV information can be very helpful in the isolation and identification of the substrate.

4. An octadecylsilane (C_{18}) reversed-phase column is the most common choice in metabolomics studies of this type. Depending on the extraction method, a C_8 reversed-phase or normal-phase column may be used for the separation of highly hydrophobic compounds.

5. For R language, go to http://www.r-project.org/ and follow the instructions for download and installation. For XCMS, go to website: http://metlin.scripps.edu/xcms/installation.php and follow the instruction to download and install XCMS into R.

6. A basic operation manual of XCMS can be downloaded at http://metlin.scripps.edu/xcms/faq.php. Click "LC/MS Preprocessing and Analysis with XCMS" and follow the instruction in the downloaded manual.

7. Even with high-quality LCMS data, false metabolic differences may be reported by XCMS. Therefore, it is very important to first check the raw LCMS data to ensure that the reported difference is valid. Second, discard insignificant differences, which may be caused by retention-time drifting, ion-source contamination, or any instrumental variation. A standard decision point is somewhat subjective. In our practice, a substrate candidate will not be selected for follow-up unless >50% of this molecule is consumed during the reaction, under prolonged conditions. If no candidate can meet the criterion, a lower setting is used until about ten candidates are picked.

References

1. Guengerich FP et al (2011) Approaches to deorphanization of human and microbial cytochrome P450 enzymes. Biochim Biophys Acta 1814:139–145

2. Guengerich FP, Cheng Q (2011) Orphans in the human cytochrome P450 superfamily: approaches to discovering functions and relevance in pharmacology. Pharmacol Rev 63: 684–699

3. Guengerich FP et al (2010) Characterizing proteins of unknown function: orphan cytochrome P450 enzymes as a paradigm. Mol Interv 10:153–163

4. Fraga CG et al (2010) Signature-discovery approach for sample matching of a nerve-agent precursor using liquid chromatography-mass spectrometry, XCMS, and chemometrics. Anal Chem 82:4165–4173

5. Smith CA et al (2006) XCMS: processing mass spectrometry data for metabolite profiling using nonlinear peak alignment, matching, and identification. Anal Chem 78:779–787

6. Cheng Q et al (2010) Cyclization of a cellular dipentaenone by *Streptomyces coelicolor* cytochrome P450 154A1 without oxidation/reduction. J Am Chem Soc 132:15173–15175

7. Hanna IH et al (1998) Role of the alanine at position 363 of cytochrome P450 2B2 in influencing the NADPH- and hydroperoxide-supported activities. Arch Biochem Biophys 350:324–332

Chapter 7

Genetic and Mass Spectrometric Tools for Elucidating the Physiological Function(s) of Cytochrome P450 Enzymes from *Mycobacterium tuberculosis*

Hugues Ouellet, Eric D. Chow, Shenheng Guan, Jeffery S. Cox, Alma L. Burlingame, and Paul R. Ortiz de Montellano

Abstract

Tuberculosis remains a leading cause of human mortality. The emergence of strains of *Mycobacterium tuberculosis* (*Mtb*), the causative agent, that are resistant to first- and second-line antitubercular drugs urges the development of new therapeutics. The genome of *Mtb* encodes 20 cytochrome P450 enzymes, at least some of which are potential candidates (CYP121, CYP125, and CYP128) for drug targeting. In this regard, we examined the specific role of CYP125 in the cholesterol degradation pathway, using genetic and mass spectrometric approaches. The analysis of lipid profiles from *Mtb* cells grown on cholesterol revealed that CYP125, by virtue of its C26-monooxygenase activity, is essential for cholesterol degradation, and, consequently, for the incorporation of side-chain carbon atoms into cellular lipids, as evidenced by an increase in the mass of the methyl-branched phthiocerol dimycocerosates (PDIM). Moreover, this work also led to the identification of cholest-4-en-3-one as a source of cellular toxicity. Herein, we describe the experimental procedures that led to elucidation of the physiological function of CYP125. A similar approach can be used to study other important *Mtb* P450 enzymes.

> **Key words** *Mycobacterium tuberculosis*, Cytochrome P450, Specialized transduction, Mycobacteriophage, Total lipid, Cholesterol, LTQFT/MS, LC–MS

---

1 Introduction

Identification of virulence factors is essential for understanding the biology of a pathogen, with the sequencing of the whole *Mtb* genome being an important step towards this objective (1). The analysis of the genome revealed that a very large proportion of the coding sequence encodes for proteins and enzymes involved in lipogenesis (2–4) and lipolysis (5–7). Structure/function studies of the enzymes participating in these key metabolic processes are expected to provide useful information for the development of new therapeutics.

Ian R. Phillips et al. (eds.), *Cytochrome P450 Protocols*, Methods in Molecular Biology, vol. 987, DOI 10.1007/978-1-62703-321-3_7, © Springer Science+Business Media New York 2013

Fig. 1 Structures of long-chain multimethyl-branched-containing lipids DAT, PAT, PDIM, PGL, and SL-1 of *Mtb*. In PAT (polyacyltrehalose), trehalose is esterified with stearic acid and multimethyl-branched mycolipenic acid. In DAT (2,3-di-*O*-acyltrehalose), trehalose is esterified with stearic acid and multimethyl-branched mycosanoic acid. In PDIM (phthiocerol dimycocerosates), the long-chain phthiocerol moiety is esterified with two mycocerosic acids; *n* = 10–11; *R* = ethyl or methyl. In PGL (phenolic glycolipids), the lipid core is composed of phenolphthiocerol esterified by mycocerosic acids; *m* = 7–8; *R* = ethyl or methyl. In SL-1 (sulfolipid-1), trehalose is sulfated at the 2′ position and esterified with palmitic acid and the multimethyl-branched phthioceranic and hydroxyphthioceranic acids. The figure is adapted from (4)

The cell envelope of pathogenic mycobacteria, including *Mtb*, is very distinct as it contains numerous lipids esterified with long-chain multi-methyl-branched acyl groups, such as the phthiocerol dimycocerosates (PDIM) and the closely related phenolic glycolipids (PGL); the trehalose ester sulfatides, including SL-1; diacyltrehaloses (DAT); and polyacyltrehaloses (PAT) (Fig. 1) (4). The biosynthesis of these lipids involves multifunctional polyketide synthases similar to the type I fatty-acid synthase (FAS-I) of eukaryotes (8, 9). However, these enzymes are distinct from FAS-I in that they introduce methyl branches into fatty-acid chains by preferentially utilizing methyl-malonyl-coenzyme A (MMCoA) instead of malonyl-CoA. In this regard, Jain et al. observed that the size and abundance of PDIM and SL-1 are metabolically coupled to the availability of MMCoA (10). Growth on odd-chain fatty acids (10, 11) and cholesterol (12–14) was shown to alter the size of methyl-branched-containing lipids, including PDIM (Fig. 1), presumably via the generation of a higher metabolic flux of propionyl-CoA, which is then converted into MMCoA.

Another class of enzymes known to be involved in lipid metabolism is the cytochromes P450, of which there are 20 isoforms in *Mtb* and even more in other pathogenic and nonpathogenic mycobacteria and related actinomycetes. The importance of P450 enzymes for *Mtb* biology is supported by the antimycobacterial activity of azole drugs that are known P450 inhibitors and are clinically used to treat fungal infections. Moreover, transposon-site hybridization (TraSH) studies revealed that CYP125 (15) and CYP128 (16) are essential for infection in mice and for growth in vitro, respectively. Recently, we and others reported that CYP125 initiates cholesterol side-chain degradation, an essential activity for growth on this carbon source. The *cyp128* gene is part of an operon with *sft3*, a sulfotransferase involved in biosynthesis of the sulfolipid S881 (17).

Although transposon mutagenesis constitutes an excellent screening tool to evaluate gene essentiality for growth under experimental conditions, gene disruption by allelic replacement remains the method of choice for understanding gene function in mycobacteria. Furthermore, in the last few years, high-resolution mass spectrometry has proven to be a powerful tool to compare lipidomic (10, 17) and metabolomic profiles (18, 19). In this work, we describe how we used a combination of genetic and mass spectrometric techniques to elucidate the physiological function of CYP125, an approach applicable to the other *Mtb* P450 enzymes as well as other potential drug targets.

2 Materials

2.1 Media, Solutions, and Commercial Kits for Generation and Growth of Mtb Gene-Disrupted Mutants

All solutions and growth media are prepared using ultrapure deionized water and analytical or cell-culture grade reagents.

2.1.1 Solid Growth Media

1. LB agar with hygromycin: To 1 L of deionized water, add 10 g of NaCl, 10 g of Bacto Tryptone (Difco), 5 g of yeast extract

(Difco), and 15 g of Bacto Agar (Difco). Autoclave for 20 min and cool to ~45°C. Add hygromycin B (Invitrogen) (to a final concentration of 150 μg/mL). Pour ~25 mL per petri dish. Let set and dry (overnight) (O/N). Wrap in foil and store at 4°C.

2. 7H10ADS: To 900 mL of deionized water, add 19 g of Middlebrook 7H10 agar base (Difco) and 10 mL of 50% glycerol. Autoclave for 20 min and cool to ~45°C. Add 100 mL of filter-sterilized ADS supplement (5% (w/v) bovine serum albumin fraction V, 2% (w/v) glucose, and 0.85% (w/v) NaCl). Pour ~25 mL per petri dish. Let set and dry O/N. Wrap in foil and store at 4°C.

3. 7H10OADC with antibiotics: To 900 mL of deionized water, add 19 g of Middlebrook 7H10 agar base and 10 mL of 50% glycerol. Autoclave for 20 min and cool to ~45°C. Add 100 mL of commercial Middlebrook OADC supplement (BBL) and hygromycin B (to a final concentration of 50 μg/mL), zeocin (Invitrogen) (to 50 μg/mL), or kanamycin sulfate (Invitrogen) (to 25 μg/mL). Pour ~30 mL per petri dish. Let set and dry O/N. Wrap in foil and store at 4°C.

4. Top agar: To 500 mL of deionized water, add 3 g of bacto-agar, 1 mL of 50% glycerol, and 1 mL of 1 M $CaCl_2$. Sterilize by autoclaving. Aliquot 4 mL of melted top agar in sterile tubes and store at 4°C until use. Melt in microwave for 30 s and keep warm at 56°C.

2.1.2 Liquid Growth Media

1. 7H9ADS + 0.05% Tween-80: To 900 mL of deionized water, add 4.7 g of Middlebrook 7H9 broth base (Difco) and 4 mL of 50% glycerol. Add 100 mL of filter-sterilized ADS (5% (w/v) bovine serum albumin fraction V, 2% (w/v) glucose, and 0.85% (w/v) NaCl) and 5 mL of 10% (v/v) Tween-80 prepared in water.

2. 7H9AS + 0.05% Tween-80: To 900 mL of deionized water, add 4.7 g of Middlebrook 7H9 broth base and 4 mL of 50% glycerol. Add 100 mL of filter-sterilized AS (5% (w/v) bovine serum albumin fraction V and 0.85% (w/v) NaCl) and 5 mL of 10% (v/v) Tween-80 prepared in water.

3. Minimal medium (MM): To 1 L of deionized water, add 0.5 g of L-asparagine, 1 g of KH_2PO_4, 2.5 g of Na_2HPO_4, 50 mg of ferric ammonium citrate, 50 mg of $MgSO_4 \cdot 7H_2O$, 0.5 mg of $CaCl_2$, and 0.1 mg of $ZnSO_4$. As a carbon source, add glucose (to a final concentration of 0.2% (w/v)) or cholesterol (to a final concentration of 0.1 mM). A 100 mM stock solution of cholesterol is prepared by dissolving 38.7 mg of cholesterol in 1 mL of ethanol-Tyloxapol (1:1). To dissolve cholesterol completely the solution is incubated for 20 min at 70°C and vor-

Genetic and Mass Spectrometric Tools for Elucidating... 83

texed frequently. Cholesterol is quickly added while still warm to filter-sterilized MM medium. The ethanol-Tyloxapol is also added to MM medium without cholesterol.

2.1.3 Solutions

1. MP (phage buffer): To 800 mL of deionized water, add 50 mL of 1 M Tris-Cl (pH 7.4), 10 mL of 1 M $MgSO_4$, 2 mL of 1 M $CaCl_2$, and 30 mL of 5 M NaCl. Complete to 1 L by addition of deionized water. Sterilize by autoclaving. Store at room temperature (RT).

2. TE: To 900 mL of deionized water, add 10 mL of 1 M Tris-Cl (pH 7.4) and 2 mL of 0.5 M EDTA (pH 8.0). Complete to 1 L by addition of water. Sterilize by autoclaving. Store at RT.

3. GTE: To 90.5 mL of deionized water, add 2.5 mL of 1 M Tris-Cl (pH 8.0), 2 mL of 0.5 M EDTA, and 5 mL of 1 M glucose. Store at RT.

4. Cetyltrimethyl ammonium bromide (CTAB): To 90 mL of deionized water, add 4.1 g NaCl and 10 g of CTAB. Complete to 100 mL by addition of water and mix by vortexing. Store at RT. Must be warmed at 65°C before use.

5. Ethanol 70%: To 300 mL of deionized water, add 700 mL of absolute ethanol.

6. Lysozyme solution, 10 mg/mL in TE.

7. DNAse I-free RNAse A solution, 10 mg/mL in TE.

8. Proteinase K solution, 10 mg/mL in TE.

9. Phenol:chloroform:isoamyl alcohol (25:24:1).

10. Vesphene IIse (Steris), diluted 1:40 in water.

2.1.4 Commercial Kits and Equipment

1. Stratagene Gigapack III XL packaging extracts (Agilent Technologies, La Jolla, CA).

2. QIAprep Spin Miniprep (Qiagen, Valencia, CA).

3. Gene Pulser electroporator (Bio-Rad) with 0.2-cm cuvettes.

2.2 Extraction of Lipids

1. 15% isopropanol in PBS.

2. Cobalt/thiocyanate solution: To 70 mL of deionized water, add 3 g of cobalt nitrate and 20 g of ammonium thiocyanate. Complete to 100 mL by addition of water. Store at RT in the dark.

3. Chloroform:methanol (17:1).

4. Hexane.

2.3 Analysis of Endogenous Lipids

1. Chloroform:methanol:isopropanol (1:2:4, $v/v/v$) containing 7.5 mM ammonium acetate.

2. 10-μL gastight syringe (Hamilton, Reno, NV) for loading samplers into nanospray tips.

3. Medium-size static nanospray tips (Proxeon, Odense, Denmark).

4. LTQ FT mass spectrometer (Thermo Fisher, Bremen, Germany).

2.4 Analysis of Steroid Metabolites

1. 25,26,26,26,27,27,27-D7-cholesterol (Avanti Polar Lipids, Alabaster, AL).

2. XTerra C_{18} column, 3.5 μm, 2.1×50 mm (Waters, Milford, MA).

3. Micromass ZQ mass spectrometer coupled to an Alliance HPLC system equipped with a 2695 separations module, and a 2487 Dual λ Absorbance detector (Waters, Milford, MA).

3 Methods

The experimental workflow shown in Fig. 2 was used to elucidate the physiological function of CYP125. A combination of genetic and mass spectrometry tools can also be used to study other Mtb P450 isoforms.

3.1 Construction of cyp Gene-Disrupted Mutant by Specialized Transduction in Mtb

The deletion mutant strains are created by homologous recombination using specialized transducing phages, as described (20). Construction of allelic-exchange substrates (AES) is made in a BSL-2 laboratory, whereas transduction of Mtb cells with mycobacteriophages is performed in a BSL-3 facility (see Note 1).

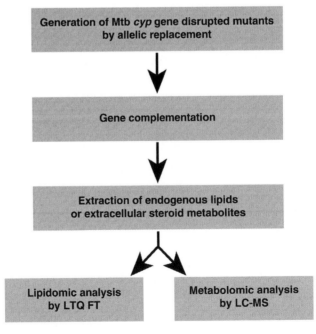

Fig. 2 Schematic diagram summarizing the steps used to elucidate the physiological function of Mtb cytochrome P450 CYP125

Genetic and Mass Spectrometric Tools for Elucidating... 85

3.1.1 Cloning of Gene Flanking Sequences into pJSC407

1. Amplify by PCR approx. 500–1,000-bp fragments from each side of the *cyp* gene to be inactivated. Restriction sites are included in the primers used for amplification. There are a number of restriction sites on both sides of the res-hyg cassette in pJSC407. *Spe*I-*Hin*d III and *Xba*I-*Asp718* sites are commonly used for the 5′ and 3′ ends, respectively.

2. Subclone the flanking fragments into a cloning vector (see Note 2) and sequence to confirm correctness.

3. Subclone the 5′ fragment into pJSC407 upstream of the res-hyg cassette, followed by insertion of the 3′ fragment downstream.

4. Digest the resulting plasmid with *Pac*I restriction endonuclease and dephosphorylate using calf-intestine phosphatase.

3.1.2 Packaging of Mycobacteriophages

1. Self-ligate a few μg of phAE87 (TM4-ts::*amp* plasmid) in as small a volume as possible (see Note 3).

2. Heat-inactivate the ligase and digest the DNA with *Pac*I (see Note 4).

3. Purify the digested phage DNA by extraction with phenol-chloroform; precipitate with 1/10th volume of sodium acetate, pH 5.2, and 2.5 volume of ethanol; and resuspend in TE.

4. Determine the approximate concentration of both phage and *Pac*I-digested and dephosphorylated plasmid, prepared above, by electrophoresis through an agarose gel.

5. Ligate the plasmid and phage DNA in as small a volume as possible (see Note 5).

6. Package the ligated DNA using the Gigapack XL kit, following the recommendation of the manufacturer.

7. Transduce *E. coli* HB101 and plate the whole packaging reaction mixtures on a few LB-agar plates containing hygromycin (150 μg/mL). Incubate O/N at 37°C.

8. Pool *hyg*-resistant colonies by scraping with 2 mL of LB medium containing hygromycin (100 mg/mL).

9. Transfer into two 1.5-mL microtubes and pellet cells by centrifugation in a microfuge at maximum speed for 30 s. Discard supernatant.

10. Resuspend the pellets with 100 μL of LB medium containing hygromycin (100 μg/mL) and expand onto a fresh LB-agar plate containing hygromycin (150 μg/mL). Incubate O/N at 32°C.

11. Scrape the lawn and prepare the cosmid DNA by taking through the neutralization step of the QIAprep Spin Miniprep kit's protocol.

12. After the centrifugation step to remove genomic DNA, proteins, and other cellular debris, extract with 1 volume of phenol:chloroform:isoamyl alcohol (25:24:1, *v/v/v*).

13. Centrifuge in a microfuge at maximum speed for 1 min and collect the aqueous phase.

14. Extract with 1 volume of chloroform:isoamyl alcohol (24:1, *v/v*).

15. Centrifuge at maximum speed for 1 min and collect the aqueous phase.

16. Precipitate cosmid DNA by adding 0.7 volume of cold isopropanol. Keep on ice for 10 min.

17. Centrifuge at maximum speed for 20 min and discard the supernatant carefully.

18. Rinse the pellet with 1 volume of 70% ethanol and centrifuge at maximum speed for 5 min.

19. Discard the supernatant carefully and air-dry.

20. Dissolve cosmid DNA with 100 µL of TE.

21. Transform 1, 2, and 5 µL of the cosmid DNA into 400 µL of electro-competent *M. smegmatis* mc^{2}155 (*Msm*), using an electroporator set at 2.5 kV, 25 µF, 1,000 Ω (see Note 6).

22. Pellet the transformed bacteria and resuspend in 200 µL of 7H9 broth base (no Tween-80) (see Note 7).

23. Combine the transformed bacteria with 3 mL of 50°C melted top agar and pour onto 7H10ADS plates (no Tween-80) prewarmed at 30°C (see Note 7).

24. Allow to solidify and then incubate at 30°C for 2 days.

25. Pick two or three single plaques into 200 mL of MP buffer by inserting the end of a sterile glass Pasteur pipette into the agar around the plaque and expelling the plug into the MP buffer.

26. Incubate at RT for 2 h. Centrifuge for a few seconds and transfer the phage solution into a clean and sterile microtube. Store at 4°C.

27. To amplify the phage, mix, in an Eppendorf tube, 3 and 10 mL of phage/MP solution with 400 mL of *Msm* cells from an O/N culture.

28. Mix the contents of each tube with 3 mL of melted (50°C) top agar, pour onto a pre-warmed 7H10 plate, and incubate at 30°C for 2 days.

29. Add 2.5 mL of MP buffer to the phage plate and incubate at RT for 2 h with gentle agitation.

30. Draw up the phage solution into a 5-mL syringe, using a 19-gauge needle. Discard the needle and filter the phage through a 0.45-µm filter. Store at 4°C.

Genetic and Mass Spectrometric Tools for Elucidating...   87

31. To titer phage, grow a dense O/N culture of *Msm* and combine 200 µL of culture with 3 mL of melted (50°C) top agar. Pour onto a 7H10 (no Tween, see Note 7) plate and allow to solidify.

32. Make serial dilutions of your phage in MP buffer and spot 5 µL of each dilution onto the solidified agar surface. Incubate 10^{-4}–10^{-8} dilutions at 30°C and 10^0–10^{-4} dilutions at 37°C. The latter dilution set is used to verify the temperature sensitivity of the phage (see Note 8).

3.1.3 Transduction of Mtb Cells

The following steps are performed in a BSL-3 laboratory (see Note 9):

1. Grow *Mtb* in a plastic 1-L roller bottle containing 7H9ADS + 0.05% (v/v) Tween-80 to an OD_{600} of ~0.8–1.0. Incubate the roller bottles at 37°C in a roll-in incubator.

2. For each transduction, centrifuge 10 mL of culture and resuspend in 10 mL of washing medium (7H9ADS without Tween-80) and incubate as a standing culture at 37°C for 24 h. This incubation is included to remove traces of the Tween-80 detergent, which can inhibit phage infection. After this incubation period again centrifuge the cells and resuspend in 1 mL of 7H9ADS broth without Tween-80, pre-warmed at 37°C, and mix with specialized transducing phage at a multiplicity of infection (moi) of 10.

3. Incubate the cell/phage mixture at the nonpermissive temperature (37°C) for 3 h, after which inoculate the mixture into 50 mL of complete 7H9ADS + 0.05% (v/v) Tween-80, pre-warmed at 37°C.

4. Perform outgrowth of the cultures for 24 h at 37°C.

5. Pellet cells by centrifugation, resuspend in 1 mL of PBS containing 0.1% (v/v) Tween-80, and plate on 7H10OADC solid medium containing hygromycin (50 µg/mL). Colonies appear after 3–5 weeks.

6. Pick a few hygromycin-resistant clones and analyze by Southern blot analysis to find the allelic-exchange mutants.

3.1.4 Preparation of Mtb Genomic DNA and Southern Blot Analysis

The following steps are performed in a BSL-3 laboratory:

1. Grow up a 15-mL culture of the parental strain and a few hygromycin-resistant clones in 7H9 + OADC + 0.05% Tween-80, containing hygromycin (50 µg/mL), to an OD of ~0.8.

2. Pellet the culture in a 15-mL Falcon tube and resuspend in 1 mL of TE.

3. Transfer to a 2-mL screw-cap microtube, then pellet again, and discard the supernatant.

4. A multi-block heater, preset to 80°C, is brought into a clean area of the biosafety cabinet and the screw-capped microtube is placed in it. The block is then covered with a piece of aluminum foil. The culture is left at 80°C for 1 h to kill the bacteria.

5. Spray the tubes and tube rack vigorously with Vesphene IIse, to decontaminate, and remove from the BSC and the BSL-3.

The following steps can be performed in a BSL-2 laboratory:

1. Rinse the pellet in 1 mL of GTE solution, by vortexing, and pellet again.

2. Resuspend the pellet in 450 μL of GTE, by vortexing.

3. Add 50 μL of lysozyme (10 mg/mL) and 5 μL of DNAse I-free RNAse A (10 mg/L), mix gently, and incubate O/N at 37°C.

4. Add 100 μL of 10% SDS and mix gently.

5. Add 50 μL of proteinase K (10 mg/mL), mix gently, and incubate at 55°C for 45 min.

6. Preheat an aliquot of CTAB solution at 65°C.

7. Add 200 μL of 5 M NaCl and mix gently.

8. Add 160 μL of CTAB, mix by inversion, and incubate for 10 min 65°C.

9. Add 1 mL of chloroform:isoamyl alcohol (24:1, v/v), shake gently to mix, and centrifuge for 5 min.

10. Transfer 900 μL of the aqueous phase to a new tube and repeat the extraction with an equal volume of chloroform:isoamyl alcohol (24:1).

11. Transfer 800 μL of the aqueous layer to a new tube, add 0.7 volume of isopropanol, and mix by inversion.

12. Incubate at RT for 5 min.

13. Pellet genomic DNA by centrifugation for 10 min at maximum speed in a microfuge.

14. Discard the supernatant gently, rinse the pellet with 500 μL of 70% ethanol, and centrifuge again for 5 min.

15. Remove the supernatant gently and air-dry the DNA pellet for 5–10 min.

16. Resuspend the DNA in 50–100 μL of TE (see Note 10).

17. Make sure always to gently mix the genomic DNA, to avoid shearing.

18. Digest approx. 1 μg of genomic DNA with the appropriate restriction enzyme.

19. Electrophorese the digested DNA through an agarose gel, transfer the DNA to a membrane, and hybridize with a DNA

Genetic and Mass Spectrometric Tools for Elucidating... 89

probe that will result in different hybridization patterns between the parental strain and deletion candidates.

3.1.5 Complementation of Mtb cyp-Disrupted Mutants

All the cloning steps required to generate the constructs will be performed in a BSL-2 laboratory, whereas the introduction of the plasmid constructs by electroporation is conducted in a BSL-3 laboratory. The *cyp* genes to be complemented are amplified by PCR from genomic DNA and subcloned into an integrative vector (pMV306), for electroporation of *Mtb* cells, and transformants are isolated at 37°C on 7H10+OADC plates containing zeocin (50 µg/mL) or kanamycin (25 µg/mL) (see Note 11):

1. Grow a 50-mL culture of *Mtb* Δcyp cells to mid-log phase ($OD_{600} = 0.6-1.0$) in 7H9+ADS+0.05% Tween-80.

2. Transfer the cells to 50-mL Falcon tubes and pellet by centrifugation at $2,000 \times g$ for 15 min.

3. Decant the supernatant and wash the cells with 50 mL of filter-sterilized 10% glycerol.

4. Pellet cells by centrifugation at $2,000 \times g$ for 15 min, decant supernatant, and repeat wash/centrifugation twice.

5. Resuspend the bacterial pellet in 1/100th of the original volume (i.e., 0.5 mL for a 50-mL culture), with 10% glycerol.

6. For each transformation, incubate 200 µL of bacteria with DNA (1 µg) for 10 min at room temperature.

7. While this incubation proceeds, bring a portable electroporation unit/platform into the biosafety cabinet, with wires trailing out of the cabinet to the pulser. Also bring the necessary number of 0.2-cm electroporation cuvettes into the cabinet.

8. When the incubation is complete, transfer the bacteria/DNA mix into a 0.2-cm electroporation cuvette, being careful that no bubbles are produced. Cap the cuvette and then gently tap it on the benchtop.

9. Place the capped cuvette into the electroporation unit and slide the handle until it locks.

10. Cover the unit with a kimtex rag that has been damped with Vesphene.

11. Shock the cells at 2.5 kV, 25 µF, 200 Ω.

12. Always shock an aliquot of cells without DNA, as a negative control.

13. Immediately add 1 mL of 7H9+ADS+0.05% Tween-80 growth medium to the cuvette and pipet the bacteria into a 15-mL Falcon tube.

14. Incubate overnight at 37°C.

15. Plate 100 µL on 7H10-OADC plates containing the appropriate selective antibiotic.

3.2 Extraction of Total Lipids and Steroids from Mtb Cells

16. Pellet the remaining cells by centrifugation at $1,600 \times g$, for 5 min at RT.

17. Resuspend the cells in 200 μL of 7H9 growth medium and spread onto 7H10 + OADC plates containing the appropriate selective antibiotic.

18. Incubate at 37°C for 2–3 weeks.

19. Pick a few colonies and check for the recovery of the wild-type phenotype.

Growth and harvest of *Mtb* cells are performed in a BSL-3 laboratory. Additional steps are performed in the BSL-2 laboratory:

1. For total lipid- or steroid-profiling experiments, grow wild-type, *cyp*-knockout, and complemented strains in 7H9AS + 0.05% (v/v) Tween-80 medium, to mid-log phase.

2. Harvest cells by centrifugation, wash in the 7H9 broth base, and synchronize cultures to an OD_{600} of 0.2 in 7H9AS + 0.01% Tween-80 containing 0.1% glycerol or 0.1 mM cholesterol.

3. Grow for an additional 24 h.

4. For analysis of secreted steroids, grow *Mtb* strains to mid-log phase in MM medium + 0.05% Tyloxapol (v/v) containing 0.2% (w/v) glucose as a sole source of carbon.

5. Harvest cells by centrifugation, wash in the MM medium + 0.01% Tyloxapol (v/v), and synchronize the cultures to an OD_{600} of 0.2 in MM medium + 0.05% Tyloxapol containing 0.1 mM cholesterol.

6. To extract total lipids or intracellular steroids, harvest cells from 25- to 50-mL cultures by centrifugation at $1,600 \times g$ for 10 min at RT.

7. If necessary, resuspend and wash cells twice with 25 mL of 15% isopropanol in PBS, to remove the excess of unused cholesterol.

8. Resuspend the cells in 1 mL of 15% isopropanol in PBS and adjust all strain at the OD_{600} or all strains.

9. Transfer 1 mL of cell suspension into a 15-mL glass tube containing 5 mL of chloroform:methanol (17:1, v/v) (see Note 12).

10. To extract secreted steroids from MM-spent media, harvest 30 mL of cultures by centrifugation at $1,600 \times g$ for 5 min at RT.

11. Transfer 3×10 mL of supernatant into 15-mL glass tubes containing 5 mL of chloroform.

12. Invert tubes several times, to ensure that the chloroform kills all *Mtb* cells.

Genetic and Mass Spectrometric Tools for Elucidating...   91

13. Spray tubes and tube rack vigorously with Vesphene IIse, to decontaminate, and remove from the BSC and the BSL-3.

14. Bring the samples to the BSL-2 laboratory and, after 24 h, collect the lower organic phase, containing apolar lipids, after centrifugation $1,600 \times g$ for 5 min at RT.

15. For the secreted steroids, collect the chloroform fractions after centrifugation at $1,600 \times g$ or 5 min at RT.

16. Evaporate chloroform from the lipid or secreted-steroid extracts under a stream of argon and resuspend the residue in 1.5 mL of methanol.

17. Remove Tween-80 or Tyloxapol from the samples by adding 0.5 mL of cobalt/thiocyanate solution, according to Shen et al. (21) (see Note 13).

18. Vortex samples for 15 s.

19. Extract the lipids by adding 4 mL of hexane.

20. Vortex the samples for 15 s.

21. Centrifuge at $1,600 \times g$ for 5 min at RT, and collect the upper (organic) phase.

22. Repeat the extraction by adding 4 mL of hexane to the lower (aqueous) phase.

23. Combine both hexane fractions and evaporate to dryness under a stream of argon.

24. Resuspend lipids in 1 mL of chloroform:methanol (2:1, v/v).

3.3 Analysis of Endogenous Lipids

1. For the analysis of the total lipid profiles, data are acquired on an LTQFT/MS instrument (ThermoFisher Scientific, Bremen, Germany).

2. Dilute the lipid samples tenfold in chloroform:methanol: isopropanol (1:2:4, $v/v/v$) containing 7.5 mM ammonium acetate.

3. Load about 5 μL of sample into a static medium nanospray tip, using a 10-μL gastight syringe (see Note 14).

4. Introduce the samples in the mass spectrometer through the loaded nanospray tip. Apply electrospray voltages of 1,000 and –700 V, for positive and negative ionization, respectively.

5. For molecular mass measurements, collect and average 10–100 scans, to produce the final spectra, with a mass resolution of 100,000 and 25,000 for lipid and cholesterol samples, respectively.

6. Acquire the final spectra in the FT mode for multistage CID MSMS (up to four stages in LTQ).

3.4 Identification of the Secreted Cholesterol-Derived Metabolites from Spent Medium

An LC/MS system equipped with a photodiode array detector can be used to isolate and identify cholesterol-derived metabolites accumulating in spent media of *Mtb* cultures.

1. For the identification of the metabolites the extracts are analyzed by LC-MS using a Waters Micromass ZQ coupled to a Waters Alliance HPLC system equipped with a 2695 separations module, a Waters 2487 Dual λ Absorbance detector, and reverse-phase C_{18} column. The mass spectrometer settings were as follows: mode, ES+; capillary voltage, 3.5 kV; cone voltage, 25 V; and desolvation temperature, 250°C.

2. For separation by LC, the samples are reconstituted in acetonitrile:water (70:30) containing 0.1% formic acid.

3. Metabolites are eluted at a flow rate of 0.5 mL/min (solvent A, H_2O + 0.1% formic acid (FA); solvent B: CH_3CN + 0.1% FA) with a gradient starting at 70% B up to 2 min and the solvents ramped up to 100% B over 12 min. The elution was kept at 100% B up to 14 min and then ramped back to 70% B by 15 min and equilibrated under the same conditions for 2 min before the next run. The absorbance of the elution was monitored at 240 nm.

4. Cholesterol and its derivatives are analyzed using the total ion current signal obtained for the specific mass with selective ion monitoring.

5. For quantification of the relative amount of cholesterol-derived metabolites, 25,26,26,26,27,27,27-D7-cholesterol is added as an internal standard at the time of extraction. The individual metabolites are thus quantified as a ratio of the integrated ion current signal of the corresponding metabolite to the known signal from the internal standard.

4 Notes

1. The introduction of the AES will confer hygromycin resistance to *Mtb* and is subjected to regulations and approvals.

2. A common commercial cloning vector system can be used (TA cloning vector, pUC, pBluescript II, or other).

3. The ligation reaction works when it gets very viscous.

4. The ligase is inactivated by incubation at 65°C for 20 min. *Pac*I digestion will eliminate the viscosity and the *amp* plasmid should be liberated.

5. To favor the formation of concatemers, a plasmid:phage DNA molar ratio of approx. 2:1 in a total reaction volume of 10–20 mL works best.

6. To prepare *Msm* electro-competent cells, a 50-mL mid-log phase culture in 7H9 + ADS + 0.05% Tween-80 is harvested by

Genetic and Mass Spectrometric Tools for Elucidating... 93

centrifugation at $1,600 \times g$ for 10 min. Cells are washed twice in 50 mL of filter-sterilized 10% glycerol. Cells are finally resuspended in 0.5 mL of 10% glycerol and kept on ice until used.

7. It is very important not to include Tween-80 in the liquid or solid growth media, as it inhibits phage i nfection.

8. The ideal titer at this stage is ~10^8 pfu/mL.

9. All work involving live *Mtb* cells must be performed in properly functioning biological safety cabinets (BSC) in a BSL-3 facility. Containment is very important, to minimize exposure to bioaerosols. Wear personal protection equipment (disposable gloves, gown, boot covers, and powered-air-purifying respirator (PAPR)). Be sure to disinfect work surfaces and tubes with Vesphene IIse when finished with an experiment and before transferring samples to the BSL-2 laboratory.

10. Always resuspend genomic DNA gently, to avoid shearing.

11. The introduction of these vectors will confer kanamycin or zeocin resistance to *Mtb* cells and is subjected to regulations and approvals.

12. Extraction with organic solvents must be done in glass tubes and PTFE-lined caps, as it may otherwise result in the extraction of plastic polymers. Also, the treatment with chloroform kills *Mtb* cells, so no heat-killing of the bacteria is required.

13. The removal of Tween-80 or Tyloxapol is essential, as the presence of these high-molecular-mass detergents interferes with the detection of endogenous lipids.

14. The lipid samples must be free of particles, as they may result in the clogging of the tip. To fix the problem, the samples can be diluted even more or clarified by a passage through silicone-coated glass wool.

Acknowledgment

This work was supported by NIH R01 grants AI074824 (to PROM) and AI51667 (to J.S.C.) and NIH NCRR grants RR01614 and RR019934 (to A.L.B).

References

1. Cole ST et al (1998) Deciphering the biology of *Mycobacterium tuberculosis* from the complete genome sequence. Nature 393: 537–544

2. Bhatt A, Fujiwara N, Bhatt K, Gurcha SS, Kremer L, Chen B, Chan J, Porcelli SA, Kobayashi K, Besra GS, Jacobs WR Jr (2007) Deletion of *kasB* in *Mycobacterium tuberculosis* causes loss of acid-fastness and subclinical latent tuberculosis in immunocompetent mice. Proc Natl Acad Sci U S A 104:5157–5162

3. Bhatt A et al (2007) The *Mycobacterium tuberculosis* FAS-II condensing enzymes: their role in mycolic acid biosynthesis, acid-fastness, pathogenesis and in future drug development. Mol Microbiol 64:1442–1454

4. Jackson M, Stadthagen G, Gicquel B (2007) Long-chain multiple methyl-branched fatty acid-containing lipids of *Mycobacterium tuberculosis*: biosynthesis, transport, regulation and biological activities. Tuberculosis (Edinb) 87:78–86

5. Mahadevan U, Padmanaban G (1998) Cloning and expression of an acyl-CoA dehydrogenase from *Mycobacterium tuberculosis*. Biochem Biophys Res Commun 244:893–897

6. Ouellet H, Johnston JB, Montellano PR (2011) Cholesterol catabolism as a therapeutic target in *Mycobacterium tuberculosis*. Trends Microbiol 19:530–539

7. Van der Geize R et al (2007) A gene cluster encoding cholesterol catabolism in a soil actinomycete provides insight into *Mycobacterium tuberculosis* survival in macrophages. Proc Natl Acad Sci U S A 104:1947–1952

8. Minnikin DE, Kremer L, Dover LG, Besra GS (2002) The methyl-branched fortifications of *Mycobacterium tuberculosis*. Chem Biol 9:545–553

9. Onwueme KC, Vos CJ, Zurita J, Ferreras JA, Quadri LE (2005) The dimycocerosate ester polyketide virulence factors of mycobacteria. Prog Lipid Res 44:259–302

10. Jain M et al (2007) Lipidomics reveals control of *Mycobacterium tuberculosis* virulence lipids via metabolic coupling. Proc Natl Acad Sci U S A 104:5133–5138

11. Savvi S et al (2008) Functional characterization of a vitamin B12-dependent methylmalonyl pathway in *Mycobacterium tuberculosis*: implications for propionate metabolism during growth on fatty acids. J Bacteriol 190:3886–3895

12. Pandey AK, Sassetti CM (2008) Mycobacterial persistence requires the utilization of host cholesterol. Proc Natl Acad Sci U S A 105:4376–4380

13. Ouellet H et al (2010) *Mycobacterium tuberculosis* CYP125A1, a steroid C27 monooxygenase that detoxifies intracellularly generated cholest-4-en-3-one. Mol Microbiol 77:730–742

14. Yang X et al (2009) Cholesterol metabolism increases the metabolic pool of propionate in *Mycobacterium tuberculosis*. Biochemistry 48:3819–3821

15. Sassetti CM, Rubin EJ (2003) Genetic requirements for mycobacterial survival during infection. Proc Natl Acad Sci U S A 100:12989–12994

16. Sassetti CM, Boyd DH, Rubin EJ (2003) Genes required for mycobacterial growth defined by high density mutagenesis. Mol Microbiol 48:77–84

17. Holsclaw CM et al (2008) Structural characterization of a novel sulfated menaquinone produced by stf3 from *Mycobacterium tuberculosis*. ACS Chem Biol 3:619–624

18. de Carvalho LP et al (2010) Metabolomics of *Mycobacterium tuberculosis* reveals compartmentalized co-catabolism of carbon substrates. Chem Biol 17:1122–1131

19. de Carvalho LP et al (2010) Activity-based metabolomic profiling of enzymatic function: identification of Rv1248c as a mycobacterial 2-hydroxy-3-oxoadipate synthase. Chem Biol 17:323–332

20. Bardarov S et al (2002) Specialized transduction: an efficient method for generating marked and unmarked targeted gene disruptions in *Mycobacterium tuberculosis*, *M. bovis* BCG and *M. smegmatis*. Microbiology 148:3007–3017

21. Shen CF, Hawari J, Kamen A (2004) Microquantitation of lipids in serum-free cell culture media: a critical aspect is the minimization of interference from medium components and chemical reagents. J Chromatogr B Analyt Technol Biomed Life Sci 810:119–127

Chapter 8

An *Escherichia coli* Expression-Based Approach for Porphyrin Substitution in Heme Proteins

Michael B. Winter, Joshua J. Woodward, and Michael A. Marletta

Abstract

The ability to replace the native heme cofactor of proteins with an unnatural porphyrin of interest affords new opportunities to study heme protein chemistry and engineer heme proteins for new functions. Previous methods for porphyrin substitution rely on removal of the native heme followed by porphyrin reconstitution. However, conditions required to remove the native heme often lead to denaturation, limiting success at heme replacement. An expression-based strategy for porphyrin substitution was developed to circumvent the heme removal and reconstitution steps, whereby unnatural porphyrin incorporation occurs under biological conditions. The approach uses the RP523 strain of *Escherichia coli*, which has a deletion of a key gene involved in heme biosynthesis and is permeable to porphyrins. The expression-based strategy for porphyrin substitution detailed here is a robust platform to generate heme proteins containing unnatural porphyrins for diverse applications.

Key words Hemoprotein, P450, Nitric oxide synthase (NOS), Heme Nitric oxide/OXygen-binding (H-NOX) domain, Porphyrin, Heme, Recombinant protein expression, Porphyrin substitution

---

1 Introduction

Heme protein scaffolds have evolved to mediate diverse chemistry such as electron transfer (1), diatomic gas binding (2), and oxygen (O_2) activation (3) with the same heme cofactor. Traditional efforts to probe the function of cytochromes P450 and other heme proteins often have relied on modification of the protein scaffold through site-directed mutagenesis (4) and use of substrate analogs, where available (5). However, these approaches provide a limited repertoire of chemical functionality. In heme proteins, unnatural porphyrins with unique properties can be introduced in place of the native heme (6–9). This powerful strategy provides new avenues to probe and modify heme protein chemistry by introducing novel and diverse functionality at the heme site.

In spite of the potential utility, porphyrin substitution has been an underutilized method to tailor heme protein properties.

Ian R. Phillips et al. (eds.), *Cytochrome P450 Protocols*, Methods in Molecular Biology, vol. 987,
DOI 10.1007/978-1-62703-321-3_8, © Springer Science+Business Media New York 2013

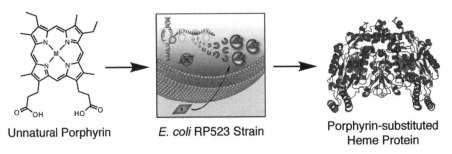

Fig. 1 Schematic for the expression-based incorporation of unnatural porphyrins in heme proteins using the RP523 strain of *E. coli*. The heme domain dimer of nitric oxide synthase (NOS) (PDB ID 1ZVL) is shown as a representative heme protein (20)

Techniques for the incorporation of unnatural porphyrins into heme proteins typically rely on harsh, denaturing conditions to remove the native heme (10–12). This has precluded the use of porphyrin substitution in most hemoproteins since denaturation often accompanies heme removal (8). Heme substitution has been a particular challenge in P450s due to complexities associated with reconstitution and a low recovery of active protein (12, 13). To address these issues, a new strategy was developed whereby unnatural porphyrin incorporation occurs during protein expression. The approach employs the RP523 strain of *Escherichia coli* (*E. coli*) as an expression host (Fig. 1) (8). The RP523 strain cannot biosynthesize heme, due to a disruption of the *hemB* (porphobilinogen synthase) gene, and has an uncharacterized mutation that renders the bacteria permeable to heme (14). These respective features limit native heme contamination and facilitate exogenous porphyrin uptake (8). Recently, other bacterial expression hosts have been introduced for porphyrin substitution (7, 15). However, these expression systems have intact heme biosynthetic pathways (7, 15) and, in one case, may not be broadly useful for recombinant protein production (15).

To evaluate the feasibility of porphyrin substitution using the RP523 strain, Fe(III) mesoporphyrin IX (Fe-MP) and Mn(III) protoporphyrin IX (Mn-PP) were incorporated into the P450-type protein nitric oxide synthase (NOS) (8). NOS is a complex, multi-cofactor enzyme that catalyzes the formation of nitric oxide (NO) through the oxidation of L-arginine to L-citrulline and NO (16). UV–visible spectra of the purified heme domain of inducible NOS (iNOS$_{heme}$) suggested proper unnatural porphyrin incorporation in the heme-binding site (Fig. 2a). Porphyrin substitution was found to occur at near stoichiometric levels with negligible heme contamination (Fig. 3a) (8). Importantly, further evaluation of iNOS$_{heme}$ bound to the unnatural porphyrins confirmed that the protein was catalytically competent, providing evidence that heme substitution with the expression-based method did not significantly perturb the protein structure (8).

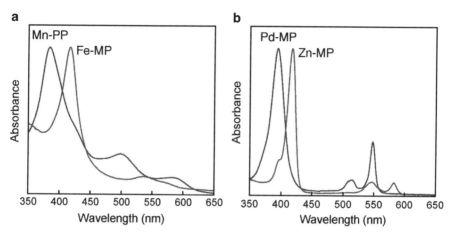

Fig. 2 Characterization of select porphyrin-substituted heme proteins with UV–visible spectroscopy. UV–visible spectra of (**a**) mammalian iNOS_heme substituted with Mn(III) protoporphyrin IX (Mn-PP) and imidazole-bound Fe(III) mesoporphyrin IX (Fe-MP) (8). UV–visible spectra of (**b**) *Tt* H-NOX substituted with Pd(II) mesoporphyrin IX (Pd-MP) and Zn(II) mesoporphyrin IX (Zn-MP)

Recently, the RP523 strategy for porphyrin substitution has been further applied to generate heme protein-based molecular sensors. Ru(II)CO mesoporphyrin IX (Ru-MP), which displays O_2-sensitive phosphorescence, was incorporated into the Heme Nitric oxide/OXygen-binding (H-NOX) domain from *Thermoanaerobacter tengcongensis* (*Tt* H-NOX) and mouse myoglobin to build new agents for biological O_2 detection (9). A crystal structure solved of Ru-MP-containing *Tt* H-NOX showed that porphyrin incorporation did not alter the protein fold and that the Ru porphyrin bound in a manner similar to the native heme (9). Other unnatural porphyrins, such as Zn(II) mesoporphyrin IX (Zn-MP) and Pd(II) mesoporphyrin IX (Pd-MP), have been successfully incorporated into the *Tt* H-NOX scaffold (Fig. 2b) without significant heme contamination (Fig. 3b). Together, these studies demonstrate the versatility of the RP523 expression-based approach.

Here, comprehensive methods are provided for expression-based porphyrin substitution using the RP523 strain of *E. coli*. The inability of RP523 cells to biosynthesize heme is an auxotrophic mutation. Therefore, aerobic growth must be carried out in the presence of hemin (8). First, *E. coli* transformation and selection are performed under aerobic conditions on a hemin-containing medium (see Subheading 3.1). Then, starter cultures (see Subheading 3.2) and large-scale cultures for protein expression (see Subheadings 3.3 and 3.4) are grown under anaerobic conditions. Heme substitution is carried out through addition of the porphyrin of interest upon induction of protein expression. Together, these methods facilitate the straightforward and robust incorporation of unnatural porphyrins in heme proteins under biological conditions.

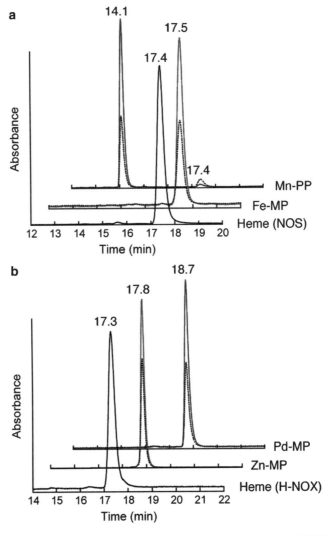

Fig. 3 Reversed-phase HPLC chromatograms of porphyrin-substituted (a) iNOS$_{heme}$ and (b) Tt H-NOX, demonstrating the absence of heme contamination in the isolated proteins. Absorbance data are reported at the λ_{max} of each porphyrin under the HPLC conditions: 371 nm (Mn-PP), 389 nm (Fe-MP), 388 nm (Pd-MP), 405 nm (Zn-MP), and 398 nm (heme) (*solid line* or *dotted line*). Wild-type NOS and Tt H-NOX were used as heme standards. Retention times are reported in minutes. Porphyrins were resolved on a C4 column with a 0–100% gradient of 0.1% formic acid to acetonitrile (containing 0.1% formic acid) over 20 mL at 1 mL/min in a manner similar to that described previously (9). All proteins were purified using established protocols (8, 21)

2 Materials

Aqueous solutions and media are prepared using ultrapure water (sensitivity of 18.2 Ω cm at 25°C). All reagents should be of the highest level of purity available and stored at room temperature unless noted. Follow all regulations for disposal of biological and chemical waste materials:

1. *E. coli* strain RP523 (*E. coli* Genetic Stock Center, Yale University). Make *E. coli* chemically competent as described (see Subheading 3.1).

2. Autoclave the following materials before use: pipette tips, pipettes (if appropriate), microfuge tubes, media, and glassware (as well as associated glassware parts) for *E. coli* growth.

3. Equipment: incubator/shaker for culture growth and UV–visible spectrophotometer for bacterial growth measurements.

4. Hemin (catalog number H651-9) (Frontier Scientific, Logan, UT). Store at 4°C in the dark.

5. Dimethyl sulfoxide (DMSO).

6. Ampicillin. Store at 4°C when dry and freeze at –20°C when in solution.

7. Carbenicillin. Store at 4°C when dry and freeze at –20°C when in solution.

8. Plastic syringes and corresponding sterile syringe filters (0.22 mm).

9. Luria Broth (LB) (catalog number L24040) (Research Products International Corp., Mount Prospect, IL).

10. Agar, granulated.

11. Disposable Petri dishes.

12. Aluminum foil.

13. Hungate tubes (16×125 mm) with septum stoppers (13 mm) (catalog number 2047-16125) (Bellco Glass, Vineland, NJ).

14. D-(+)-glucose.

15. Needles (PrecisionGlide™, 22 G × 1½ in.) (Becton, Dickinson and Company, Franklin Lakes, NJ).

16. Ultra high purity argon (99.999%) (Praxair, Inc., Danbury, CT).

17. Terrific Broth (TB) (catalog number T15000) (Research Products International Corp.) or other optimized medium.

18. Antifoam (e.g., polypropylene glycol P2,000) (catalog number 81380) (Sigma-Aldrich Co. LCC, St. Louis, MO).

19. N_2 gas, house or prepurified grade (99.998%) (Praxair, Inc.).

100   Michael B. Winter et al.

20. Gas-tight tubing.

21. Sterile filter (Millex-FG$_{50}$, 0.2 μm) (catalog number SLFG05010) (Millipore, Billerica, MA).

22. Disposable cuvettes for bacterial growth measurements.

23. Unnatural porphyrins: Fe(III) mesoporphyrin IX chloride (catalog number FeM658) and Mn(III) protoporphyrin IX chloride (catalog number MnP562-9) (Frontier Scientific). Pd(II) mesoporphyrin IX (Pd-MP) was prepared according to Cowan et al. (6). Zn(II) mesoporphyrin IX (Zn-MP) was prepared using a procedure similar to Ito et al. (17). Experimental details for the preparation of Pd-MP and Zn-MP are provided (see Note 1). Store porphyrins at –20°C or 4°C in the dark.

24. Isopropyl β-D-1-thiogalactopyranoside (IPTG). Store at 4°C when dry and freeze at –20°C when in solution.

25. 2-L glass bottles (Corning Incorporated, Corning, NY).

26. VICI-Cap (GL45, 2 Ports) (catalog number JR-S-11001) (Lab Unlimited, UK), or similar HPLC caps for glass bottles, with associated HPLC tubing and fittings.

27. 4-L Erlenmeyer glass flasks.

3   Methods

3.1   Day 1: Transformation of E. coli (RP523) and Growth on Hemin Plates

Hemin and other porphyrins are prone to photodegradation. Therefore, care should be taken throughout to limit exposure to light:

1. Prepare a 200× solution of hemin in DMSO (6 mg/mL).

2. Prepare a 1,000× solution of appropriate antibiotic (e.g., 75–100 mg/mL ampicillin or carbenicillin), and sterile filter the solution with a syringe.

3. Transfer the hemin (see Note 2) and antibiotic to autoclaved LB agar.

4. Pour the LB agar into Petri dishes and allow it to solidify under aluminum foil. Store the plates at 4°C wrapped in aluminum foil.

5. Transform *E. coli* (RP523) with the appropriate expression plasmid (see Note 3). Chemically competent RP523 cells can be prepared using standard techniques with slight modifications (see Note 4).

6. Transfer *E. coli* to the hemin plates. Incubate the plates overnight (approximately 12 h) in the dark until colonies appear.

3.2   Day 2: Anaerobic E. coli (RP523) Starter Cultures

1. Transfer sterile LB to four autoclaved Hungate tubes (approximately 14 mL of medium each) (see Note 5).

2. Prepare a 100× solution of 20% D-(+)-glucose (w/v in water), and sterile filter the solution with a syringe (see Note 6).

Expression-Based Porphyrin Substitution 101

3. Prepare a 1,000× solution of antibiotic as described (see Subheading 3.1, step 2).

4. Transfer the sterile glucose (140 μL) and antibiotic solutions (14 μL) to each of the Hungate tubes.

5. Cap the Hungate tubes with sterile stoppers and carefully invert to mix.

6. Pick individual RP523 colonies from the hemin plates and inoculate each tube.

7. Pierce the septum of each Hungate tube with an entrance and exit needle.

8. Sparge the headspace of the tubes with ultra high purity argon for ≥30 min (see Note 7).

9. Incubate the cultures at 37°C for 12 h (or until dense *E. coli* growth is achieved) while shaking at approximately 250 rpm.

3.3 Days 2–4: Expression-Based Porphyrin Incorporation (Fermentor)

The fermentor setup (see Note 8) used in this section facilitates continuous degassing of larger RP523 *E. coli* cultures with N_2. Alternatively, cultures can be grown anaerobically in sealed bottles (see Subheading 3.4).

Day 2: Preparation of the fermentor

1. Transfer approximately 9 L of sterile TB (see Note 9) to the autoclaved fermentation vessel (volumes given below assume a 9-L scale).

2. Add antifoam to the medium to a final concentration of approximately 0.005% (v/v).

3. Connect the fermentor to N_2 gas through gas-tight tubing in-line with a sterile filter. Adjust the N_2 flow to achieve gentle bubbling. Titrate the medium with antifoam as needed to prevent excessive foaming.

4. Sparge the fermentor overnight with N_2 (>12 h) to ensure full deoxygenation. To limit O_2 contamination, continuously sparge the fermentor with N_2 throughout subsequent steps unless otherwise noted.

Days 3–4: *E. coli* growth and protein expression

5. Prepare a 100× solution (90 mL) of D-(+)-glucose as described (see Subheading 3.2, step 2), and sterile filter the solution into the medium with a syringe.

6. Prepare a 1,000× solution (9 mL) of antibiotic as described (see Subheading 3.1, step 2), and sterile filter the solution into the medium with a syringe.

7. Thoroughly mix the medium.

8. Remove a sample of medium and save as a reference for *E. coli* growth.

9. Warm the medium to 37°C.

102   Michael B. Winter et al.

10. Uncap each Hungate tube (see Note 10) and immediately inoculate the medium with the starter cultures.

11. Grow the expression culture until an OD_{600} of 0.8–1.0 is reached (approximately 4–8 h).

12. If needed, cool the medium to the desired expression temperature (e.g., 25°C). Expression temperatures may be adjusted to optimize protein expression and/or porphyrin incorporation.

13. Cover the fermentor with aluminum foil to prevent porphyrin photodecomposition during *E. coli* growth.

14. Prepare a 200× stock of porphyrin in DMSO (see Note 11) and pour the solution into the culture (3–30 µg/mL final concentration). Final porphyrin concentrations may need to be optimized depending on porphyrin toxicity, solubility, efficiency of *E. coli* uptake, protein expression levels, and/or porphyrin incorporation levels (see Notes 12 and 13).

15. Prepare a 1,000× solution (9 mL) of IPTG (in water), and sterile filter the solution into the culture with a syringe. Final IPTG concentrations of 1 mM are typically used but can be adjusted to optimize protein expression levels.

16. Let the culture grow under a constant flow of N_2 (see Note 14) for 10–22 h in the absence of light at the desired induction temperature. Expression times can be optimized to improve protein yield and porphyrin incorporation levels (see Note 12).

17. Harvest the expression culture (see Note 15).

18. Pellet the *E. coli* by centrifugation and store the pellets at −80°C prior to lysis and protein purification.

19. Follow recommended purification procedures for the protein of interest (see Note 16) but in the absence of light.

3.4 Days 3–4: Expression-Based Porphyrin Incorporation (Anaerobic Bottles)

Anaerobic protein expression can also be carried out in sealed bottles as an alternative to the fermentor approach:

1. Prepare sterile 1,000× antibiotic (see Note 17) and 100× glucose solutions as described (see Subheading 3.1, step 2 and Subheading 3.2, step 2, respectively).

2. Transfer the antibiotic (1.5 mL) and glucose (15 mL) solutions to each of four autoclaved 2-L bottles containing 1.5 L of TB (volumes given assume a 6-L expression scale).

3. Thoroughly mix the medium.

4. Inoculate the medium with the starter cultures.

5. Fit the bottles with HPLC caps containing HPLC tubing (see Note 18).

6. Degas the cultures for ≥15 min with ultra high purity argon through the HPLC tubing.

Expression-Based Porphyrin Substitution   103

7. Incubate the cultures at 37°C for 10–12 h while shaking at 120 rpm.

8. For aerobic expression (see Note 14), transfer each culture to a 4-L expression flask.

9. Prepare a 200× solution of porphyrin and a 1,000× solution of IPTG as described (see Subheading 3.3, steps 14 and 15).

10. Transfer the porphyrin (7.5 mL) and IPTG (1.5 mL) solutions to each of the cultures.

11. For anaerobic expression, immediately degas the bottles, as described in step 6, after porphyrin and IPTG addition.

12. Incubate the gas-tight anaerobic bottles or aerobic expression flasks at the desired induction temperature for 10–22 h in the dark while shaking.

13. Harvest, pellet, and store *E. coli* as described (see Subheading 3.3, steps 17 and 18).

14. Follow recommended purification procedures for the protein of interest as described (see Subheading 3.3, step 19).

4 Notes

1. Synthesis and subsequent manipulations of Pd-MP and Zn-MP should be carried out in low light. For Pd-MP, reflux $PdCl_2$ (80 mg, 0.45 mmol) and mesoporphyrin IX dihydrochloride (100 mg, 0.156 mmol) (Frontier Scientific) under N_2 in DMF for several hours. To prepare Zn-MP, stir $Zn(OAc)_2 \cdot 2H_2O$ (35 mg, 0.16 mmol) with mesoporphyrin IX dihydrochloride (10 mg, 0.016 mmol) at room temperature under N_2 in $CHCl_3/MeOH$ for several hours. Monitor reaction progress by UV–visible spectroscopy. Cool the products in an ice bath and precipitate through the dropwise addition of chilled ddH_2O. Collect the products on Whatman grade 42 filter paper (55 mm), wash with several volumes of chilled ddH_2O, and dry overnight under house vacuum. Electrospray ionization-mass spectrometry (ESI-MS) calculated and (found) masses: $[M–H]^-$ 669.2 (669.2) and 627.2 (627.2) for Pd-MP and Zn-MP, respectively.

2. The inability to synthesize heme in the RP523 strain is an auxotrophic mutation (14). Therefore, supplementation with hemin is required for growth under aerobic conditions (14). Media will appear amber brown after the addition of hemin.

3. Expression in the RP523 strain can only be carried out with plasmids utilizing endogenous *E. coli* promoters because the strain is not a T7 lysogen. The ampicillin-resistant pCW plasmid, which is under the control of a *Taq* promoter, has been reliably used for robust protein expression.

4. To generate competent RP523 cells, follow a standard protocol for preparation of chemically competent *E. coli* (18). However, grow *E. coli* cultures in the presence of 30 µg/mL hemin (prepared from a 200× stock in DMSO) and limit exposure to light. Competent *E. coli* (RP523) stocks will appear light green due to the presence of hemin. No further addition of hemin is required during transformation with the expression plasmid.

5. Leave approximately 1 cm of headspace in the tubes. This will facilitate sparging without excessive foaming of the medium.

6. Slowly add glucose to water while stirring. A compact crystalline layer will form at the bottom of the container if water is added directly to glucose.

7. To avoid excessive foaming and prevent loss of medium, do not put the needles directly in the culture. Following sparging, remove the exit needle first to limit O_2 contamination.

8. The 10-L fermentor used here was built in-house following the design of Prof. B. Krantz, which has been made publically available (19). However, commercial fermentors with sparging capabilities can be readily substituted. The fermentor vessel used here has the following general features: an inlet port connected to a sparger, outlet valve (for venting), harvest tube, temperature control probe and cooling coil, mixing propeller, and access point (for addition of medium and other materials).

9. Depending on the fermentor setup, sterilized medium can be effectively transferred to the fermentor vessel via a funnel that has been sterilized with ethanol. Transfer of warm, freshly autoclaved medium decreases the likelihood of contamination. However, caution should be taken to avoid injury with warm medium. Opening the outlet valve will decrease the splashing back of medium upon pouring it into the fermentor vessel.

10. Caution: the overnight cultures may become slightly pressurized during cell growth. Particularly dense cultures will make a popping noise upon opening.

11. Porphyrin stock solutions prepared in DMSO can be diluted fivefold into base (100 mM NaOH and 200 mM Na_2PO_4, pH 12) to enhance porphyrin solubility in the medium (8). Alternative solvents must be selected carefully to not affect *E. coli* growth.

12. Mn and Zn porphyrins were prone to demetalation and replacement with Fe during expression conditions. Adding a solution of the appropriate metal salt before inducing protein expression, increasing porphyrin concentrations, and/or decreasing expression times limited Fe contamination in the isolated proteins. Care should be taken to avoid increasing metal and metalloporphyrin concentrations beyond toxic levels for *E. coli*.

13. Attempts have only been made to incorporate mesoporphyrin IX and protoporphyrin IX derivatives. Success at incorporating

other porphyrin macrocycles will depend on solubility in the expression medium, permeability across the *E. coli* membrane, and ability to bind to the protein of interest.

14. For nitric oxide synthase (NOS), better porphyrin incorporation was obtained during aerobic conditions. Therefore, porphyrin-substituted NOS was obtained in two steps: initial anaerobic *E. coli* growth, to achieve a suitable cell density, followed by aerobic expression (8). For new proteins, expression should be compared under both anaerobic and aerobic conditions.

15. For the fermentor setup used here, *E. coli* can be harvested through the harvest tube by pressurization of the fermentor upon closing the outlet valve.

16. The genes for endogenous *E. coli* proteases are not disrupted in the RP523 strain. Therefore, additional protease inhibitors may be required if proteolysis of the protein of interest is observed.

17. Carbenicillin is recommended as an alternative antibiotic to ampicillin in Subheading 3.4 because of the longer growth times.

18. To prepare the HPLC caps, fit two ports in the cap with HPLC tubing (seal the remaining port(s) if appropriate). HPLC tubing from one port is used to directly sparge the medium, whereas the tubing from the remaining port is used as an outlet. Put the outlet tubing in the headspace to avoid loss of medium during sparging. Cap the tubing as appropriate following sparging to keep the bottles anaerobic.

Acknowledgements

We thank Karla M. Ramos-Torres (UC Berkeley Amgen Scholars Program) for assistance with porphyrin synthesis and protein expression. We also thank Prof. Susan Marqusee for use of equipment and Prof. Bryan A. Krantz for guidance with fermentor construction. Finally, we thank members of the Marletta laboratory for continued helpful discussions. This work was supported by the Aldo DeBenedictis Fund (M.B.W. and M.A.M.) and NIH grant GM070671 (M.A.M.).

References

1. Battistuzzi G, Borsari M, Cowan JA et al (2002) Control of cytochrome c redox potential: axial ligation and protein environment effects. J Am Chem Soc 124:5315–5324
2. Olson JS, Phillips GN (1997) Myoglobin discriminates between O_2, NO, and CO by electrostatic interactions with the bound ligand. J Inorg Biochem 2:544–552
3. Ortiz de Montellano PR, De Voss JJ (2005) Substrate oxidation by cytochrome P450

enzymes. In: Ortiz de Montellano PR (ed) Cytochrome P450: structure, mechanism, and biochemistry, 3rd edn. Kluwer Academic/Plenum, New York, pp 183–244
4. Jung ST, Lauchli R, Arnold FH (2011) Cytochrome P450: taming a wild type enzyme. Curr Opin Biotechnol 22:1–9
5. Jung C, Hoa GH, Schroder KL et al (1992) Substrate analogue induced changes of the CO-stretching mode in the cytochrome

P450cam-carbon monoxide complex. Biochemistry 31:12855–12862

6. Cowan JA, Gray HB (1989) Synthesis and properties of metal-substituted myoglobins. Inorg Chem 28:2074–2078

7. Lelyveld VS, Brustad E, Arnold FH et al (2010) Metal-substituted protein MRI contrast agents engineered for enhanced relaxivity and ligand sensitivity. J Am Chem Soc 133:649–651

8. Woodward JJ, Martin NI, Marletta MA (2007) An *Escherichia coli* expression-based method for heme substitution. Nat Methods 4:43–45

9. Winter MB, McLaurin EJ, Reece SY et al (2010) Ru-porphyrin protein scaffolds for sensing O_2. J Am Chem Soc 132:5582–5583

10. Teale F (1959) Cleavage of the haem-protein link by acid methylethylketone. Biochim Biophys Acta 35:543

11. Schmidt P, Schramm M, Schroder H et al (2003) Preparation of heme-free soluble guanylate cyclase. Protein Expr Purif 31:42–46

12. Yu C, Gunsalus IC (1974) Cytochrome P-450cam. III. Removal and replacement of ferriprotoporphyrin IX. J Biol Chem 249:107–110

13. Wagner GC, Perez M, Toscano WA Jr et al (1981) Apoprotein formation and heme reconstitution of cytochrome P-450cam. J Biol Chem 256:6262–6265

14. Li JM, Umanoff H, Proenca R et al (1988) Cloning of the *Escherichia coli* K-12 *hemB* gene. J Bacteriol 170:1021–1025

15. Brugna M, Tasse L, Hederstedt L (2010) In vivo production of catalase containing haem analogues. FEBS J 277:2663–2672

16. Marletta MA, Hurshman AR, Rusche KM (1998) Catalysis by nitric oxide synthase. Curr Opin Chem Biol 2:656–663

17. Ito S, Uno H, Murashima T et al (2001) Synthesis of benzoporphyrins functionalized with octaester groups. Tetrahedron Lett 42:45–47

18. Chung CT, Niemela SL, Miller RH (1989) One-step preparation of competent *Escherichia coli*: transformation and storage of bacterial cells in the same solution. Proc Natl Acad Sci U S A 86:2172–2175

19. Krantz BA (2008) 10L Fermentor vessel. http://ebookbrowse.com/10l-vessel-model-3-pdf-d101805732. Accessed 26 Nov 2011

20. Matter H, Kumar HS, Fedorov R et al (2005) Structural analysis of isoform-specific inhibitors targeting the tetrahydrobiopterin binding site of human nitric oxide synthases. J Med Chem 48:4783–4792

21. Weinert EE, Plate L, Whited CA et al (2010) Determinants of ligand affinity and heme reactivity in H-NOX domains. Angew Chem Int Ed Engl 49:720–723

Chapter 9

Expression in *Escherichia coli* of a Cytochrome P450 Enzyme with a Cobalt Protoporphyrin IX Prosthetic Group

Wesley E. Straub, Clinton R. Nishida, and Paul R. Ortiz de Montellano

Abstract

Unlike many hemoproteins, the prosthetic heme group of most cytochrome P450 enzymes cannot be extracted and replaced by modified heme groups. Here, we describe a procedure for generating a cytochrome P450 enzyme (CYP119) with cobalt protoporphyrin IX as its prosthetic group. This is achieved by expressing the protein in *Escherichia coli* in iron-limited medium and adding cobalt to the medium at the moment that inducible protein expression is initiated.

Key words Cobalt protoporphyrin IX, Cobalt heme, CYP119, Heme modification, Bacterial ferrochelatase

1 Introduction

Hemoproteins with a modified heme group are valuable in biophysical, structural, and mechanistic studies. Heme modifications can involve changes in the porphyrin framework, such as replacement of the peripheral substituents by other groups, or changes in the central metal atom. In many hemoproteins, including the peroxidases and globins (1, 2), the introduction of modified heme groups is readily achieved by extracting the heme group from the protein and reconstituting the resulting apoprotein with the modified heme. However, except for cytochrome P450cam (CYP101) (3–7), this approach has not succeeded with P450 enzymes. In situ reconstitution of mammalian cytochrome P450 enzymes has been shown to occur when the heme is destroyed by a mechanism-based inhibitor and is then replaced by a normal heme group (8, 9). However, this approach is not applicable to the preparation of purified enzymes.

An alternative approach to the incorporation of a modified heme group into a cytochrome P450 is to express the desired

Ian R. Phillips et al. (eds.), *Cytochrome P450 Protocols*, Methods in Molecular Biology, vol. 987,
DOI 10.1007/978-1-62703-321-3_9, © Springer Science+Business Media New York 2013

protein in *Escherichia coli* under conditions where normal heme is absent, but the modified heme is present. Expression of P450BM3 (CYP102) with a manganese protoporphyrin IX (MnPPIX) prosthetic group has recently been reported by co-expressing the P450BM3 heme domain with the bacterial heme transporter ChuA in medium supplemented with MnPPIX (10). In another approach, *E. coli* strain RP523, which bears a heme biosynthesis defect and an unknown permeability mutation that allows easy entry of exogenous heme, was used to incorporate manganese protoporphyrin IX into inducible nitric oxide synthase (iNOS) (11). Here, we report a related, but different, approach for the introduction of cobalt heme into a P450 enzyme. Our protocol involves growing *E. coli* bearing an IPTG-inducible P450 construct in iron-limited medium to the point where the iron is severely depleted. At this point cobalt is added to the medium and expression of the P450 enzyme is induced by IPTG. This approach depends on, and is limited by, both the ability of the *E. coli* to take up the metal ion and of its ferrochelatase to incorporate it into protoporphyrin IX. In general, uptake and incorporation of iron, cobalt, and zinc occurs readily, although some ferrochelatases can utilize other metals (12).

Using the present procedure, we have expressed the thermophilic cytochrome P450 enzyme CYP119 with a cobalt prosthetic group that is shown by mass spectrometric and electronic absorption analyses to be contaminated by no more than 2% of the iron species (13).

2 Materials

Prepare all aqueous solutions using ultrapure water (prepared by purifying deionized water to attain a resistivity of 18 MΩ cm at 25°C) and analytical grade reagents. Prepare and store all reagents at room temperature (unless indicated otherwise). Autoclave for 20 min. When sterilizing by filtration, use 0.2-µm filters.

2.1 Protein Expression Components

1. BL21 competent cells.
2. pCWori/CYP119 plasmid DNA encoding for CYP119 expression.
3. CHELEX™ 100 cation-exchange resin (see Note 1): Handle the CHELEX 100 according to the manufacturer's instructions.
4. 2.8-L Fernbach flasks (see Note 2).
5. Magnesium sulfate solution: 1 M solution in water. Dissolve 24.6 g of $MgSO_4 \cdot 7H_2O$ in water and dilute to 100 mL. Sterilize by autoclaving.

Expression in *Escherichia coli* of a Cytochrome P450 Enzyme... 109

6. Calcium chloride solution: 1 M in water. Dissolve 14.7 g of $CaCl_2 \cdot 2H_2O$ in water and dilute to 100 mL. Sterilize by autoclaving.

7. Glucose solution: 20% (w/v) solution in water. Dissolve 20.0 g of D-glucose in water and dilute to 100 mL. Add 2 g of CHELEX 100 and incubate for 1 h at room temperature with shaking. Remove CHELEX and sterilize by filtration.

8. Ampicillin solution: 100 mg/mL solution in water. Dissolve 4 g of ampicillin, sodium salt, in water and dilute to 40 mL. Add 0.8 g of CHELEX 100 and incubate for 1 h at 4°C with shaking. Remove CHELEX and sterilize by filtration. Store at −20°C.

9. Ferrous sulfate solution: 25 mM solution (500,000×) in water. Dissolve 0.35 g of $FeSO_4 \cdot 7H_2O$ in water and dilute to 50 mL. Make fresh and sterilize by filtration.

10. Cobalt (II) chloride solution: 0.3 M solution (1,000×) in water. Dissolve 0.71 g of $CoCl_2 \cdot 6H_2O$ in water and dilute to 10 mL. Sterilize by filtration.

11. δ-Aminolevulinic acid solution: 0.1 M solution (1,000×) in water. Dissolve 0.34 g of δ-aminolevulinic acid hydrochloride in water and dilute to 20 mL. Add 0.4 g of CHELEX 100 and incubate for 1 h with shaking. Remove CHELEX and sterilize by filtration. Store at −20°C.

12. IPTG (isopropyl-beta-D-thiogalactopyranoside) solution: 1 M solution (1,000×) in water. Dissolve 4.77 g of IPTG in water and dilute to 20 mL. Add 0.4 g CHELEX 100 and incubate for 1 h at 4°C with shaking. Remove CHELEX and sterilize by filtration. Store at −20°C.

13. Metal-depleted M9-based expression medium: To 100 mL of water, add 6 g of $Na_2HPO_4 \cdot 7H_2O$, 3 g of KH_2PO_4, 0.5 g of NaCl, and 1.5 g of NH_4Cl. Adjust the pH to 7.5. Add 2 g of CHELEX 100, and incubate the suspension for 1 h at room temperature with shaking. Remove the resin by filtration, and add water to a final volume of 1 L. Sterilize by autoclaving.

2.2 Protein Purification Components

1. Sonicator.
2. Ultracentrifuge.
3. UV–visible spectrophotometer.
4. Anion-exchange DEAE-agarose resin.
5. Hydrophobic affinity Toyopearl Ether-650 M agarose resin.
6. PMSF (phenylmethanesulfonylfluoride) solution: 50 mM (1,000×) in isopropanol. Dissolve 44 mg in isopropanol and dilute to 5 mL. Make fresh.

7. Antipain solution: 1 mg/mL (1,000×) in water. Dissolve 10 mg of antipain in water and dilute to 10 mL. Store at –80°C.

8. Leupeptin hemisulfate solution: 10 mM (10,000×) in water. Dissolve 10 mg of leupeptin hemisulfate in water and dilute to 2.1 mL. Store at –80°C.

9. DTT (dithiothreitol) solution: 100 mM (1,000×) in water. Dilute 15 mg of DTT in 1 mL of water. Make fresh immediately before use.

10. DNase (deoxyribonuclease) I, solid.

11. RNase (ribonuclease) A, solid.

12. Lysozyme, solid.

3 Methods

3.1 Expression of Metal-Substituted CYP119

1. Transform BL21 competent cells with pCWori/CYP119 expression plasmid.

2. Prepare 30 mL of starter culture for every 1.5 L of expression culture. For 30–60 mL, use a 125-mL Erlenmeyer flask. For 90–120 mL or 150–240 mL, use a 250-mL or 500-mL flask, respectively. With a single colony, inoculate LB medium containing ampicillin (100 µg/mL). Grow overnight at 37°C.

3. To 1.5 L of metal-depleted M9-based expression medium, add 1.5 mL of magnesium sulfate solution, 15 µL of calcium chloride solution (see Note 3), 15 mL of glucose solution, 1.5 mL of ampicillin solution, and 1.5 mL of 1/500 diluted (50 µM) ferrous sulfate solution (see Note 4). Mix the solution with gentle shaking.

4. Harvest the cells from the 30-mL starter culture by centrifugation for 10 min at 5,000×g. Resuspend the cells with 20 mL of minimal medium and add to the expression flask. Incubate at 42°C with shaking at 225 rpm.

5. At an OD_{600} of 0.7–0.8 (see Note 4), add 1.5 mL of cobalt(II) chloride solution and 1.5 mL of δ-aminolevulinic acid solution. Continue incubation with shaking.

6. After 1 h, add 1.5 mL of IPTG solution. Lower the temperature to 32°C and incubate with shaking at 180 rpm.

7. After 24 h, harvest the cells by centrifugation for 10 min at 5,000×g. Collect the cell paste and store at –80°C (see Note 5).

Expression in *Escherichia coli* of a Cytochrome P450 Enzyme... 111

3.2 Purification of Metal-Substituted CYP119

Perform all steps at 4°C.

1. Break the cell paste into smaller pieces (see Note 5) and add to ice-cold 50 mM Tris–HCl, pH 7.5 (see Note 6) containing DNase I (32 units/mL), RNase A (3 units/mL), lysozyme (2 mg/mL), 500 μM PMSF, antipain (1 μg/mL), and 1 μM leupeptin. Use 5 mL of buffer per gram of cells. Stir 30 min, to aid suspension of the cells and allow for the action of lysozyme, DNase, and RNase.

2. Sonicate three cycles for 1 min each while the solution is kept on ice. Monitor the temperature of the solution to avoid a rise in temperature beyond 8°C (see Note 7).

3. Collect the supernatant by centrifugation at $100,000 \times g$ for 30 min.

4. Load the supernatant onto a DEAE weak anion-exchange column (see Note 8), and wash with three column volumes of 50 mM Tris–HCl, pH 7.2, 50 mM KCl, 500 μM PMSF, antipain (1 μg/mL), and 1 μM leupeptin.

5. Elute with a KCl gradient from 50 to 300 mM over five column volumes, collecting fractions.

6. Pool fractions with an Rz value of 1 or higher, where Rz is the ratio of the absorbance of the Soret band to that at 280 nm $(A_{\text{Soret}}/A_{280})$.

7. Concentrate the pooled fractions to 20 mL (see Note 9).

8. Elute and fractionate the protein by ammonium sulfate precipitation between 58 and 85% saturated ammonium sulfate: To the pooled fractions, slowly add ammonium sulfate to 58% saturation. Centrifuge at $20,000 \times g$ for 10 min. Retain the supernatant and add ammonium sulfate to 85% saturation. Centrifuge at $20,000 \times g$ for 10 min to recover the pellet containing CYP119.

9. Resuspend the pellet in a minimal volume of 50 mM Bis-Tris (Bis(2-hydroxyethyl)-amino-tris(hydroxymethyl)-methane), pH 6.8. Add ammonium sulfate to 1.4 M.

10. Load onto an ether-substituted Toyopearl® Ether-650 M hydrophobic column (Tosoh Bioscience) and wash with three column volumes of 50 mM Bis-Tris, pH 6.8, 1.4 M ammonium sulfate.

11. Elute and fractionate the protein with a gradient from 1.4 to 0 M ammonium sulfate over 5 column volumes.

12. Analyze the fractions by SDS-PAGE and pool fractions containing the desired purity.

13. Concentrate the protein to approximately 200 μM, supplement with glycerol to 15% (v/v), add DTT to 100 μM, and flash freeze.

4 Notes

1. CHELEX resin is available in different mesh sizes. We use CHELEX 100, 100–200 mesh, as it can be removed by passage of the treated solution through 0.2-μm filter devices without clogging. Treatment of solutions involves addition of excess equivalents of CHELEX 100 to bind metal ions. The metal content of chemicals should be considered and compared with the metal-binding capacity of the resin as described by Bio-Rad, the manufacturer of CHELEX. Based on metal assays reported by manufacturers of the reagents we use, 2 g of CHELEX per 100 mL is an ample amount. Solutions are treated with CHELEX by incubation for 1 h with gentle shaking.

2. Cell growth and therefore protein yields are influenced by aeration. Thus, the type of culture flask can dramatically affect protein yield. We routinely use Fernbach flasks without baffles.

3. It is normal to observe the visible formation of calcium phosphate precipitate when calcium chloride is added to the phosphate-containing M9 medium.

4. Ideally, each investigator will determine the minimum amount of supplemented iron that is required to obtain cell growth to an OD_{600} of 1.5. The amount of added iron can vary depending on the small amount of contaminating iron present in magnesium sulfate and calcium chloride reagents. If cells stop growing and fail to reach the desired OD_{600} for induction, the iron content might be too low. In such a case, perform empirical testing of the minimum required amount of supplemented iron, because $CaCl_2$ and $MgSO_4$ can contain varying amounts of iron contamination. A range of 0–1,000 nM added iron can be examined.

5. Flattening the cell paste within a plastic freezer bag before freezing allows the frozen paste to be more easily pulverized, which will (1) speed the thawing process during later processing and (2) provide a means to portion the cells. Care should be taken to avoid cracking the plastic bag, which is more brittle when first removed from extremely low temperature (−80°C).

6. The pH of Tris–HCl and Bis-Tris solutions are temperature dependent. Adjust the pH of solutions at the temperature of use.

7. Sonication will raise the temperature of the solution. To maximize cooling, use a metal or glass beaker and avoid plastic ones. Because sonication becomes less efficient with larger volume, we use at most 150 mL of solution in a 250-mL beaker.

8. The nature of the anion exchanger is important. Use of the strong anion exchanger Q Sepharose instead of DEAE Sepharose leads to substantial loss of cobalt porphyrin.

9. We use centrifugal concentrator devices such as Amicon Ultracel units.

Acknowledgment

This work was supported by National Institutes of Health Grant GM25515.

References

1. Scholler DM, Wang MY, Hoffman BM (1978) Metal-substituted hemoglobin and other hemoproteins. Methods Enzymol 52:487–493
2. Wang MY, Hoffman BM, Hollenberg PF (1977) Cobalt-substituted horseradish peroxidase. J Biol Chem 252:6268–6275
3. Dolphin D, James BR, Welborn HC (1980) Oxygenation, and carbonylation, of a reduced P450cam enzyme and derivatives reconstituted with mesodeutero-, dibromodeutero, and diacetyldeuteroheme. J Mol Catal 7:201–213
4. Wagner GC et al (1981) Apoprotein formation and heme reconstitution of cytochrome P-450cam. J Biol Chem 256:6262–6245
5. Wagner GC et al (1981) Cobalt-substituted cytochrome P-450cam. J Biol Chem 256:6266–6273
6. Furukawa Y, Ishimori K, Morishima I (2000) Electron transfer reactions in Zn-substituted cytochrome P450cam. Biochemistry 39:10996–11004
7. Gelb MH, Toscano WA Jr, Sligar SG (1982) Chemical mechanisms for cytochrome P-450 oxidation: spectral and catalytic properties of a

manganese-substituted protein. Proc Natl Acad Sci U S A 79:5758–5762
8. Correia MA et al (1979) Incorporation of exogenous heme into hepatic cytochrome P-450 *in vivo*. J Biol Chem 254:15–17
9. Farrell GC et al (1979) Exogenous heme restores *in vivo* functional capacity of hepatic cytochrome P-450 destroyed by allylisopropylacetamide. Biochem Biophys Res Commun 89:456–463
10. Lelyveld VS et al (2011) Metal-substituted protein MRI contrast agents engineered for enhanced relaxivity and ligand sensitivity. J Am Chem Soc 133:649–651
11. Woodward JJ, Martin NI, Marletta MA (2007) An *Escherichia coli* expression-based method for heme substitution. Nat Methods 4:43–45
12. Dailey HA, Dailey TA (2011) Ferrochelatase. In: Kadish KM, Smith KM, Guilard R (eds) The porphyrin handbook, vol 12. Academic, New York, pp 93–156
13. McGinn-Straub WE (2003) Biosynthetic metal substitution of P450 enzymes with cobalt and its applications. Ph.D. Thesis, University of California, San Francisco

Chapter 10

Nanodiscs in the Studies of Membrane-Bound Cytochrome P450 Enzymes

A. Luthra, M. Gregory, Y.V. Grinkova, I.G. Denisov, and S.G. Sligar

Abstract

Cytochromes P450 from eukaryotes and their native redox partners cytochrome P450 reductases both belong to the class of monotopic membrane proteins containing one transmembrane anchor. Incorporation into the lipid bilayer significantly affects their equilibrium and kinetic properties and plays an important role in their interactions. We describe here the detailed protocols developed in our group for the functional self-assembly of mammalian cytochromes P450 and cytochrome P450 reductases into Nanodiscs with controlled lipid composition. The resulting preparations are fully functional, homogeneous in size, composition and oligomerization state of the heme enzyme, and show an improved stability with respect to P420 formation. We provide a brief overview of applications of Nanodisc technology to the biophysical and biochemical mechanistic studies of cytochromes P450 involved in steroidogenesis, and of the most abundant xenobiotic-metabolizing human cytochrome P450 CYP3A4.

Key words Cytochrome P450, Nanodiscs, CYP3A4, CYP17, CYP19A1, Eukaryotic P450, Membrane mimetics

1 Introduction

Cytochromes P450 constitute a broad superfamily of heme enzymes with more than 20,000 isozymes identified in genomes of organisms from all biological kingdoms (1, 2). Prokaryotic cytochromes P450 are soluble. However, eukaryotic enzymes are monotopic integral membrane proteins with the N-terminal transmembrane fragment and the hydrophobic fragment between F and G helices inserted into the lipid bilayer with an estimated depth of insertion of 1–2 nm (3–9). As with other membrane proteins, isolation, purification, and biochemical studies of eukaryotic cytochromes P450 are mostly based on the use of detergents to improve solubilization of hydrophobic constituents of the membrane. However, in most cases, the resulting reconstitution systems of mixed detergents, lipid micelles, and solubilized membrane proteins do not contain a lipid bilayer and are poor mimetics of

Ian R. Phillips et al. (eds.), *Cytochrome P450 Protocols*, Methods in Molecular Biology, vol. 987, DOI 10.1007/978-1-62703-321-3_10, © Springer Science+Business Media New York 2013

biological membrane protein systems. Lipid vesicles represent a better approximation to the real in vivo arrangement of membrane protein assemblies, but working with small and large unilamellar vesicles is difficult because of limited stability and the inability to control the local environment of the membrane protein incorporated into the lipid bilayer. Therefore, there is a need for a new model system for the functional reconstitution of membrane proteins and their complexes (10).

The development of Nanodiscs, self-assembled nanoscale lipid bilayers solubilized by an amphipathic scaffold protein (11–14), has created a new list of methods and experimental approaches that can be used in studying the biophysical chemistry of membrane proteins (15–20). Since its development, the Nanodisc technology has proved to be the method of choice for detailed mechanistic studies of G-protein-coupled receptor systems (21–26), cytochromes P450 (16, 20, 27–37), ion channels (38–41), transporters (42–47), chemotactic receptors (48, 49), and other biological systems (50–53).

There are several important advantages for using Nanodiscs in biochemical and biophysical studies of membrane proteins. First is the precise control of oligomerization state of the target protein and subsequent separation of Nanodiscs containing populations of monomers and oligomers (15, 21, 27, 50–54). This approach offers an unprecedented opportunity to study the functional properties of monomers and oligomers with no need to deal with the monomer–oligomer equilibria and simultaneous presence of multiple species with different properties. Nanodiscs allowed the first direct measurement of the binding and native photochemical properties of monomeric rhodopsin (21, 24–26) and green proteorhodopsin (55), activation of G-proteins by monomeric β2 receptors (22, 56), activation of unclustered integrins (53) and the identification and characterization of the functional unit of chemotactic receptors (49).

In addition, highly homogeneous stable and reproducible preparations of solubilized membrane proteins allow for the application of a plethora of modern biophysical methods and precise quantitative analysis of acquired experimental data. Examples include the application of global analysis to the deconvolution of binding constants and functional properties of binding intermediates during cooperative catalysis of testosterone hydroxylation by CYP3A4 (29), single-molecule experiments describing the allosteric effects in CYP3A4 substrate binding (37), stabilization of transient intermediates in CYP3A4 and CYP19 for spectroscopic studies at cryogenic temperatures (16, 34), and the functional immobilization of CYP3A4 on gold nanoparticles for sensing and screening using local surface plasmon resonance methods (57).

In many cases Nanodiscs improve the stability of membrane proteins as compared with detergent-solubilized preparations.

This was shown for CYP3A4 at room temperature (by the decreased rate of formation of inactive P420 form), for ATP synthase from *E. coli*, for human monoamine oxidase A, and for epidermal growth factor receptor (15, 58–60). Although in most cases the stabilization mechanisms have not been identified, the general effect may be attributed to the native-like incorporation of membrane proteins into lipid bilayers and to the absence or reduced concentrations of detergents.

Control of the local lipid composition provides new opportunities for detailed microscopic studies of protein binding to membranes, influence of lipids on protein–protein interactions mediated by the membrane, and certain specific functional properties of membrane proteins (61, 62). For P450 systems this is especially important, because not only cytochrome P450 enzymes but their redox partner proteins, hepatic cytochrome P450 reductase (CPR) and mitochondrial adrenodoxin reductase can also be incorporated into the membrane. Therefore, mutual orientations and matching of protein–protein interfaces between interacting cytochrome P450 and reductase are defined by the depth of insertion and orientation of each protein with respect to the lipid bilayer, electrostatic interactions and metal ions at the bilayer interface, as well as the mobility and dynamics of all components. Equilibrium and kinetic parameters of hydrophobic substrates binding to membrane cytochromes P450 can be significantly modulated by presence of the lipid bilayer, since some substrates can access the active site of the heme enzyme directly from the membrane (7, 63, 64). In addition, the electrostatic field created by membrane lipids may significantly affect the redox potentials of cytochromes P450 (65) and of the flavoprotein redox partners (66). Such effects demand a platform where one can perform mechanistic studies of cytochromes P450 reconstituted in the lipid bilayers, with an ability to control and vary lipid composition, oligomerization state of the proteins, and the stoichiometric ratio between protein components. Nanodiscs provide such a platform, making available a wide range of lipid–membrane protein systems that can be probed experimentally.

2 Materials

Prepare all solutions using ultrapure water with a sensitivity of 18 MΩ at 25°C. Unless a specific vendor or product is listed, all reagents should be of the highest grade available.

2.1 Materials

Amberlite XAD-2 (Supelco)

Ni-NTA resin (Invitrogen)

2′-5′ ADP agarose resin

Size-exclusion column: Superdex 200 10/300 (GE Life Sciences)

Microcentrifuge filters, 0.22 μm (Millipore)

118   A. Luthra et al.

2.2 Buffers

Column buffer: 100 mM phosphate buffer, pH 7.4

Disc buffer: 100 mM phosphate buffer, pH 7.4, 50 mM NaCl

Buffer with cholate: 100 mM phosphate buffer, pH 7.4, 50 mM NaCl, 100 mM sodium cholate. 10 mM 2-mercaptoethanol

Buffer without cholate: 100 mM phosphate buffer, pH 7.4, 50 mM NaCl, 10 mM 2-mercaptoethanol

Buffer A: 100 mM phosphate buffer pH 7.4, 300 mM NaCl

Buffer W: Buffer A containing 15 mM imidazole

Buffer E: Buffer A containing 300 mM imidazole

2.3 Other Solutions

Lipid solubilization solution: 100 mM sodium cholate, 100 mM NaCl

100 mM $NiSO_4$

Triton X-100 stock: 20% Triton X-100 (w/v) in ultrapure water

Palmitoyloleoylphosphatidylcholine (POPC) stock in chloroform

2.4 Protein Stocks

P450 target in buffer of choice.

His-tagged membrane scaffold protein (MSP) in disc buffer. MSP1D1 for P450 Nanodiscs and MSP1E3D1 for P450-CPR Nanodiscs.

3 Methods

3.1 Self-Assembly of P450 in Nanodiscs Using Cholate

1. Using a glass Hamilton syringe, dispense 427 μl of 75 mM chloroform–POPC stock into a disposable 18 ml glass culture tube (see Notes 1–3).

2. In a fume hood, dry the chloroform using a gentle stream of nitrogen gas. A thin, uniform film of lipid on the lower walls eases lipid solubilization in an aqueous medium later, and can be obtained by rotating the tube while holding it at an angle.

3. Place the tube in a vacuum desiccator for at least 4 h to remove any residual solvent (see Note 4).

4. Add 640 μl of solution containing 100 mM sodium cholate and 100 mM NaCl to the tube (see Note 5).

5. Solubilize lipids by immersing the tube in warm water (50–70°C) for 30–60 s and then vortexing. Repeat this until no lipid remains on the walls of the tube.

6. Sonicate the tubes in an ultrasonic bath for 5–10 min (see Note 6).

   Steps 7–13 are done at 4°C, or on ice unless otherwise mentioned.

7. Add 640 μl of buffer with cholate to the tube (see Note 7).

Nanodiscs in the Studies of Membrane-Bound Cytochrome P450 Enzymes   119

8. Add 160 μl of buffer without cholate to the tube (see Note 7).

9. Add 3,333 μl of 150 μM MSP1D1(−) to the tube (see Notes 8, 9).

10. Add 500 μl of 50 μM P450 to the tube (see Note 10).

11. Let the mixture incubate on ice for 20–30 min.

12. Add 6 ml equivalent of wet Amberlite XAD-2 beads of the disc reconstitution mixture (see Note 11).

13. Place the tubes in an ice bucket secured on a shaker, to gently agitate the mixture. Make sure all the beads are suspended in solution. Alternatively, one may put the tube on a shaker platform placed inside a cold cabinet (see Note 12).

14. After at least 6 h of incubation with Amberlite XAD-2, filter the solution to remove the hydrophobic beads with the help of plastic disposable filter columns (see Note 13).

15. Load the solution on to a 1 ml Ni-NTA column (see Note 14), pre-equilibrated with Buffer A.

16. Wash the column with 5 ml of Buffer W and elute with Buffer E (see Note 15).

17. Concentrate the eluate from the Ni-NTA column and filter it using disposable microcentrifugal filters (0.22 μm).

18. Load the filtrate onto a calibrated Superdex-200 10/300 column (GE Life Sciences), pre-equilibrated with the column buffer.

19. Collect fractions and pool those containing P450 Nanodiscs and having the desired level of enzymatic activity, substrate-binding characteristics, or with the anticipated size to obtain a homogenous, monodisperse, stable membrane–P450 preparation (see Note 16).

20. P450–POPC Nanodiscs can be stored at 4°C for 1–2 weeks. For long-term storage, add 15–20% of ultrapure glycerol, flash-freeze, and store at −80°C.

3.2 Self-Assembly of P450 in Nanodiscs Without Using Cholate

1. Perform steps 1–3 of Subheading 3.1.

2. Add 1,960 μl of disc buffer to the tube containing dried lipids, and seal with laboratory film. Resuspend the lipids by three cycles of vortexing for 30 s, immersing the tube in liquid N_2 and thawing the frozen solution at room temperature (see Note 17).

3. Deliver 115 μl of 20% Triton X-100 to the tube (see Note 18).

4. Seal the tube with laboratory film and sonicate for 30 min in a heated bath sonicator set to ~60°C. Transfer the solution to an ice bucket and allow 5 min of cooling before proceeding. The

120   A. Luthra et al.

solution should now be slightly more translucent than observed in step 2.

5. Add 2,000 μl of 250 μM MSP and 500 μl of 50 μM detergent-solubilized P450.

6. If the solution remains cloudy after 1 min, add 5–10 μl of 20% Triton X-100 and gently mix with a transfer pipette. Repeat this step until the solution has completely clarified.

7. Perform steps 11–20 of Subheading 3.1.

3.3 Co-incorporation of P450 and Cytochrome P450 Reductase into Nanodiscs

1. In order to self-assemble Nanodiscs containing both P450 and CPR, mix the two proteins in a ratio of 1:2, preserving the detergent concentration that the membrane P450 is stored in.

2. Incubate this mixture for 1 h at (room temperature).

3. Add this mixture to Nanodisc components consisting of cholate-solubilized POPC and MSP1E3D1(−). For MSP1E3D1 POPC Nanodiscs, the optimal ratio of lipids to MSP is 120:1 per Nanodisc.

4. Incubate the mixture at room temperature for 45 min and then on ice for an additional hour.

5. Perform steps 12–16 of Subheading 3.1.

6. Load the Ni-NTA eluent onto a 2′-5′-ADP-agarose column equilibrated with buffer of choice (see Note 19).

7. Wash the column with 3–5 column volumes of equilibration buffer.

8. Elute the protein with 2–3 column volumes of equilibration buffer containing 2.5 mM 2′-AMP.

9. Dialyze the eluent against 1,000-fold volume of disc buffer, with three exchanges every 3–4 h.

10. Load the solution on to a calibrated Superdex-200 10/300 column (GE Life Sciences) pre-equilibrated with column buffer.

11. Perform steps 20–21 of Subheading 3.1.

Functional CPR: P450 complexes with different CPR to P450 ratios can also be prepared by adding full-length CPR to the solution of P450 preassembled into Nanodiscs at 37°C. The complexes thus formed are stable and can be separated by size-exclusion chromatography. Functional properties for P450–CPR complexes assembled in this manner depend on the molar ratio of P450 and CPR used. For CYP3A4, NADPH oxidation and testosterone hydroxylation have been found to depend on the CPR:CYP3A4 ratio peaking at tenfold molar excess of CPR (67).

4 Notes

1. Starting with the desired amount of P450, calculate the required amount of lipid. We have used POPC as an example. Other lipids such as Dimyristoylphosphatidylcholine (DMPC) and Dipalmitoylphosphatidylcholine (DPPC) can also be used to make Nanodiscs (12, 13). Other membrane-scaffolding proteins can be used (we have used MSP1D1 as an example), using the following ratios: P450: MSP = 1:20, and POPC: MSP1D1 = 1:64.

2. The size of the membrane-scaffolding protein variant (MSP1D1, MSPE3D1, etc.) and the lipid molecule determine the optimal ratio of lipids to MSP to be used for a Nanodisc preparation (12–14, 18).

3. Lipid stocks are maintained in chloroform, usually in concentrations between 25 and 100 mM. Lipids stocks are stored in 4 ml glass vials with Teflon-lined screw caps, to minimize changes in concentration between usages. The stock concentration is determined by analyzing the total phosphorus, by a colorimetric assay (68, 69).

4. For best results, leave the tube in the vacuum desiccator overnight.

5. A final cholate concentration of 50 mM ensures quick and efficient solubilization.

6. The lipid solution should be clear at this point. If not, repeat steps 5 and 6 until the solution becomes transparent. For hard-to-solubilize lipid films sonicate in short cycles of 2–3 min, vortexing the tube between cycles.

7. Given the amount of lipids, calculate the total volume of the disc reconstitution mixture that would result in a final lipid concentration of 5–8 mM. Using the calculated final volume and the amount of sodium cholate used to solubilize the lipids, calculate the amount of buffer with cholate that would need to be added to the tube to achieve a final cholate concentration of 14–20 mM. Add the required amount of MSP and P450 and adjust the volume of the mixture to equal that calculated earlier using Buffer without cholate.

8. MSP1D1(−) denotes MSP1D1 without the histidine tag. The histidine tag on membrane-scaffold proteins can be cleaved with tobacco etch virus (TEV) protease. It is best to prepare His-tag-cleaved MSP beforehand and have it ready in a concentrated stock of 150–250 mM in disc buffer. Use of a His-tag-cleaved variant of MSP and a His-tagged target protein allows a simple isolation of Nanodiscs with embedded P450 from bare discs, using Ni-NTA chromatography.

9. Measure the concentration of MSP every time a frozen stock solution is thawed. The extinction coefficients of more routinely used MSP variants are (18):

	ε_{280} (in M^{-1} cm^{-1})	
MSP variant	With His-tag	Without His-tag
MSP1D1	21,000	18,200
MSP1E3D1	29,400	26,600

10. A stock solution of concentration 25–50 μM of the protein works best.

11. For best results, use 0.8–1 ml equivalent of wet BioBeads per ml of sample. Remove all excess water from the beads before use. This can be done by spreading the beads on a stack of KimWipes on a glass plate.

12. Choice of incubation temperature depends on the type of lipid molecules, with the goal of doing the assembly at a temperature slightly higher than the phase transition of lipids in use. For POPC, this is 270 K (70).

13. For efficient recovery of protein solution wash spent beads with 2–3 equal volumes of cold buffer and add it to the filtrate recovered earlier.

14. Use 1 ml of Ni-NTA resin for every 5–6 mg of protein sample.

15. Wash until there is no protein in the flow-through. To verify, mix 2 μl of Commassie reagent with 100 μl of the flow-through. The appearance of a blue color indicates the presence of a protein. Typically, washing with 4–6 column volumes of the wash buffer is sufficient.

16. The retention time of P450-embedded Nanodiscs is only 0.3–0.5 min less (depending on the sample volume) than that for empty Nanodiscs of same MSP-lipid pair.

17. Ensure that after vortexing all POPC has dislodged from the surface of the tube. At the end of this process the solution will possess a cloudy white appearance.

18. The amount of Triton X-100 in this preparation is an important determinant of the final yield of Nanodisc-associated P450. Ideally, only enough detergent to fully solubilize the POPC will be present in the mixture before addition of Amberlite XAD-2. We have observed that detergent excesses result in the formation of large aggregates and result in extremely low yields. However, failure to completely solubilize the lipids will similarly interfere with the assembly of Nanodiscs. For protein stock solutions containing ~50 μM P450 and

0.02–0.2% Emulgen 913, we have determined that the optimal molar ratio of Triton X-100 to POPC is 1.15:1 at this step.

19. Nanodiscs containing both P450 and CPR will bind to the 2′,5′-ADP column, whereas those containing only P450 would flow-through.

20. Following is an example of how the reconstitution mixture is prepared, showing all the calculations involved. The example is based on 25 nmol of P450.

Assume the following values for the concentration of the stock solutions that would be used in this mock disc assembly:

$(POPC)_{stock}$	$= 75$ mM
$(MSP1D1(-))_{stock}$	$= 150$ μM
$(P450)_{stock}$	$= 75$ μM
POPC: MSP1D1(−): P450	$= 1,280{:}20{:}1$

Note: The ratio between lipid and MSP would depend on the choice of lipid (POPC, POPS, DMPC, DPPC, etc.) and membrane-scaffold protein type (MSP1D1, MSP1E3D1, etc.) pair. The ratio between P450 and the number of Nanodiscs is kept high (typically 1:10) to ensure a high yield of incorporation.

Amount of P450 to be incorporated into discs	$= 25$ nmol	
Amount of MSP1D1(−) needed	$= 25 \times 20$	$= 500$ nmol
Amount of POPC needed	$= 500 \times 64$	$= 32{,}000$ nmol
Volume of POPC–chloroform stock to be dried	$= 32{,}000/75$	$= 426.67$ μl, i.e., 427 μl
Amount of cholate needed to solubilize POPC	$= 32{,}000/50$	$= 640$ μl
For a final POPC concentration of 6 mM		
Total volume of the solution should be	$32{,}000/6$	$= 5{,}333$ μl
To achieve a final cholate concentration of 15 mM		
Volume of buffer with cholate to be added	$= ((5{,}333 \times 15) - (640 \times 100))/100$	$= 160$ μl
Volume of 150 μM MSP1D1(−) to be added	$= 500/150$	$= 3{,}333$ μl
Volume of 50 μM P450 to be added	$= 25/50$	$= 500$ μl
Volume of buffer with cholate to be added	$= 5{,}333 - (640 + 160 + 3{,}333 + 500)$	$= 700$ μl

More specific details can be found in original publications, including supplemental materials in references (11, 13, 30, 65).

In conclusion, we provide a list of cytochrome P450 studies that used Nanodiscs and a brief summary of results. Most of them were done with the most abundant xenobiotic-metabolizing human CYP3A4; however, recently, we initiated systematic mechanistic studies of steroidogenic human enzymes CYP19 and CYP17. Different aspects of substrate binding have been addressed in (20, 27, 32, 36, 37, 71–73), dynamics and kinetics in (20, 28, 31, 35, 37, 74, 75), catalytic activity in (29, 33, 67), and redox properties of cytochrome P450 (28, 65, 74) and CPR in Nanodiscs (66). We also applied cryogenic spectroscopy to study unstable oxygenated intermediates in CYP19 (34, 75) and in CYP3A4 (16), and documented the EPR spectra and annealing of the peroxo-intermediate in CYP19, obtained using cryoradiolytic reduction (34).

Acknowledgments

We gratefully acknowledge the valuable contribution to development of these methods made by T. H. Bayburt, B. J. Baas, M. A. McLean, and other members of the Sligar lab, whose works are cited in the list of references. This work was supported by NIH grants GM31756 and GM33775 to S.G. Sligar.

References

1. Nelson DR (2009) The cytochrome p450 homepage. Human Genomics 4:59–65
2. Nelson DR (2011) Progress in tracing the evolutionary paths of cytochrome P450. Biochim Biophys Acta 1814:14–18
3. Annalora AJ, Goodin DB, Hong WX, Zhang Q, Johnson EF, Stout CD (2010) Crystal structure of CYP24A1, a mitochondrial cytochrome P450 involved in vitamin D metabolism. J Mol Biol 396:441–451
4. Bayburt TH, Sligar SG (2002) Single-molecule height measurements on microsomal cytochrome P450 in nanometer-scale phospholipid bilayer disks. Proc Natl Acad Sci U S A 99:6725–6730
5. Berka K, Hendrychova T, Anzenbacher P, Otyepka M (2011) Membrane position of ibuprofen agrees with suggested access path entrance to cytochrome P450 2C9 active site. J Phys Chem A 115:11248–11255
6. Cojocaru V, Balali-Mood K, Sansom MS, Wade RC (2011) Structure and dynamics of the membrane-bound cytochrome P450 2C9. PLoS Comput Biol 7:e1002152
7. Denisov IG, Shih AY, Sligar SG (2012) Structural differences between soluble and membrane bound cytochromes P450. J Inorg Biochem 108:150–158
8. Mast N, Liao WL, Pikuleva IA, Turko IV (2009) Combined use of mass spectrometry and heterologous expression for identification of membrane-interacting peptides in cytochrome P450 46A1 and NADPH-cytochrome P450 oxidoreductase. Arch Biochem Biophys 483:81–89
9. Zhao Y, White MA, Muralidhara BK, Sun L, Halpert JR, Stout CD (2006) Structure of microsomal cytochrome P450 2B4 complexed with the antifungal drug bifonazole: insight into P450 conformational plasticity and membrane interaction. J Biol Chem 281:5973–5981
10. Popot J-L (2010) Amphipols, nanodiscs, and fluorinated surfactants: three nonconventional approaches to studying membrane proteins in aqueous solutions. Annu Rev Biochem 79:737–775
11. Bayburt TH, Grinkova YV, Sligar SG (2002) Self-assembly of discoidal phospholipid bilayer nanoparticles with membrane scaffold proteins. Nano Lett 2:853–856
12. Denisov IG, McLean MA, Shaw AW, Grinkova YV, Sligar SG (2005) Thermotropic phase tran-

sition in soluble nanoscale lipid bilayers. J Phys Chem B 109:15580–15888

13. Denisov IG, Grinkova YV, Lazarides AA, Sligar SG (2004) Directed self-assembly of monodisperse phospholipid bilayer nanodiscs with controlled size. J Am Chem Soc 126:3477–3487
14. Grinkova YV, Denisov IG, Sligar SG (2010) Engineering extended membrane scaffold proteins for self-assembly of soluble nanoscale lipid bilayers. Protein Eng Des Sel 23:843–848
15. Bayburt TH, Sligar SG (2010) Membrane protein assembly into Nanodiscs. FEBS Lett 584:1721–1727
16. Denisov IG, Sligar SG (2011) Cytochromes P 450 in nanodiscs. Biochim Biophys Acta 1814:223–229
17. Nath A, Atkins WM, Sligar SG (2007) Applications of phospholipid bilayer nanodiscs in the study of membranes and membrane proteins. Biochemistry 46:2059–2069
18. Ritchie TK, Grinkova YV, Bayburt TH, Denisov IG, Zolnerciks JK, Atkins WM, Sligar SG (2009) Reconstitution of membrane proteins in phospholipid bilayer nanodiscs. Methods Enzymol 464:211–231
19. Sligar SG (2003) Finding a single-molecule solution for membrane proteins. Biochem Biophys Res Commun 312:115–119
20. Nath A, Trexler AJ, Koo P, Miranker AD, Atkins WM, Rhoades E (2010) Single-molecule fluorescence spectroscopy using phospholipid bilayer nanodiscs. Methods Enzymol 472:89–117
21. Bayburt TH, Leitz AJ, Xie G, Oprian DD, Sligar SG (2007) Transducin activation by nanoscale lipid bilayers containing one and two rhodopsins. J Biol Chem 282:14875–14881
22. Whorton MR, Bokoch MP, Rasmussen SG, Huang B, Zare RN, Kobilka B, Sunahara RK (2007) A monomeric G protein-coupled receptor isolated in a high-density lipoprotein particle efficiently activates its G protein. Proc Natl Acad Sci U S A 104:7682–7687
23. Whorton MR, Jastrzebska B, Park PS, Fotiadis D, Engel A, Palczewski K, Sunahara RK (2008) Efficient coupling of transducin to monomeric rhodopsin in a phospholipid bilayer. J Biol Chem 283:4387–4394
24. Tsukamoto H, Sinha A, DeWitt M, Farrens DL (2010) Monomeric rhodopsin is the minimal functional unit required for arrestin binding. J Mol Biol 399:501–511
25. Bayburt TH, Vishnivetskiy SA, McLean MA, Morizumi T, Huang C-C, Tesmer JJG, Ernst OP, Sligar SG, Gurevich VV (2011) Monomeric rhodopsin is sufficient for normal rhodopsin kinase (GRK1) phosphorylation and arrestin-1 binding. J Biol Chem 286:1420–1428
26. Tsukamoto H, Szundi I, Lewis JW, Farrens DL, Kliger DS (2011) Rhodopsin in nanodiscs has

native membrane-like photointermediates. Biochemistry 50:5086–5091

27. Baas BJ, Denisov IG, Sligar SG (2004) Homotropic cooperativity of monomeric cytochrome P450 3A4 in a nanoscale native bilayer environment. Arch Biochem Biophys 430:218–228
28. Davydov DR, Fernando H, Baas BJ, Sligar SG, Halpert JR (2005) Kinetics of dithionite-dependent reduction of cytochrome P450 3A4: heterogeneity of the enzyme caused by its oligomerization. Biochemistry 44:13902–13913
29. Denisov IG, Baas BJ, Grinkova YV, Sligar SG (2007) Cooperativity in cytochrome P450 3A4: linkages in substrate binding, spin state, uncoupling, and product formation. J Biol Chem 282:7066–7076
30. Denisov IG, Grinkova YV, Baas BJ, Sligar SG (2006) The ferrous-dioxygen intermediate in human cytochrome P450 3A4: substrate dependence of formation and decay kinetics. J Biol Chem 281:23313–23318
31. Denisov IG, Grinkova YV, McLean MA, Sligar SG (2007) The one-electron autoxidation of human cytochrome P450 3A4. J Biol Chem 282:26865–26873
32. Frank DJ, Denisov IG, Sligar SG (2009) Mixing apples and oranges: analysis of heterotropic cooperativity in cytochrome P450 3A4. Arch Biochem Biophys 488:146–152
33. Frank DJ, Denisov IG, Sligar SG (2011) Analysis of heterotropic cooperativity in cytochrome P450 3A4 using α-naphthoflavone and testosterone. J Biol Chem 286:5540–5545
34. Gantt SL, Denisov IG, Grinkova YV, Sligar SG (2009) The critical iron-oxygen intermediate in human aromatase. Biochem Biophys Res Commun 387:169–173
35. Grinkova YV, Denisov IG, Waterman MR, Arase M, Kagawa N, Sligar SG (2008) The ferrous-oxy complex of human aromatase. Biochem Biophys Res Commun 372:379–382
36. Mak PJ, Denisov IG, Grinkova YV, Sligar SG, Kincaid JR (2011) Defining CYP3A4 structural responses to substrate binding. Raman spectroscopic studies of a nanodisc-incorporated mammalian cytochrome P450. J Am Chem Soc 133:1357–1366
37. Nath A, Koo PK, Rhoades E, Atkins WM (2008) Allosteric effects on substrate dissociation from cytochrome P450 3A4 in nanodiscs observed by ensemble and single-molecule fluorescence spectroscopy. J Am Chem Soc 130:15746–15747
38. Raschle T, Hiller S, Etzkorn M, Wagner G (2010) Nonmicellar systems for solution NMR spectroscopy of membrane proteins. Curr Opin Struct Biol 20:471–479
39. Shenkarev ZO, Lyukmanova EN, Solozhenkin OI, Gagnidze IE, Nekrasova OV, Chupin VV,

Tagaev AA, Yakimenko ZA, Ovchinnikova TV, Kirpichnikov MP, Arseniev AS (2009) Lipid-protein nanodiscs: possible application in high-resolution NMR investigations of membrane proteins and membrane-active peptides. Biochemistry (Mosc) 74:756–765

40. Shenkarev ZO, Paramonov AS, Lyukmanova EN, Shingarova LN, Yakimov SA, Dubinnyi MA, Chupin VV, Kirpichnikov MP, Blommers MJJ, Arseniev AS (2010) NMR structural and dynamical investigation of the isolated voltage-sensing domain of the potassium channel KvAP: implications for voltage gating. J Am Chem Soc 132:5630–5637

41. Yu TY, Raschle T, Hiller S, Wagner G (2011) Solution NMR spectroscopic characterization of human VDAC-2 in detergent micelles and lipid bilayer nanodiscs. Biochim Biophys Acta DOI: 10.1016/j.bbamem.2011.11.012

42. Alami M, Dalal K, Lelj-Garolla B, Sligar SG, Duong F (2007) Nanodiscs unravel the interaction between the SecYEG channel and its cytosolic partner SecA. EMBO J 26:1995–2004

43. Alvarez FJD, Orelle C, Davidson AL (2010) Functional reconstitution of an ABC transporter in nanodiscs for use in electron paramagnetic resonance spectroscopy. J Am Chem Soc 132:9513–9515

44. Dalal K, Duong F (2010) Reconstitution of the SecY translocon in nanodiscs. Methods Mol Biol 619:145–156

45. Kawai T, Caaveiro JM, Abe R, Katagiri T, Tsumoto K (2011) Catalytic activity of MsbA reconstituted in nanodisc particles is modulated by remote interactions with the bilayer. FEBS Lett 585:3533–3537

46. Ritchie TK, Kwon H, Atkins WM (2011) Conformational analysis of human ATP-binding cassette transporter ABCB1 in lipid nanodiscs and inhibition by the antibodies MRK16 and UIC2. J Biol Chem 286:39489–39496

47. Zou P, McHaourab HS (2010) Increased sensitivity and extended range of distance measurements in spin-labeled membrane proteins: Q-band double electron–electron resonance and nanoscale bilayers. Biophys J 98:L18–20

48. Boldog T, Grimme S, Li M, Sligar SG, Hazelbauer GL (2006) Nanodiscs separate chemoreceptor oligomeric states and reveal their signaling properties. Proc Natl Acad Sci U S A 103:11509–11514

49. Li M, Hazelbauer GL (2011) Core unit of chemotaxis signaling complexes. Proc Natl Acad Sci U S A A108:9390–9395

50. Frauenfeld J, Gumbart J, van der Sluis EO, Funes S, Gartmann M, Beatrix B, Mielke T, Berninghausen O, Becker T, Schulten K, Beckmann R (2011) Cryo-EM structure of the ribosome-SecYE complex in the membrane environment. Nat Struct Mol Biol 18:614–621

51. Katayama H, Wang J, Tama F, Chollet L, Gogol EP, Collier RJ, Fisher MT (2010) Three-dimensional structure of the anthrax toxin pore inserted into lipid nanodiscs and lipid vesicles. Proc Natl Acad Sci U S A 107:3453–3457, S3453/3451–S3453/3453

52. Pandit A, Shirzad-Wasei N, Wlodarczyk LM, van Roon H, Boekema EJ, Dekker JP, de Grip WJ (2011) Assembly of the major light-harvesting complex II in lipid nanodiscs. Biophys J 101:2507–2515

53. Ye F, Hu G, Taylor D, Ratnikov B, Bobkov AA, McLean MA, Sligar SG, Taylor KA, Ginsberg MH (2010) Recreation of the terminal events in physiological integrin activation. J Cell Biol 188:157–173

54. Bayburt TH, Sligar SG (2003) Self-assembly of single integral membrane proteins into soluble nanoscale phospholipid bilayers. Prot Sci 12:2476–2481

55. Ranaghan MJ, Schwall CT, Alder NN, Birge RR (2011) Green proteorhodopsin reconstituted into nanoscale phospholipid bilayers (nanodiscs) as photoactive monomers. J Am Chem Soc 133:18318–18327

56. Leitz AJ, Bayburt TH, Barnakov AN, Springer BA, Sligar SG (2006) Functional reconstitution of beta 2-adrenergic receptors utilizing self-assembling nanodisc technology. BioTechniques 40:601–602, 604, 606, 608, 610, 612

57. Das A, Zhao J, Schatz GC, Sligar SG, Van Duyne RP (2009) Screening of type I and II drug binding to human cytochrome P450-3A4 in nanodiscs by localized surface plasmon resonance spectroscopy. Anal Chem 81:3754–3759

58. Cruz F, Edmondson DE (2007) Kinetic properties of recombinant MAO-A on incorporation into phospholipid nanodisks. J Neural Transm 114:699–702

59. Ishmukhametov R, Hornung T, Spetzler D, Frasch WD (2010) Direct observation of stepped proteolipid ring rotation in E. coli F0F1-ATP synthase. EMBO J 29:3911–3923

60. Mi L-Z, Grey MJ, Nishida N, Walz T, Lu C, Springer TA (2008) Functional and structural stability of the epidermal growth factor receptor in detergent micelles and phospholipid nanodiscs. Biochemistry 47:10314–10323

61. Morrissey JH, Pureza V, Davis-Harrison RL, Sligar SG, Ohkubo YZ, Tajkhorshid E (2008) Blood clotting reactions on nanoscale phospholipid bilayers. Thromb Res 122:S23–S26

62. Shaw AW, Pureza VS, Sligar SG, Morrissey JH (2007) The local phospholipid environment modulates the activation of blood clotting. J Biol Chem 282:6556–6563

63. Johnson EF, Stout CD (2005) Structural diversity of human xenobiotic-metabolizing cytochrome P450 monooxygenases. Biochem Biophys Res Commun 338:331–336

64. Mast N, White MA, Bjorkhem I, Johnson EF, Stout CD, Pikuleva IA (2008) Crystal structures of substrate-bound and substrate-free cytochrome P450 46A1, the principal cholesterol hydroxylase in the brain. Proc Natl Acad Sci U S A 105:9546–9551

65. Das A, Grinkova YV, Sligar SG (2007) Redox potential control by drug binding to cytochrome P 450 3A4. J Am Chem Soc 129:13778–13779

66. Das A, Sligar SG (2009) Modulation of the cytochrome P450 reductase redox potential by the phospholipid bilayer. Biochemistry 48:12104–12112

67. Grinkova YV, Denisov IG, Sligar SG (2010) Functional reconstitution of monomeric CYP3A4 with multiple cytochrome P450 reductase molecules in Nanodiscs. Biochem Biophys Res Commun 398:194–198

68. Chen PS, Toribara TY, Warner H (1956) Microdetermination of phosphorus. Anal Chem 28:1756–1759

69. Fiske CH, Subbarow Y (1925) The colorimetric determination of phosphorus. J Biol Chem 66:375–400

70. Cevc G (1993) Phospholipids handbook. Marcel Dekker, Inc., New York, p 988

71. Chougnet A, Grinkova Y, Ricard D, Sligar S, Woggon W-D (2007) Fluorescent probes for rapid screening of potential drug-drug interactions at the CYP3A4 level. ChemMedChem 2:717–724

72. Davydov DR, Baas BJ, Sligar SG, Halpert JR (2007) Allosteric mechanisms in cytochrome P450 3A4 studied by high-pressure spectroscopy: pivotal role of substrate-induced changes in the accessibility and degree of hydration of the heme pocket. Biochemistry 46:7852–7864

73. Nath A, Grinkova YV, Sligar SG, Atkins WM (2007) Ligand binding to cytochrome P450 3A4 in phospholipid bilayer nanodiscs: the effect of model membranes. J Biol Chem 282:28309–28320

74. Davydov DR, Sineva EV, Sistla S, Davydova NY, Frank DJ, Sligar SG, Halpert JR (2010) Electron transfer in the complex of membrane-bound human cytochrome P450 3A4 with the flavin domain of P450BM-3: the effect of oligomerization of the heme protein and intermittent modulation of the spin equilibrium. Biochim Biophys Acta 1797:378–390

75. Luthra A, Denisov IG, Sligar SG (2011) Temperature derivative spectroscopy to monitor the autoxidation decay of cytochromes P450. Anal Chem 83:5394–5399

Chapter 11

Rapid LC-MS Drug Metabolite Profiling Using Bioreactor Particles

Linlin Zhao, Besnik Bajrami, and James F. Rusling

Abstract

Enzyme-coated magnetic bioreactor particles enable a fast, convenient approach to metabolic screening. A semi-automated metabolite-profiling technique using these particles in a 96-well plate with liquid chromatography (LC)-mass spectrometry (MS)/MS detection is described. Reactions can be investigated over 1- to 2-min periods, and 96 or more reactions or reaction time points can be processed in parallel. Incorporation of DNA in the particle films facilitates determination of rates of DNA damage and metabolite-DNA adduct structures.

Key words Cytochrome P450, Metabolism, Sample preparation, High throughput, LC-MS/MS, Magnetic separation

1 Introduction

Rapidly and accurately predicting in vivo metabolism, pharmacokinetics, and toxicity of drug candidates early in the development process is a major challenge in drug discovery. This fact has driven the development of high-throughput in vitro bioanalytical methodologies in pharmaceutical and academic laboratories. We developed a high-throughput sample preparation approach using bioreactor particles coated with thin enzyme films coupled with LC-MS/MS for metabolite identification and quantification (1). These bioreactor particles feature densely packed enzyme coatings (with or without DNA) fabricated by electrostatic layer-by-layer (LbL) adsorption assembly (2). Enzymes and DNA on the particle surface are in close proximity, creating high local concentrations to facilitate fast reactions using small amounts of materials. Faster enzyme turnover rates compared with those obtained using conventional microsomal enzyme dispersions result from better access to the enzymes. Significantly decreased reaction times, often several minutes, are achieved for reactive metabolite-DNA reactions

Ian R. Phillips et al. (eds.), *Cytochrome P450 Protocols*, Methods in Molecular Biology, vol. 987, DOI 10.1007/978-1-62703-321-3_11, © Springer Science+Business Media New York 2013

due to very high surface concentrations of DNA close to sites on the particles where the metabolites are generated (1–3).

The rapid drug metabolite-profiling method features parallel processing of reactions through the use of the bioreactor particles in a 96-well plate format, in which up to 96 reactions (or more if a 384-well plate is used) can be processed simultaneously in relatively short times. This approach was optimized by using magnetic particles to simplify sample-handling procedures and introduce partial automation (3), although silica particles were used as the initial platform (1). This chapter describes the rapid metabolite-profiling methodology using microsomal enzyme magnetic bioreactor particles in a parallel-processing format illustrated for two example reactants, diclofenac and troglitazone.

2 Materials

Prepare all the solutions using twice-deionized (DI) water and analytical grade reagents. All the prepared solutions and reagents should be stored at 4°C, unless otherwise noted. Comprehensively follow all waste-disposal regulations when disposing of solutions and other reagents.

2.1 Magnetic Bioreactor Film Assembly

1. Tris Buffer: 5 mM Tris–HCl (pH 7.0), 5 mM NaCl. Dissolve and mix well 0.61 g of Trizma base and 0.29 g of NaCl in 1.0 L of DI water. Add concentrated HCl drop-wise as needed to adjust to pH 7.0.

2. Carboxylated magnetic particles (Polysciences, Warrington, PA, average diameter 1 μm, 20 mg/mL). These particles are diluted by a factor of 5 using 5 mM Tris buffer (pH 7.0, 5 mM NaCl) in a 2-mL polypropylene Eppendorf tube or 15-mL Falcon tube (BD®) in a batch preparation before film assembly.

3. The positively charged polyion poly(diallyldimethylammonium chloride) (PDDA) is used as a molecular glue to facilitate the electrostatic assembly of negatively charged species, e.g., microsomes or DNA. PDDA solution: 2 mg/mL + 50 mM NaCl in water.

4. Rat liver microsomes (pooled, Fischer) and human liver microsomes are from BD biosciences (Woburn, MA), and are used for film assembly. Microsomes contain important metabolizing enzymes, such as cytochromes P450, and are overall negatively charged due to lipid components.

2.2 Reaction Components

1. A 96-well filtration plate (AcroPrep™, 500 μL, 3 kDa mass cutoff) and a multi-well plate vacuum manifold (part no. 5017) from Pall Life Sciences are used together with a 96-well receiver plate (Axygen Scientific). A 12.5 cm×6 cm rectangular

Rapid LC-MS Drug Metabolite Profiling Using Bioreactor Particles 131

magnetic separation unit (BioMag® flask separator, Polysciences, Warrington, PA) is placed on top of the 96-well filtration plate cover to facilitate the filtration.

2. Diclofenac (Sigma): stock solution of 5 mM in dimethyl sulfoxide, store at 4°C.

3. Troglitazone (Sigma): stock solution of 5 mM in water, store at 4°C.

4. NADPH (β-nicotinamide adenine dinucleotide phosphate reduced, Sigma): 0.2 M stock solution is freshly prepared in water before addition to reaction mixes.

5. Caffeine is used as an internal standard for LC-MS analysis: 1 mM stock solution in water.

2.3 LC-MS Components

1. A Capillary LC (Waters, Capillary LC-XE, Milford, MA) equipped with a C18 (Atlantis®, 23.5 mm) trap column and a C18 reverse-phase analytical column (Atlantis®, 150 mm, 300 μm I.D., 5-μm particle size).

2. Any mass spectrometer (e.g. QTRAP, AB Sciex, Foster City, CA, with Analyst 1.4 software) operated in the electrospray ionization positive mode. Total ion scans are first used for identification, and selected reaction monitoring and production scans are used later for structural confirmation. Any spectrometer with equivalent capabilities can be used.

3. A binary mobile phase was used for capillary LC, which is composed of A, ammonium acetate buffer (10 mM, pH 4.5 with 0.1% formic acid), and B, 100% acetonitrile (HPLC grade). Mobile-phase solvent was filtered (0.2 μm) and degassed before use.

3 Methods

3.1 Film Fabrication

The following protocol is described in terms of the amount of particles used per reaction well, but is scalable. Particles are generally prepared in batch depending on the enzymes and reactions desired (see Note 1).

1. Disperse 20 μL of magnetic-particle dispersion (20 mg/mL) in 80 μL of 5 mM Tris buffer (pH 7.0, 5 mM NaCl) in a 2-mL Eppendorf tube.

2. Add 100 μL of PDDA dispersion, mix gently, and allow 20 min for assembly on the negatively charged magnetic-particle surface.

3. Use a neodymium disc rare-earth magnet to pull magnetic particles to the side or the bottom of the 2-mL Eppendorf tube.

132   Linlin Zhao et al.

Depending on the amount of particles used, they are usually separated from the solution within 1 min.

4. Carefully remove the supernatant without disturbing the magnetic particles being held in place by the magnet, and then wash particles with 200 µL of 5 mM Tris buffer (pH 7.0, 5 mM NaCl) twice to remove loosely bound polyions (follow the same procedure to remove the buffer each time).

5. Re-disperse the particles into 100 µL of Tris buffer. Add 40 µL of microsomes drop-wise with gentle shaking and allow 30 min for assembly of the microsomes, followed by washing twice with buffer and magnetic-particle separation as described above.

6. Disperse the bioreactor particles with PDDA/microsome films in 100 µL of Tris buffer (pH 7.0, 5 mM NaCl) and store at 4°C until use.

3.2 Sample Workup for Diclofenac and Troglitazone

Safety note: All procedures are done under closed hoods while wearing gloves.

1. Add 100 µL bioreactor particle dispersion coated with PDDA/microsome films into each well in a 96-well plate.

2. Add 1 µL of 5 mM diclofenac or troglitazone stock solutions to reaction well to achieve a final concentration of 50 µM.

3. Reactions are carried out in triplicate at multiple reaction time points, i.e., 1, 3, 5, and 7 min. Initiate the reaction by adding 1 µL of 0.2 M NADPH solution (a final concentration of 2 mM in the reaction mixture) and terminate the reaction by adding 15 µL of cold acetonitrile containing 6% (v/v) formic acid at the appropriate time. A multichannel pipette can be used to facilitate precise control of time for a series of wells.

4. After the reaction, transfer the 96-well plate to a vacuum manifold while applying a rectangular magnetic plate on top of the 96-well plate cover. Particles are suspended on the surface of the solution to facilitate filtration. The metabolite-containing solution is collected as filtrate, leaving the particles and most of the enzyme components in the filtration plate (see Notes 2 and 3).

5. A final concentration of 10 µM caffeine is added as an internal standard before LC-MS analysis.

3.3 LC-MS Analysis

General strategy: It is important to develop suitable LC separation conditions for the metabolites. In the absence of authentic standards, a surrogate LC-MS scan is preformed to identify the target metabolite in the total ion chromatogram. Retention times of ions of interest are matched to the LC-diode array chromatogram to ensure they are adequately resolved. In the absence of isotope-labeled standards, a semiquantitative method (peak area ratio of analyte/internal standard) can be used. Using integrated peak areas of interest from LC-diode array chromatograms can also minimize the difference in ionization efficiencies of different metabolites. Product-ion and selected reaction monitoring scans are used for structural confirmatory purposes for a specific metabolite.

1. Load a 10-μL sample onto the trap column, and flush at a flow rate of 5 μL/min with water (containing 0.1% formic acid) to pre-concentrate and desalt the sample. After 1 min, sample is back-flushed to the analytical column at a flow rate of 9.25 μL/min using a binary separation gradient composed of ammonium acetate buffer (10 mM, pH 4.5 with 0.1% formic acid) and 100% of acetonitrile.

2. The solvent gradient is adjusted to suit the analysis of different metabolites. The following acetonitrile composition is used: for troglitazone, 10% for 5 min, 10–50% for 3 min, 50% for 12 min, 50–90% for 10 min, 90% for 10 min, 90–10% for 1 min, 10% for 2 min; for diclofenac, 10% for 5 min, 10–40% for 3 min, 40% for 20 min, 40–10% for 1 min, 10% for 5 min.

3. The following parameters are used in 4000 QTRAP mass spectrometer: ion-spray voltage 4500 V, declustering potential 40 V, collision energy 20–40 eV, and 0.15-s dwell time for different mass transitions. LC-MS batch analysis can be programmed using MassLynx software to interface with mass spectrometer; automation of the process is facilitated by the auto-sampler equipped with capLC.

4 Notes

1. Film assembly described in Subheading 3.1 can be easily modified to suit the desired experiment. Other enzyme sources can be used instead of microsomes (e.g., supersomes, liver cytosolic fractions, pure bioconjugation enzymes). The absorption efficiency may vary when different enzyme sources are used; appropriate buffer pH should be adjusted according to the pI of the protein to be assembled in the protein assembly step. The amount of the enzyme on the surface can be increased by repeating the adsorption cycles for particle coating, as described above. However, the amount of electrostatically absorbed components saturate at about 3–4 bilayers in this case and particle aggregation is often observed as more component layers are added.

2. Negatively charged DNA can be included as a component in the film in order to determine metabolite-DNA reaction rates and metabolite-nucleobase adduct structures (3). Particles with

positively charged PDDA on the outermost layer can be mixed with an equal volume of salmon testis DNA (0.2 mg/mL) in 5 mM Tris buffer (pH 7.0), 5 mM NaCl, then washed.

3. The processing of particles with DNA films is slightly different after the enzyme reaction. The DNA film on the particle surface must be hydrolyzed to release metabolite-nucleobase adducts before LC-MS/MS analysis, and this is also done in the 96-well plate. Different hydrolysis procedures can be employed and one more step involving sample transfer is needed. Interested readers should refer to ref. (3) for details.

References

1. Bajrami B, Zhao L, Schenkman JB, Rusling JF (2009) Rapid LC-MS drug metabolite profiling using microsomal enzyme bioreactors in a parallel processing format. Anal Chem 81: 9921–9929

2. Bajrami B, Hvastkovs EG, Jensen GC, Schenkman JB, Rusling JF (2008) Enzyme-DNA biocolloids for DNA adduct and reactive metabolite detection by chromatography-mass spectrometry. Anal Chem 80:922–932

3. Zhao L, Schenkman JB, Rusling JF (2010) High-throughput metabolic toxicity screening using magnetic biocolloid reactors and LC–MS/MS. Anal Chem 82:10172–10178

Chapter 12

Fluorescence-Based Screening of Cytochrome P450 Activities in Intact Cells

M. Teresa Donato and M. José Gómez-Lechón

Abstract

Fluorimetric methods to assess cytochrome P450 (P450) activities that do not require metabolite separation have been developed. These methods make use of non- or low-fluorescent P450 substrates that produce highly fluorescent metabolites in aqueous solutions. The assays are based on the direct incubation of intact cells in culture with appropriate fluorogenic probe substrates, followed by fluorimetric quantification of the product formed and released into incubation medium. We describe a battery of fluorescence assays for rapid measurement of the activity of nine P450s involved in drug metabolism. For each individual P450 activity the probe showing the best properties (highest metabolic rates, lowest background fluorescence) has been selected. Fluorescence-based assays are highly sensitive and allow the simultaneous activity assessments of cells cultured in 96-well plates, using plate readers, with notable reductions in costs, time, and cells, thus enhancing sample throughput.

Key words Cytochromes P450, Fluorogenic substrate, Cell-based assays

1 Introduction

Drug metabolism is one of the key determinants of a drug's disposition in the body, interindividual pharmacokinetic differences, and an indirect determinant of the clinical efficacy and toxicity of drugs. Cytochrome P450 (P450) enzymes are a superfamily of enzymes that constitutes the major players in the oxidative metabolism of a wide range of structurally diverse xenobiotics including drugs (1). To speed up the selection of new drug candidates, pharmaceutical companies increasingly make use of different in vitro systems to investigate drug metabolism at the early preclinical stages. As a result of this, it is now possible to identify the metabolic profile of drug candidates, potential drug interactions, and the role of polymorphic enzymes.

A critical point for measuring P450 activities is the selection of probe substrates. Only a few molecules are useful as selective probes for individual P450s, and most of them involve chromatographic

Ian R. Phillips et al. (eds.), *Cytochrome P450 Protocols*, Methods in Molecular Biology, vol. 987,
DOI 10.1007/978-1-62703-321-3_12, © Springer Science+Business Media New York 2013

separations for metabolite detection, which severely limits sample throughput (2, 3). Alternatively, fluorimetric methods that do not require metabolite separation have been developed for P450 activity measurements (4). These methods are based on the quantification of formation rates of highly fluorescent metabolites produced from non- or low-fluorescent P450 substrates. Different O-alkyl derivatives of resorufin (5), fluorescein (6), 7-hydroxycoumarins (7–9), 6-hydroxyquinolines (6), and 4-methylsulfonylphenyl furanones (10) have been proposed as useful probes for this purpose. Fluorimetric measurements are rapid and highly sensitive, and enable the use of 96-well-plate formats, thereby markedly increasing assay throughput in comparison with conventional analytical procedures.

In this chapter a battery of fluorescence assays for rapid measurement of the activity of nine P450 enzymes in intact monolayers of cells is described. Although most of the assays make use of fluorogenic substrates that are not selective for an individual P450, some are selective and can also be used for measuring the activity of a particular enzyme in cells expressing multiple P450s (e.g., hepatocytes) (9–11). Cultured cells are directly incubated with the selected substrate, and the corresponding metabolite formed and released into incubation medium is fluorimetrically quantified (4). Major advantages of this procedure over P450 activity measurements using subcellular fractions are that, as living cells are used, manual handling and enzyme damage are minimized, the endoplasmic reticulum of the cells remains intact, exogenous cofactors or NADPH-regenerating systems are not required, and transport processes are maintained. These assays offer the possibility of easily selecting cells showing P450 activities as well as screening inhibitory or inductive effects of new drugs on individual P450 enzymes.

2 Materials

2.1 Cell Cultures

1. Seeding culture medium for hepatocytes: Ham's F-12/Williams (1:1) medium (Gibco BRL, Paisley, Scotland) supplemented with 2% (v/v) newborn calf serum (NBCS; Gibco BRL, Paisley, Scotland), 0.2% (w/v) bovine serum albumin fraction V (BSA), 10 nM insulin (Novo Nordisk, A/S Bagsvaerd, Denmark), transferrin (25 μg/mL), 0.1 μM sodium selenite, 65.5 μM ethanolamine, 7.2 μM linoleic acid, 7 mM glucose, 6.14 mM ascorbic acid (Sigma), penicillin (50 mU/mL), and streptomycin (50 μg/mL) (Gibco BRL, Paisley, Scotland).

2. 24-h chemically defined medium for hepatocytes: The same culture medium composition as above, but NBCS-free and supplemented with 10 nM dexamethasone.

3. Non-tumorigenic SV40-immortalized human liver epithelial (THLE) cell lines expressing CYP1A2, CYP2A6, CYPB6, CYP2D6, CYP2E1, or CYP3A4 genes were kindly supplied

Fluorescent P450 Assays in Cells   137

by Dr. A. Pfeifer from Nestec Ltd. (Basel, Switzerland) (12). THLE cell lines expressing CYP2C8, CYP2C9, or CYP2C19 were generated as described elsewhere (13).

4. Culture medium for THLE cell lines: Pasadena Foundation for Medical Research Medium 4 (PMFR-4 medium) supplemented with 200 nM hydrocortisone, 640 nM insulin, 500 nM triiodothyronine, 200 mM L-glutamine, epidermal growth factor (5 ng/mL), transferrin (10 µg/mL), bovine pituitary extract (15 µg/mL), 0,33 nM retinoic acid, 500 nM phosphorylethanolamine/ethanolamine (Biofluids, Rockville, USA), 3% (v/v) fetal bovine serum, and gentamicin (40 µg/mL) (Gibco BRL, Paisley, Scotland).

5. Coating mixture for culture plates: Dissolve 1 mg of human fibronectin in 97 mL of Dulbecco Minimum Essential Medium (DMEM) (Gibco BRL, Paisley, Scotland) supplemented with 0.1% (w/v) BSA. Add to this solution 3 mL of 0.1% (w/v) collagen Type I (from calf skin) in 0.1 M acetic acid (Sigma).

2.2 P450 Activity Assays

1. Substrate stock solutions: 4 mM dibenzylfluorescein (DBF) (BD Gentest, Woburn, MA) in acetonitrile;

   10 mM 3[2-(N,N-diethyl-N-methylammonium)ethyl]-7-methoxy-4-methylcoumarin (AMMC) (BD Gentest) in acetonitrile;

   12.5 mM 7-benzyloxy-4-trifluoromethylcoumarin (BFC) (BD Gentest) in dimethyl sulfoxide (DMSO);

   12.5 mM 7-methoxy-4-trifluoromethylcoumarin (MFC) (BD Gentest) in DMSO;

   12.5 mM 7-ethoxy-4-trifluoromethylcoumarin (EFC) (Ultrafine Chemicals, Manchester, UK) in DMSO;

   20 mM 3-Cyano-7-ethoxycoumarin (CEC) (Molecular Probes Europe BV, Leiden, The Netherlands) in acetonitrile;

   2 mM 7-ethoxyresorufin;

   4 mM 7-pentoxyresorufin;

   5 mM 7-methoxyresorufin;

   4 mM 7-benzoxyresorufin;

   100 mM coumarin (all from Sigma-Aldrich) in DMSO.

   All solutions are stored in aliquots at –80°C.

2. Metabolite stock solutions: 10 mM 3-[2-(diethylamino)ethyl]-7-hydroxy-4-methylcoumarin (AHMC) (BD Gentest) in DMSO;

   50 mM 7-hydroxycoumarin (Sigma-Aldrich) in DMSO;

   1 mM resorufin (Sigma-Aldrich) in methanol;

   10 mM 7-hydroxy-4-trifluoromethylcoumarin (HFC) (Ultrafine Chemicals) in water;

10 mM 3-cyano-7-hydroxycoumarin (CHC) (Molecular Probes Europe BV) in water;

10 mM fluorescein (Sigma-Aldrich) in water.

All solutions are stored in aliquots at –80°C.

3. Incubation medium: pH 7.4-buffered saline solution containing 1 mM Na_2HPO_4, 137 mM NaCl, 5 mM KCl, 0.5 mM $MgCl_2$, 10 mM glucose, and 10 mM Hepes (Sigma-Aldrich) and stored at –20°C. Just before use add $CaCl_2$ (1 mM final concentration) (see Note 1).

4. Conjugated hydrolysis solution: Just before use prepare a β-glucuronidase/arylsulfatase (Roche, Barcelona, Spain) solution in 0.1 M acetate, pH 4.5 (14).

5. Incubation medium with substrate: Just before use substrate stock solution is diluted in incubation medium plus 1 mM $CaCl_2$ to obtain the final incubation concentrations of each substrate indicated in Table 1.

6. Chemical inhibitor stock solutions: 10 mM furafylline; 0.5 mM ketoconazole; 10 mM methoxsalen; 10 mM quercetin; 2 mM quinidine; 10 mM sulfaphenazole; 10 mM tranylcypromine; and 50 mM diethyldithiocarbamate (all from Sigma-Aldrich) are each prepared as aqueous solutions or in organic solvent (methanol or DMSO) and stored in aliquots at –80°C (see Note 2).

7. P450 inducers: 1 mM dexamethasone (DX) and 6 M ethanol (ET) (Merck Chemicals) are prepared as aqueous solutions; 1 mM 3-methylcholanthrene (MC); 400 mM phenobarbital (PB); 1 mM clofibric acid (CL); 20 mM rifampicin (RF) (all from Sigma-Aldrich) stock solutions are each prepared in DMSO and stored in aliquots at –20°C.

3 Methods

Primary human hepatocytes are recognized as the closest in vitro model to human liver and are currently considered a valuable tool for determining drug metabolism and for assessing the risk of drug hepatotoxicity in humans (15–17). However, the well-known phenotypic instability of hepatocytes throughout culture, the scarce and irregular availability of fresh human livers for cell-harvesting purposes, and the high batch-to-batch functional variability of fresh and cryopreserved hepatocyte preparations obtained from different human liver donors seriously hinder their use in routine testing (18). To overcome these limitations, different cell-line models have been proposed for drug metabolism screening. Human liver-derived cell lines would be ideal models for this purpose given their availability, unlimited life span, stable phenotype, and the fact that they are easy to handle. However, most of them

Table 1
Fluorescence-based assays for P450 activities

Substrate[a]	CYP	Concentration in assay (μM)	Incubation time (min)	Metabolite	Standard curve (nM)	Ex/Em (nm)[b]	Quenching solution
7-MR[c]	1A2	10	45	Resorufin[c]	5–200	530/580	Methanol
7-ER	1A2	10	45	Resorufin	5–200	530/580	Methanol
CEC	1A2	30	90	CHC	20–500	408/455	0.1 M phosphate potassium buffer pH 7.4
Coumarin[c]	2A6	50	45	7-Hydroxycoumarin	200–8000	355/460	0.1 M Tris pH 9
7-BR[c]	2B6	10	60	Resorufin	5–200	530/580	Methanol
7-PR	2B6	10	60	Resorufin	5–200	530/580	Methanol
EFC	2B6	30	60	HFC	5–100	410/510	0.25 M Tris in 60% (v/v) acetonitrile
DBF	2C8	10	120	Fluorescein	5–60	485/538	2 N sodium hydroxide
MFC	2C9	150	120	HFC	5–100	410/510	0.25 M Tris in 60% (v/v) acetonitrile
CEC	2C19	60	90	CHC	5–80	408/455	0.1 M phosphate potassium buffer pH 7.4
EFC	2C19	100	120	HFC	5–60	410/510	0.25 M Tris in 60% (v/v) acetonitrile
AMMC[c]	2D6	100	120	AHMC	5–60	390/460	0.25 M Tris in 60% (v/v) acetonitrile

(continued)

Table 1
(continued)

Substrate[a]	CYP	Concentration in assay (μM)	Incubation time (min)	Metabolite	Standard curve (nM)	Ex/Em (nm)[b]	Quenching solution
MFC	2E1	10	60	HFC	5–250	410/510	0.25 M Tris in 60% (v/v) acetonitrile
BFC	3A4	100	60	HFC	5–100	410/510	0.25 M Tris in 60% (v/v) acetonitrile

[a]7-*MR* 7-methoxyresorufin, 7-*ER* 7-ethoxyresorufin, *CEC* 3-Cyano-7-ethoxycoumarin, *CHC* 3-Cyano-7-hydroxycoumarin, 7-*BR* 7-benzoyresorufin, 7-*PR* 7-pentoxyresorufin, *EFC* 7-ethoxy-4-trifluoromethylcoumarin, *HFC* 7-hydroxy-4-trifluoromethylcoumarin, *DBF* dibenzylfluorescein, *MFC* 7-methoxy-4-trifluoromethylcoumarin, *AHMC* 3-[2-(diethylamino)ethyl]-7-hydroxy-4-methylcoumarin, *AMMC* 3-[2-(*N*,*N*-diethyl-*N*-methylammonium)ethyl]-7-methoxy-4-methylcoumarin, *BFC* 7-benzyloxy-4-trifluoromethylcoumarin
[b]Ex/Em: excitation and emission wavelengths
[c]Enzyme-specific assay able to be used to determine an individual activity in cells expressing multiple P450s (e.g., hepatocytes)

do not constitute a real alternative to primary cultured hepatocytes, given their very low/partial expression of P450 enzymes (19). Advances in molecular biology have enabled the possibility of developing genetically manipulated cells stably or transiently expressing human P450s as new tools for investigating drug metabolism, drug-drug interactions, and metabolism-based toxic bioactivation.

3.1 Cell Culture

1. Coat culture plates (BD Falcon™ tissue culture dishes, New Jersey USA) with 10 µL/cm² of the fibronectin/collagen coating mixture and leave to stand at 37°C for 1 h.

2. Remove excess coating mixture.

3.1.1 Culturing of Human Hepatocytes

1. Suspend freshly isolated or cryopreserved human hepatocytes (5×10^5 viable cells/mL) in seeding culture medium and plate on fibronectin/collagen-coated plates at a density of 8×10^4 cells/cm² (see Note 3).

2. 1 h later, discard unattached cells and debris and renew the medium (see Note 4).

3. After 24 h, replace medium with serum-free chemically defined medium.

4. Replace medium every 24 h.

3.1.2 Culturing P450-Expressing THLE Cells

1. For subculturing, harvest cells after trypsin/EDTA (0.25%/0.02%) (Gibco, BRL, Paisley, Scotland) treatment for 5 min at 37°C.

2. Suspend in PMFR-4 culture medium THLE cells genetically manipulated to express individual P450s and plate on fibronectin/collagen-coated flasks.

3. For P450 activity assays, use cells 24 h after seeding (at 75% monolayer confluence).

3.2 Incubation of Cultured Cells with P450 Substrates

P450 activity measurements are key tools for the evaluation of metabolic capacity of cultured cells (hepatocytes, hepatoma cell lines, engineered cells) and the exploration of potential metabolism-based drug-drug interactions (16, 20). In the search of rapid and reliable methods for medium- or high-throughput screenings, we propose a battery of sensitive assays for evaluating the activity of major P450 enzymes responsible of drug metabolism in human liver. The methods consist in the direct incubation of intact cell monolayers with appropriate probe substrates, which are oxidized by P450s, to produce highly fluorescent metabolites in aqueous solution. In selecting fluorogenic probes, substrates with limited solubility, poor metabolite formation rates, high background fluorescence, low signal/noise ratio, or excitation wavelength in the UV range were discarded (4). A notable advantage of the assays is their sensitivity, which allows activity assessments in cells cultured in 96-well plates, with notable reductions in costs (smaller amounts of substrates and reagents), time (microtiter plate readers, potential automatization), and cells (a key issue for cells with limited availability such as human hepatocytes). Moreover, because living cells are used, substrate incubations can be extended for several hours, which can be useful for activity measurements in cells with low metabolic rates.

1. Remove culture medium from culture plates and add an appropriate volume of incubation medium containing the substrate (recommended volumes: 150 μl, 300 μl, or 750 μl for 96-well, 24-well, or 3.5-cm diameter plate formats, respectively).

2. Incubate culture plates in the presence of the appropriate substrate in a cell culture incubator ($37°C$ 5% CO_2) for the indicated time (see Table 1). Then, transfer incubation medium to new 96-well plates or tubes. At this step, samples can be immediately processed or kept frozen at $-80°C$ until used (see Note 5).

3. At this stage, cell monolayers are washed out with phosphate-buffered saline solution, frozen in liquid N_2, and kept at $-20°C$ for further cell protein quantification.

4. Samples of incubation medium with substrate not incubated with cells (blank samples) are also processed.

3.3 Sample Preparation

1. Before metabolite analysis, thaw incubation samples (see Subheading 3.2) at room temperature and add β-glucuronidase/arylsulfatase in 0.1 M acetate pH 4.5, to hydrolyze metabolite conjugates (see Note 6).

2. After 120 min in a shaking incubator ($37°C$), stop hydrolysis reactions by diluting (1:2) in the appropriate quenching solution (Table 1).

3. If sample turbidity is observed after adding quenching solution, centrifuge for 5 min at $10,000 \times g$ at $4°C$ and transfer clean supernatants to new 96-well plate or tubes.

3.4 Preparation of Standard Calibration Curves

1. Dilute (1:100) an aliquot of thawed metabolite stock solution in incubation medium (saline solution plus Ca^{2+}).

2. Make serial dilutions in incubation medium (saline solution plus Ca^{2+}) to prepare a calibration curve (up to 8 different concentrations) in the range of concentrations indicated in Table 1.

3. Add an equal volume (1:2 dilution) of appropriate quenching solution (Table 1).

3.5 Fluorescence Measurements

1. Transfer 100 µl of each sample (cell incubation medium, blank, or standard) to 96-well plates.

2. Determine fluorescence intensity by using a Spectra Max Geminis XS fluorescence microplate reader (Molecular Devices Co, Sunnyvale, CA). See Table 1 for details (excitation and emission wavelengths).

3. Quantify metabolite formation by interpolation of fluorescence intensity in the standard curve. For each P450 enzyme, activity is expressed as pmol of the corresponding metabolite formed per minute and per mg of cell protein.

3.6 Screening of P450 Inhibitors

The early detection of potential P450 inhibitors is an important application of activity assays (21). Inhibition of P450 enzymes resulting in impaired hepatic clearance of co-administered drugs is a major cause of drug-drug interactions. Although P450 inhibitory effects of new chemicals are usually tested in subcellular fractions (recombinant P450 enzymes or human liver microsomes), inhibition studies can also be performed in living cells (4, 22, 23). A comparison of the effects of model inhibitors on P450 enzymes using the fluorescence-based assays in intact cells or conventional probe assays in subcellular systems showed highly comparable results (4). Assays in intact cells more closely reflect the environment to which drugs are exposed in the liver (transport mechanisms, cytosolic enzymes, binding to intracellular proteins), which can be determinant of the actual concentration of substrate and inhibitor available to the P450 enzyme. Obviously, nonselective fluorogenic substrates cannot be used for in vitro systems expressing several P450s (e.g., hepatocytes), and their application is limited to those models expressing an individual enzyme (e.g., P450-expressing THLE cells). Moreover, inhibitory effects of fluorescent compounds or of compounds metabolized to fluorescent products cannot be tested.

1. Add the chemical to be screened (together with substrate) to incubation medium. Several concentrations of the inhibitor covering at least a 100-fold concentration range is recommended (see Note 7). Low substrate concentrations proximate to K_m values are used for inhibition assays (see details in Fig. 1) (see Note 8).

2. Incubate cultured cells with an appropriate volume of incubation medium containing substrate and inhibitor. In parallel, incubate control cells with incubation medium containing substrate but no inhibitor, but containing an equivalent concentration of the solvent (<0.5%, v/v) used to prepare the inhibitor (usually DMSO or acetonitrile).

3. After sample processing and metabolite quantification (see Subheadings 3.3–3.5), P450 activity values in cells incubated

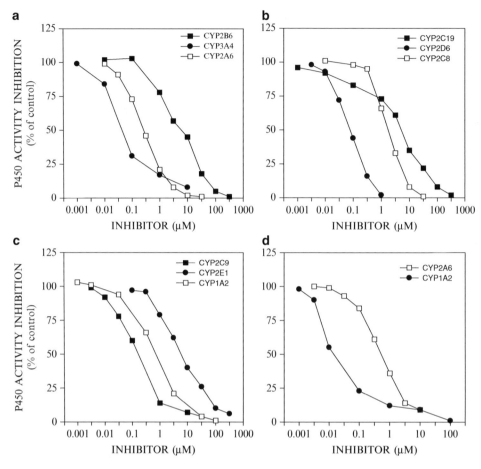

Fig. 1 Inhibition of P50 activities in intact cells. THLE cells expressing individual P450 enzymes (**a–c**) or human hepatocytes in primary culture (**d**) were incubated with substrates in the presence of different concentrations of model chemical inhibitors. (**a**) Inhibition of CYP2B6 (10 μM EFC) by tranylcypromine, CYP3A4 (30 μM BFC) by ketoconazole, and CYP2A6 (15 μM coumarin) by methoxalen; (**b**) inhibition of CYP2C19 (10 μM EFC) by tranylcypromine, CYP2D6 (30 μM AMMC) by quinidine, and CYP2C8 (3 μM DBF) by quercetin; (**c**) inhibition of CYP2C9 (50 MFC) by sulfaphenazole, CYP2E1 (3 μM MFC) by diethyldithiocarbamate, and CYP1A2 (3 μM CEC) by furafylline; (**d**) inhibition of CYP2A6 (15 μM coumarin) by methoxalen and CYP1A2 (5 μM 7-MR) by furafylline. Results are expressed as percentage of control activity (assayed in the absence of inhibitor)

in the presence of inhibitors are expressed as a percentage of activity in cells incubated with substrate in the absence of the inhibitor (control) (Fig. 1). IC50 values (concentration of inhibitor causing a 50% reduction in activity relative to the control) are calculated by plotting the linear regression analysis of the log inhibitor concentration versus the percentage of the control activity.

4. Time-dependent inactivation or mechanism-based inhibition (produced by formation of metabolite intermediates that bind tightly and irreversibly to the enzyme) can easily be assessed

Fluorescent P450 Assays in Cells 145

using fluorogenic P450 probes. To this end, cultured cells are preincubated for variable times with incubation medium containing the inhibitor (several concentrations). After preincubation, P450 substrate is added and activity assay performed as described above.

3.7 Induction Studies

Drug-drug interactions due to P450 induction are far less frequent than those caused by inhibition; however, their consequences can be clinically relevant. A drug with inductive properties can accelerate its own metabolism or the metabolism of co-administered drugs. Therapeutic inefficacy, due to a faster clearance, or toxic effects, resulting from generation of reactive metabolites by the induced enzyme(s), are potential consequences of P450 induction. At present, primary cultures of human hepatocytes are the most reliable system to study induction potential of new drugs (2, 16). Quantification of mRNA by RT-PCR or enzyme activity measurements are recommended assays for the screening of potential P450 inducers (16, 24). The high variations in the induction response shown by cultured hepatocytes from different donors recommend the use of at least three cell preparations obtained from different individuals. The inclusion of prototypical inducers (positive controls) in each experiment contributes to standardization of the induction response and helps categorize the tested compounds as potent, weak/mild, or non-inducers (24, 25).

1. Stock solutions of chemical inducers are conveniently diluted in serum-free chemically defined medium to reach desired concentrations: 2 μM MC, 1 mM PB, 100 mM ET, 1 mM CL, 50 μM RF, and 1 μM DX (see Note 9). Concentration of the organic solvent (DMSO) in the cultured medium does not exceed 0.5% (v/v).

2. Incubate hepatocytes in the presence of inducers for 48 or 72 h. Start treatments 24 h after cell plating by removing cultured medium and adding fresh medium containing the inducer or solvent (control cells). Inducers are re-added 24 h later with medium change.

3. After treatments, remove culture medium and incubate cells with incubation medium containing substrate, as described in Subheading 3.2.

4. Activity values in treated cells are compared to those in control (untreated) cells (Fig. 2). Results are usually expressed as fold-induction over vehicle-treated cells (negative control) or cells treated with prototypical P450 inducers (positive controls).

3.8 Notes

1. Substrate can be prepared in other pH 7.4 buffered saline solutions or in culture media.

2. Many compounds are able to inhibit reversibly or irreversibly P450 activities. The chemicals included in this list are considered

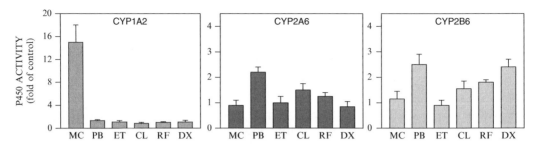

Fig. 2 Effects of model inducers on the P450 activities in primary cultured human hepatocytes. After 24 h in culture, hepatocytes were exposed to 2 μM 3-methylcholanthrene (MC), 1 mM phenobarbital (PB), 100 mM ethanol (ET), 1 mM clofibrate (CL), 50 μM rifampicin (RF), or 1 μM dexamethasone (DX). After 48 h of treatment, P450 activities were determined by incubation of cell monolayers with selective fluorogenic probes: 7-MR (CYP1A2), 7-BR (CYP2B6), or coumarin (CYP2A6). Data are mean ± SD of three independent cell cultures and are expressed as fold increase over activity measured in control cells

as the preferred inhibitors for in vitro experiments, although some of them are not absolutely specific for a particular P450 (24).

3. Cell viability is determined by the trypan blue exclusion method. An aliquot of a mixture of hepatocyte suspension and 0.4% trypan blue solution in saline (1:1 v/v) is immediately loaded in a counter chamber and viable (non-blue) cells counted in five different fields under the optical microscope. Cell preparations with viability below 70% are generally inadequate for further cultivation.

4. The attachment efficiency of viable hepatocytes to fibronectin/collagen-coated plates 1–2 h after cell seeding is usually 80%.

5. Longer incubation times than those recommended in Table 1 could be required for the assessment of P450 activities in cells with low metabolic capacity (e.g., hepatoma cell lines).

6. Because living cells are used, all enzymatic systems and cofactors are present, including those involved in conjugating (phase II) reactions. Therefore, to avoid an underestimation of fluorescent metabolite formed by P450 enzymes, any potential conjugated metabolite is hydrolyzed before fluorimetric quantification.

7. Potential toxicity of the chemicals to cultured cells should be tested and only non-cytotoxic concentrations of the inhibitors used for inhibition assays.

8. Typical experiments for determining IC50 values involve incubating the substrate (if the metabolic rate is sufficient) at concentrations below its Km, to more closely relate the inhibitor IC50 to its Ki. For Ki determinations, both the substrate and inhibitor concentrations should be varied, to cover ranges

above and below the Km for the substrate and the Ki for the inhibitor.

9. In addition to prototypical inducers of P450 enzymes, other chemicals can be used for P450 induction studies. For a general induction screening at least three different concentrations of the compound are recommended. Cytotoxicity to hepatocytes should be previously tested and only sub-cytotoxic concentrations of the chemical used for P450 induction assays.

References

1. Cytochrome P450 home page (http://drnelson.utmem.edu/CytochromeP450.html)
2. Hewitt NJ, Gomez-Lechón MJ, Houston JB et al (2007) Primary hepatocytes: Current understanding of the regulation of metabolic enzymes, transporter proteins and how they are used in pharmaceutical practice. Drug Metab Rev 39:159–234
3. Lahoz A, Donato MT, Castell JV, Gómez-Lechón MJ (2008) Strategies to in vitro assessment of major human CYP enzyme activities by using liquid chromatography tandem mass spectrometry. Curr Drug Metab 9:12–19
4. Donato MT, Jiménez N, Castell JV, Gómez-Lechón MJ (2004) Fluorescence-based assays for screening nine cytochrome P450 (P450) activities in intact cells expressing individual human P450 enzymes. Drug Metab Dispos 32:699–706.5
5. Burke MD, Thompson S, Elcombe CR, Halpert J, Haaparanta T, Mayer RT (1995) Ethoxy-, pentoxy-, and benzyloxyphenoxasones and homologues: a series of substrates to distinguish between different induced cytochromes P-450. Biochem Pharmacol 34:3337–3345
6. Stresser DM, Turner SD, Blanchard AP, Miller VP, Crespi CL (2002) Cytochrome P450 fluorometric substrates: identification of isoform-selective probes for rat CYP2D2 and human CYP3A4. Drug Metab Dispos 30:845–852
7. White INH (1988) A continuous fluorimetric assay for cytochrome P-450-dependent mixed function oxidases using 3-cyano-7-ethoxycoumarin. Anal Biochem 172:304–310
8. Venhorst J, Onderwater RCA, Meerman JHN, Vermeulen NPE, Commandeur JNM (2000) Evaluation of a novel high-throughput assay for cytochrome P450 2D6 using 7-methoxy-4-(aminophenyl)-coumarin. Eur J Pharmaceut Sci 12:151–158
9. Chauret N, Dobss B, Lackman RL, Bateman K, Nicoll-Griffith DA, Stresser DM, Ackermann JM, Turner SD, Miller V, Crespi CL (2001) The use of 3-[2-N, N-diethyl-N-methylammonium)

ethyl]-7-methoxy-4-methylcoumarin (AMMC) as a specific CYP2D6 probe in human liver microsomes. Drug Metab Dispos 29:1196–1200
10. Chauret N, Tremblay N, Lackman RL, Gauthier JY, Silva JM, Marois J, Yergey JA, Nicoll-Griffith DA (1999) Description of a 96-well plate assay to measure cytochrome P4503A inhibition in human liver microsomes using a selective fluorescent probe. Anal Biochem 276:215–226
11. Nerurkar PV, Park SS, Thomas PE, Nims RW, Lubet RA (1993) Methoxyresorufin and benzyloxyresorufin: substrates preferentially metabolized by cytochromes P4501A2 and 2B, respectively, in the rat and mouse. Biochem Pharmacol 46:933–943
12. Soars MG, Mcginnity DF, Grime K, Riley RJ (2007) The pivotal role of hepatocytes in drug discovery. Chem Biol Interact 168:2–15
13. Gómez-Lechón MJ, Castell JV, Donato MT (2008) An update on metabolism studies using human hepatocytes in primary culture. Expert Opin Drug Metab Toxicol 4:837–854
14. Gómez-Lechón MJ, Lahoz A, Gombau L, Castell JV, Donato MT (2010) In vitro evaluation of potential hepatotoxicity induced by new drugs. Curr Pharm Design 16:1963–1977
15. Gómez-Lechón MJ, Donato MT, Castell JV, Jover R (2004) Human hepatocytes in primary culture: the choice to investigate drug metabolism in man. Curr Drug Metab 5:443–462
16. Donato MT, Lahoz A, Castell JV, Gómez-Lechón MJ (2008) Cell lines: a tool for in vitro drug metabolism studies. Curr Drug Metab 9:1–11
17. Pfeifer AMA, Cole KE, Smoot DT, Weston A, Groopman JD, Shields PG, Vignaud JM, Juillerat A, Lipsky MM, Trump BF, Lechner JF, Harris CC (1993) Simian virus 40 large tumor antigen-immortalized normal human liver epithelial cells express hepatocyte characteristics and metabolise chemical carcinogens. Proc Natl Acad Sci 90:5123–5127
18. Bort R, Castell JV, Pfeifer A, Gómez-Lechón MJ, Macé K (1999) High expression of human

CYP2C in immortalized human liver epithelial cells. Toxicol In Vitro 13:633–638

19. Donato MT, Castell JV (2003) Strategies and molecular probes to investigate the role of cytochrome P450 in drug metabolism: focus on in vitro studies. Clin Pharmacokinet 42:153–178

20. Donato MT, Gómez-Lechón MJ, Castell JV (1993) A microassay for measuring cytochrome P450IA1 and P450IIB1 activities in intact human and rat hepatocytes cultured on 96-well plates. Anal Biochem 213:29–33

21. Lahoz A, Gombau L, Donato MT, Castell JV, Gómez-Lechón MJ (2006) In vitro ADME medium/high-throughput screening in drug preclinical development. Mini Rev Med Chem 6:1053–1062

22. Di Marco A, Yao D, Laufer R (2003) Demethylation of radiolabelled dextromethorphan in rat microsomes and intact hepatocytes. Eur J Biochem 270:376837–376877

23. Xu L, Chen Y, Pan Y, Skiles GL, Shou M (2009) Prediction of human drug-drug interactions from time-dependent inactivation of CYP3A4 in primary hepatocytes using a population-based simulator. Drug Metab Dispos 37:2330–2339

24. Guidance for industry: drug interaction studies—study design, data analysis, and implications for dosing and labelling. Guidance draft, 2006. Available at: www.fda.gov/cber/gdlns/interactstud.htm

25. Hewitt NJ, de Kanter R, LeCluyse E (2007) Induction of drug metabolizing enzymes: a survey of in vitro methodologies and interpretations used in the pharmaceutical industry—do they comply with FDA recommendations? Chem Biol Interact 168:51–65

Chapter 13

Screening for Cytochrome P450 Reactivity with a Reporter Enzyme

Kersten S. Rabe and Christof M. Niemeyer

Abstract

The identification of novel substrates of cytochrome P450 enzymes by high-throughput screening assays is of utmost importance to further increase the scope of these enzymes for future applications. Most screens are either confined to individual substrate analogues or hampered by low throughput due to elaborate analysis techniques. Here we describe a general high-throughput screening assay that interrogates the activity of P450 enzymes with the aid of catalase as a reporter enzyme.

Key words High-throughput screening, P450 enzyme, Substrate independent, Reporter enzyme, Catalase, Amplex Red, Resorufin

1 Introduction

Cytochrome P450 enzymes constitute one of the most widespread enzyme families throughout all classes of life on earth, and they can be found in all of the three domains of life, archaea, bacteria, and eukaryota, the latter including examples from fungi, plants, and animals. Owing to the broad range of reactivities displayed by P450 enzymes (1–5), several screening approaches have been established, often based on specific fluorogenic or liquid chromatography/mass spectrometry (LC/MS)-detectable substrates (6); reactive cleavage products (7); or the consumption of cofactors, such as NAD(P)H (8). Here we describe a generic assay to analyze the enzymatic activity of P450 enzymes through the use of the reporter enzyme catalase. This assay is based on the observation that, in the presence of organic peroxides like tert-butylhydroperoxide (TBHP) and cumene hydroperoxide (CHP), catalase can readily convert the fluorogenic substrate Amplex Red to the strongly fluorescent product resorufin (9). Hence, it is possible to analyze conversion of a substrate by a P450 enzyme by analyzing the amount of remaining peroxide, using the catalase reaction.

Ian R. Phillips et al. (eds.), *Cytochrome P450 Protocols*, Methods in Molecular Biology, vol. 987,
DOI 10.1007/978-1-62703-321-3_13, © Springer Science+Business Media New York 2013

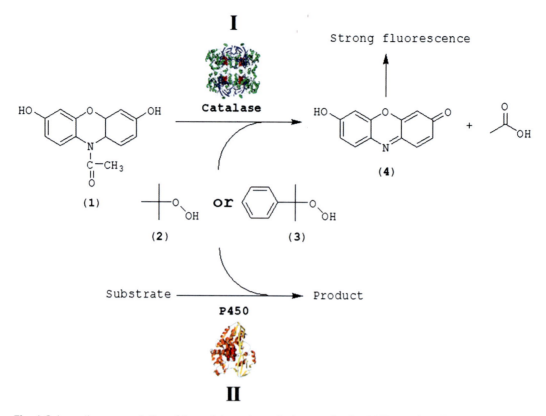

Fig. 1 Schematic representation of the catalase-dependent screening for P450 reactivity. The assay is based on the quantification of organic peroxides (**2, 3**), which are co-substrates required for putative P450 reactivity against potential substrates (reaction **II**). Depletion of the hydroperoxide is measured in reaction **I** by addition of Amplex Red (**1**) and catalase, and subsequent fluorescence detection of resorufin (**4**)

The overall reaction scheme of the assay is illustrated in Fig. 1. Subsequent to this screen, any potential P450 enzyme/substrate pairs identified can be further analyzed by more elaborate techniques, preferably LC-MS (9).

2 Materials

All solutions should be prepared using ultrapure water and can be stored at room temperature if not indicated otherwise. All enzymes, chemicals, and reagents are used without further purification. All materials should be discarded after the experiments following the corresponding regulations for disposing waste materials.

2.1 Buffers and Stock Solutions

1. Tenfold stock standard buffer (500 mM potassium phosphate, 3 M NaCl, pH 7.4), referred to as 10× PBS: 15.6 g $NaH_2PO_4 \times H_2O$, 103.7 g $Na_2HPO_4 \times 7H_2O$, and 175.3 g NaCl dissolved in H_2O to a volume of 1 L (see Note 1). The pH should be adjusted with either 500 mM NaH_2PO_4 or Na_2HPO_4, if necessary.

Reporter Enzyme-based P450 Screen 151

2. Resorufin (Invitrogen) (10 mM in DMSO): store at –20°C until use (see Note 2). This stock solution is then diluted by a factor of 1,000 in 1× PBS. The concentration of this working solution can be determined using the extinction coefficient of resorufin at 571 nm (ε_{571}=54,000 M^{-1} cm^{-1}) supplied by the manufacturer.

3. Amplex Red (Sigma): dissolve in anhydrous DMSO (see Note 3) to a concentration of 10 mM and store at –20°C until use (see Note 4).

4. Catalase (CAT, from bovine liver, Sigma, see Note 5): dissolve in 1× PBS to a concentration of 15 mg/ml (60 µM, see Note 6) and store at –20°C until use.

5. Organic peroxide stock solutions: should be prepared fresh for every experiment by dilution of the commercially available peroxide to 1 mM in 1× PBS (see Note 7).

6. Substrates: prepare as 20 mM stock in either acetonitrile or dimethylformamide (see Note 8) and store at –20°C until use.

7. P450 enzymes: prepare fresh stock solutions (2 µM in 1× PBS) for every experiment.

2.2 Equipment

1. Black microtiter plates (96 wells, nontreated, flat bottom, Nunc).

2. Adhesive sealing tape for well plates (Nunc).

3. Unimax 1010 shaker equipped with Incubator 1000 (Heidolph).

4. Synergy 2 microtiter-plate reader (Biotek).

5. Multipipette for handling volumes from 10 to 100 µl (Finnpipette).

6. Pipetting reservoirs.

7. Eppendorf centrifuge 5430 R equipped with Eppendorf A-2-DWP rotor for microtiter plates.

3 Methods

Carry out all steps at room temperature unless indicated otherwise.

3.1 Calibration Curve for Determination of Resorufin Concentration

1. In order to be able to convert the fluorescence readings into the amount of resorufin produced, prepare dilutions of the resorufin stock in 1× PBS, ranging from 1 to 1,000 µM. At least four dilutions (1, 10, 100, 1,000 µM) should be analyzed.

2. Transfer 100 µl of these solutions into a black microtiter plate and record their fluorescence. The corresponding filters are set

to $\lambda_{ex} = 530/25$ nm and $\lambda_{em} = 590/35$ nm (see Note 9). This should be independently reproduced at least in triplicate.

3. Plot the logarithm of the fluorescence against the logarithm of the resorufin concentrations and calculate a linear regression curve (see Note 10). The formula of the resulting regression curve will be used in all subsequent experiments to determine the amount of resorufin produced.

3.2 Substrate Conversion Assay

Every time an experiment is carried out, calibration samples containing varying concentrations of the organic peroxide, ranging from 100 µM down to no organic peroxide in 1× PBS, should be included in the microplate design. These samples need to be incubated and analyzed the same way as all other reaction samples. That way the peroxide concentration can be linked to the amount of resorufin generated for each individual experiment, excluding the possibility of errors due to changes in concentration, activity of the catalase, or differing reaction conditions.

Furthermore, each substrate/P450 pair should be analyzed at least in triplicate along with control reactions lacking the enzyme, the substrate, or the peroxide. A layout for one such pair is suggested in Fig. 2.

1. Set up three multichannel pipette-compatible reservoirs containing 1× PBS, the 2 µM solution of the enzyme and the 1 mM solution of the organic peroxide.

2. Transfer 158 µl of the 1× PBS solution to all wells that will contain all components (black wells in Fig. 2) using a multipipette. To wells that will lack the enzyme (striped wells in Fig. 2) and wells that will lack the peroxide (gray wells in Fig. 2) transfer 178 µl 1× PBS. To all wells that will lack the substrate (dotted wells in Fig. 2) transfer 160 µl of 1× PBS.

3. Add 2 µl of the 20 mM substrate stock solution individually to all wells, except those control wells that will lack the substrate (dotted circles in Fig. 2).

4. Add 20 µl of the 1 mM peroxide stock solution to all wells, except those controls that should lack the peroxide (gray circles in Fig. 2).

5. Start the reactions by adding 20 µl of the 2 µM enzyme stock solution to all wells, except those that should lack the enzyme (striped circles in Fig. 2).

6. Seal the microtiter plate with sealing foil and incubate in an orbital shaker at 37°C overnight at 150 rpm (see Note 11) to allow for P450 enzyme-catalyzed substrate conversion.

7. The next day prepare a solution containing 10 µM Amplex Red and 1 µM catalase (see Note 12).

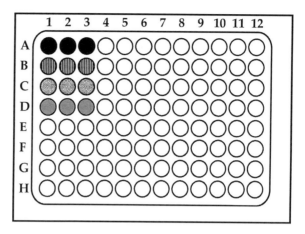

Fig. 2 Layout for one substrate/P450 pair analyzed in triplicate including all controls. *Black wells* contain all components, *striped wells* lack the enzyme, *dotted wells* lack the substrate, and *gray wells* lack the peroxide

8. Centrifuge the microtiter plate at $1,000 \times g$ for 5 min to remove any condensate from the sealing foil. Subsequently, carefully remove the sealing foil.

9. Transfer 50 µl of each overnight reaction mixture into a fresh black microtiter plate (analysis plate) with a multipipette and place the analysis plate in close proximity to the microtiter-plate reader.

10. Place a reservoir with the Amplex Red/catalase solution in close proximity to the microtiter-plate reader.

11. To each well of the analysis plate add 50 µl of the Amplex Red/catalase solution, with a multipipette, and record the development of fluorescence signals immediately after addition (see Note 13).

3.3 Data Analysis

1. For all wells calculate the change in fluorescence per minute.

2. Applying the calibration curves detailed in Subheading 3.1, convert all fluorescence values to amounts of resorufin produced per minute.

3. For each experiment calculate the average and the standard deviation of the triplicates.

4. By comparison with the data obtained from the wells with known organic peroxide concentrations, the remaining peroxide can be calculated as a percent value.

5. By comparing the results obtained from reaction mixes that contained peroxide, P450 enzyme, and substrate with those obtained from ones lacking either enzyme or substrate (see Note 14), five different results can be observed (Fig. 3). Case A represents clear evidence of a peroxide-dependent P450 substrate conversion,

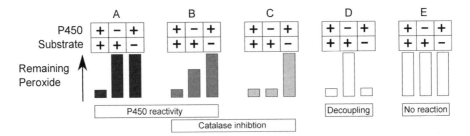

Fig. 3 Classification of the various reactivity patterns which can be observed in the screening assay (cases a–e). The heights of the bars represent the amount of remaining peroxide

because considerable amounts of peroxide are consumed only when both the substrate and the enzyme are present in solution. In case B, low amounts of peroxide were consumed even in the absence of the enzyme, whereas in its presence peroxide depletion was almost complete. This reaction profile indicates either that in the absence of the P450 the peroxide reacts with the substrate to a lower extent than in the P450-catalyzed reaction or that the catalase-catalyzed reaction (reaction **I** in Fig. 1) is inhibited by the substrate. As an augmentation of case B, almost the entire amount of peroxide is consumed in case C, even when no enzyme is present. This case indicates either efficient reaction between the substrate and the peroxide, or else strong inhibition of the catalase. Case D, in which peroxide is consumed by the enzyme even in the absence of a substrate, indicates the decoupling of peroxide consumption and substrate oxygenation. This decoupling has been observed previously in H_2O_2-dependent P450 reactions (10). Finally, in case E no consumption of peroxide is detected in any of the reaction vessels, thus indicating that no reaction occurred.

4 Notes

1. Other buffers may be applied according to the specific requirements of individual proteins under investigation. When the buffer has to be changed, the conversion of the standard concentrations of peroxides should be checked first, to ensure that the catalase is still active in that buffer.

2. Resorufin solutions should be kept in the dark because the dye is light sensitive.

3. Amplex Red can hydrolyze if the solvent is not dry. Anhydrous dimethyl sulfoxide can be purchased from Sigma and is being handled in a glove box under argon to ensure the solvent stays dry.

4. Amplex Red solutions should be kept in the dark because the dye is light sensitive.

5. In general, other catalase preparations might be applicable as well. It is recommended to experimentally verify that particular catalase preparations do in fact catalyze the detection reactions (Fig. 1, bottom half).

6. The concentration is calculated based on the molecular weight of the catalase homotetramer.

7. Organic peroxides are hydrophobic. They tend to stick to plastic tips if such are used. By wiping the tip on the wall of the dilution tube the experimenter can ensure that the peroxide stays in the tube. Extensive vortexing leads to a homogeneous solution. If the solution is left unstirred for periods longer than 10 min, it should be vortexed again before continued usage.

8. If a substrate is not soluble or stable in acetonitrile or DMF other solvents might be used. In this case the conversion of the standard concentrations of peroxides in the presence of that particular solvent should be checked first, to ensure that catalase remains active under these conditions.

9. Most microtiter-plate readers utilize filters for emission and excitation; therefore other models can be used. In that case the best available filter set should be determined based on the maximal excitation of resorufin at 571 nm and the maximal emission at 585 nm (according to the manufacturer).

10. The individual calibration curve is dependent on different factors related to the particular plate reader used. In addition to the filter settings (see Note 9), settings like gain and the mirror used to deflect the emission can modulate the fluorescence readings. The experimenter should make sure that with the individual settings used a linear regression curve describes the correlation of fluorescence and concentration over the whole concentration range from 1 to 1,000 μM.

11. The reaction time and temperature can be adjusted according to the particular enzyme tested.

12. The Amplex Red/catalase solution has to be prepared immediately before use and cannot be stored because the catalase will react with the Amplex Red over time.

13. As the reaction of Amplex Red, catalase, and the peroxide can be rather fast, it is important to keep the time interval between addition of the Amplex Red/catalase solution and the beginning of the measurement as short as possible. The microplate reader should be programmed to perform a shaking step before the first measurement in order to ensure complete mixing of the reaction mixture.

14. The reaction mix lacking the peroxide is only needed as a background control. Because no peroxides are present, very little or no Amplex Red conversion should be observed.

Acknowledgments

We are most grateful to Deutsche Forschungsgemeinschaft (DFG), Bundesministerium für Bildung und Forschung (BMBF), Alexander-von-Humboldt Stiftung, the European Union, and the Max-Planck Society for financial support of our work. KSR thanks the Deutscher Akademischer Austauschdienst (DAAD) for a postdoctoral fellowship.

References

1. Jung ST, Lauchli R, Arnold FH (2011) Cytochrome P450: taming a wild type enzyme. Curr Opin Biotechnol. (6):809–817. doi: 10.1016/j.copbio.2011.02.008. Epub 2011 Mar 14

2. Guengerich FP (2002) Cytochrome P450 enzymes in the generation of commercial products. Nat Rev Drug Discov 1:359–366

3. Urlacher V, Schmid RD (2002) Biotransformations using prokaryotic P450 monooxygenases. Curr Opin Biotechnol 13:557–564

4. Kelly SL, Lamb DC, Kelly DE (2006) Cytochrome P450 biodiversity and biotechnology. Biochem Soc Trans 34:1159–1160

5. Rabe KS, Gandubert VJ, Spengler M, Erkelenz M, Niemeyer CM (2008) Engineering and assaying of cytochrome P450 biocatalysts. Anal Bioanal Chem 392(6):1059–1073. doi: 10.1007/s00216-008-2248-9. Epub 2008 Jul 13

6. Delaporte E, Rodrigues AD (2001) Cytochrome P450 assays. Curr Protoc Pharmacol Chapter 3: Unit 3.9 doi: 10.1002/0471141755.ph0309s15.

7. Peters MW, Meinhold P, Glieder A, Arnold FH (2003) Regio- and enantioselective alkane hydroxylation with engineered cytochromes P450 BM-3. J Am Chem Soc 125:13442–13450

8. Tsotsou GE, Cass AE, Gilardi G (2002) High throughput assay for cytochrome P450 BM3 for screening libraries of substrates and combinatorial mutants. Biosens Bioelectron 17:119–131

9. Rabe KS, Spengler M, Erkelenz M, Muller J, Gandubert VJ, Hayen H, Niemeyer CM (2009) Screening for cytochrome p450 reactivity by harnessing catalase as reporter enzyme. Chembiochem 10:751–757

10. Xu F, Bell SG, Rao Z, Wong LL (2007) Structure-activity correlations in pentachlorobenzene oxidation by engineered cytochrome P450cam. Protein Eng Des Sel 20:473–480

Chapter 14

High-Throughput Fluorescence Assay of Cytochrome P450 3A4

Qian Cheng and F. Peter Guengerich

Abstract

Microtiter plate-based fluorescence assays allow rapid measurement of the catalytic activities of cytochrome P450 oxygenases (P450s). We describe a high-throughput fluorescence assay of P450 3A4, one of the key enzymes involved in xenobiotic metabolism. The assay involves the oxidative debenzylation of 7-hydroxy-4-trifluoromethyl coumarin, producing an increase in fluorescence.

Key words P450, Fluorescence, High-throughput assay, Enzyme inhibition

1 Introduction

Cytochrome P450 oxygenases (P450s) are the major enzymes involved in the oxidative metabolism of drugs and other xenobiotics (1). Inhibition of P450s catalytic activities is a principal mechanism for in vivo drug-drug interactions (2–4). Various in vitro assays have been developed to measure the activity or inhibition of P450s (5). Historically, the most commonly applied method is to use recombinant P450s with probe substrates, with the products identified by high-performance liquid chromatography (HPLC) coupled with mass spectrometry (MS) (6–8). However, this method is relatively time-consuming and labor intensive and cannot be readily adapted to high-throughput formats (e.g., testing the inhibitory properties of a chemical library). Thus, microtiter plate-based fluorescence assays are more efficient and cost effective than HPLC-MS methods, allowing rapid in vitro testing of many samples in parallel (9, 10), if they provide accurate reports of the function of the enzymes. In a fluorescent assay, a "pro-fluorescent" molecule is oxidized by P450 to a fluorescent product, which can be directly measured using a fluorescence microplate reader. P450 3A4 is, in most individuals, the most abundant P450 in the liver and small intestine and is involved in the metabolism of about one-half of the drugs on the market. Here we describe a high-throughput

Ian R. Phillips et al. (eds.), *Cytochrome P450 Protocols*, Methods in Molecular Biology, vol. 987,
DOI 10.1007/978-1-62703-321-3_14, © Springer Science+Business Media New York 2013

158 Qian Cheng and F. Peter Guengerich

fluorescence assay for measuring the activity of P450 3A4. A number of pro-fluorescent substrates (mainly benzyl ethers) have been used in P450 3A4 fluorescence assays (11, 12). 7-Benzoyloxy-4-trifluoromethyl coumarin (BFC) is selected in this protocol because BFC metabolism by P450 3A4 is linear with respect to BFC concentration up to 100 μM (13). A typical P450 3A4 inhibitor, ketoconazole, is used to demonstrate the use of microtiter plate-based fluorescence assays to assess the activity of P450 3A4 and characterize the IC_{50} of this inhibitor.

2 Materials

2.1 P450 Assay

1. Human P450 3A4 "bicistronic" membranes (14) containing both P450 3A4 and NADPH-P450 reductase (concentration of both enzymes is 1 μM). Similar commercial products are available from BD Biosciences (see Note 1).

2. 10 mM NADP+ stock solution: 382 mg NADP+ in 50 ml Milli-Q water. Store at 4°C.

3. 100 mM glucose-6-phosphate: 3.4 g in 100 ml Milli-Q water. Store at 4°C.

4. Yeast glucose-6-phosphate dehydrogenase (10^3 IU/ml): 1.0 mg in 1 ml 10 mM Tris-acetate buffer (pH 7.4), containing 20% (v/v) glycerol and 1 mM EDTA. Store at 4°C.

5. 4 mM BFC stock solution: 2.56 mg in 2 ml methanol. Store in Teflon-sealed amber glass vial at 4°C.

6. 1 mM ketoconazole stock solution: 5.31 mg in 10 ml methanol. Store in Teflon-sealed glass vial at 4°C.

7. NADPH-generating system: combine 50 parts 10 mM NADP+, 100 parts 100 mM glucose-6-phosphate, and 1 part yeast glucose 6-phosphate dehydrogenase (1 mg/ml). Prepare fresh daily and store on ice when not in use (see Note 2).

8. Stop buffer: 80% acetonitrile and 20% 0.5 M Tris-base (v/v), stored at ambient temperature.

9. P450-BFC 2× mix: mix P450 3A4 bicistronic membranes and 4 mM BFC stock solution in 100 mM potassium phosphate buffer (pH 7.4) so that the final concentration of P450 is 20 nM and of BFC is 40 μM. Prepare the mix fresh daily and keep on ice.

2.2 Equipment

1. Microplate reader. Polarstar microplate reader is used in this protocol (BMG Labtech).

2. 96-well black microtiter plate (Corning Costar).

3. Microchannel pipette (Gilson).

4. Software to process the data. Prism (Graphpad.com) is used in this protocol (see Note 3).

High-Throughput Fluorescence Assay of Cytochrome P450 3A4   159

3 Methods

3.1 Plate Setup

The assay is performed in duplicate and the plate setup is summarized in Table 1.

1. Dispense 60 μl of 100 mM potassium phosphate buffer into the wells (columns 2–12) of a 96-well microplate using a multichannel pipette.

2. Dispense 118 μl of 100 mM potassium phosphate and 2 μl of 1 mM ketoconazole into the wells (column 1).

3. Serially dilute 60 μl of inhibitor solution from the wells in column 1 to the other wells (2–10). Discard the extra 60 μl of solution in the wells in column 10.

4. Dispense 100 μl of P450-BFC 2× mix into all of the wells. Add 75 μl of stop buffer to the wells in column 11.

5. Preincubate the plate at 37°C for 5 min (see Note 4).

6. Add 40 μl of NADPH generating system into each well to initiate the reaction.

7. Incubate the plate at 37°C for 20 min.

8. Stop the reaction by adding 75 μl of stop buffer to each well (except wells in column 11).

3.2 Data Acquirement and Processing

1. Scan the plate using the Polarstar plate reader. Average the replicates of the reading for each column.

2. Subtract the blank from the mean value of all other columns and calculate the percentage of inactivated enzyme for each ketoconazole concentration (designated as I%). $I = (1 - (\text{mean of individual column} - \text{mean of column } 11)/(\text{mean of column } 12 - \text{mean of column } 11)) \times 100$.

3. Convert the concentration of inhibitor into log format (designated as X).

4. Fit the data with the 4-parameter logistic fit: $I = b + (a - b)/(1 + 10^{\wedge}((\text{LogIC}_{50} - X) \times \text{HillSlope}))$ using Prism (see Note 5).

5. The processed data is shown in Table 2 and the fitted plot is shown in Fig. 1.

Table 1
The scheme of plate setup

	1	2	3	4	5	6	7	8	9	10	11	12
I	5,000	2,500	1,250	625	312	156	78	39.1	19.5	9.76	Blank	No inhibitor
II	5,000	2,500	1,250	625	312	156	78	39.1	19.5	9.76	Blank	No inhibitor

The assays are performed in duplicate, as designated by row I and row II. The concentration of ketoconazole (nM) in each well is listed

Table 2
Fluorescence reading of each well and data processing

	1	2	3	4	5	6	7	8	9	10	11	12
I	7,845	8,072	8,229	8,729	9,536	11,753	15,018	18,213	22,589	23,408	7,489	24,886
II	7,571	7,647	7,850	8,861	9,827	10,940	14,914	16,373	20,459	21,789	7,223	23,760
Mean	7,708	7,859	8,039	8,795	9,681	11,346	14,966	17,293	21,524	22,598	7,356	24,323
Mean blank	424	503	683	1,439	2,325	3,990	7,610	9,937	14,168	15,242	0	16,967
I (%)	**97.5**	**97**	**96**	**91.5**	**86.3**	**76.5**	**55.1**	**41.4**	**13.8**	**9.1**		
Inhibitor concentration (nM)	5,000	2,500	1,250	625	312	156	78	39.1	19.5	9.76		
X		**3.7**	**3.4**	**3.1**	**2.8**	**2.5**	**2.2**	**1.9**	**1.6**	**1.3**	**1**	

Data listed in bold font are used for 4-parameter logistic fit: $I = b + (a-b)/(1 + 10^{\wedge}((\mathrm{LogIC}_{50} - X) \times \mathrm{HillSlope}))$

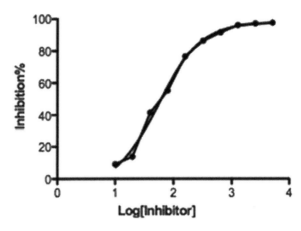

Fig. 1 Calculated IC$_{50}$ of ketoconazole with P450 3A4. The response (inhibition percentage) is plotted against the log$_{(10)}$ of ketoconazole concentration

6. Subtract the blank from the mean value of all other columns and calculate the percentage of inactivated enzyme for each ketoconazole concentration (designated as I%). I = (1 − (mean of individual column − mean of column 11)/(mean of column 12 − mean of column 11)) × 100.

7. The calculated log IC$_{50}$ is 1.57–1.92. Thus, the measured IC$_{50}$ in this experiment is 37–83 nM (see Note 6).

4 Notes

1. The activities of P\450 3A4 "bicistronic" membranes may decrease over the time. Avoid extensive thaw-freeze cycles.

2. Determine that the NADPH-generating system is active by monitoring the increase in absorbance at 340 nm.

3. Other statistical software (than GraphPad Prism) that can perform nonlinear regression can be used for this experiment. Web-based curve-fitting programs are also available, e.g., www.chanbioscience.com/stat/ec50.html.

4. Preincubation before the addition of NADPH, to equilibrate temperature, is a critical step. Insufficient preincubation can result in inconsistent readings among replicates.

5. Although simple linear regression may be used in certain situations, the 4-parameter logistic fit is more appropriate in most cases. Parameter *a* is the highest response and *b* is the lowest response.

6. The raw data used in this protocol were reported previously (15).

References

1. Isin EM, Guengerich FP (2007) Complex reactions catalyzed by cytochrome P450 enzymes. Biochim Biophys Acta 1770:314–329
2. Lamb DC et al (2007) Cytochromes P450 and drug discovery. Curr Opin Biotechnol 18:504–512
3. Di L et al (2007) Comparison of cytochrome P450 inhibition assays for drug discovery using human liver microsomes with LC-MS, rhCYP450 isozymes with fluorescence, and double cocktail with LC-MS. Int J Pharm 335:1–11
4. Wienkers LC, Heath TG (2005) Predicting in vivo drug interactions from in vitro drug discovery data. Nat Rev Drug Discov 4:825–833
5. Fowler S, Zhang H (2008) In vitro evaluation of reversible and irreversible cytochrome P450 inhibition: current status on methodologies and their utility for predicting drug-drug interactions. AAPS J 10:410–424
6. Youdim KA et al (2008) An automated, high-throughput, 384 well cytochrome P450 cocktail IC_{50} assay using a rapid resolution LC-MS/MS end-point. J Pharm Biomed Anal 48:92–99
7. Lin T et al (2007) In vitro assessment of cytochrome P450 inhibition: strategies for increasing LC/MS-based assay throughput using a one-point IC(50) method and multiplexing high-performance liquid chromatography. J Pharm Sci 96:2485–2493
8. Turpeinen M et al (2005) Multiple P450 substrates in a single run: rapid and comprehensive in vitro interaction assay. Eur J Pharm Sci 24:123–132
9. Schaeffner I et al (2005) A microtiterplate-based screening assay to assess diverse effects on cytochrome P450 enzyme activities in primary rat hepatocytes by various compounds. Assay Drug Dev Technol 3:27–38
10. Yamamoto T, Suzuki A, Kohno Y (2002) Application of microtiter plate assay to evaluate inhibitory effects of various compounds on nine cytochrome P450 isoforms and to estimate their inhibition patterns. Drug Metab Pharmacokinet 17:437–448
11. Nakamura K et al (2001) Coumarin substrates for cytochrome P450 2D6 fluorescence assays. Anal Biochem 292:280–286
12. Crespi CL, Miller VP, Penman BW (1997) Microtiter plate assays for inhibition of human, drug-metabolizing cytochromes P450. Anal Biochem 248:188–190
13. Stresser DM et al (2000) Substrate-dependent modulation of CYP3A4 catalytic activity: analysis of 27 test compounds with four fluorometric substrates. Drug Metab Dispos 28:1440–1448
14. Parikh A, Gillam EM, Guengerich FP (1997) Drug metabolism by *Escherichia coli* expressing human cytochromes P450. Nat Biotechnol 15:784–788
15. Cheng Q, Sohl CD, Guengerich FP (2009) High-throughput fluorescence assay of cytochrome P450 3A4. Nat Protoc 4:1258–1261

Chapter 15

Targeted Protein Capture for Analysis of Electrophile-Protein Adducts

Rebecca E. Connor, Simona G. Codreanu, Lawrence J. Marnett, and Daniel C. Liebler

Abstract

Proteomic analyses of protein-electrophile adducts generally employ affinity capture of the adduct moiety, which enables global analyses, but is poorly suited to targeted studies of specific proteins. We describe a targeted molecular probe approach to study modifications of the molecular chaperone heat-shock protein 90 (Hsp90), which regulates diverse client proteins. Noncovalent affinity capture with a biotinyl analog of the Hsp90 inhibitor geldanamycin enables detection of the native protein isoforms Hsp90α and Hsp90β and their phosphorylated forms. We applied this probe to map and quantify adducts formed on Hsp90 by 4-hydroxynonenal (HNE) in RKO cells. This approach was also applied to measure the kinetics of site-specific adduction of selected Hsp90 residues. A protein-selective affinity capture approach is broadly applicable for targeted analysis of electrophile adducts and their biological effects.

Key words Electrophile, Hsp90, Affinity probe, Geldanamycin, Kinetic analysis

1 Introduction

Covalent modification of proteins by electrophilic metabolites is a key step in drug toxicity mediated by cytochrome P-450 enzymes. A growing body of literature describes the identification of proteins that are modified by electrophiles (1). Progress in this field has been driven by the combination of affinity capture for electrophile adducts and mass spectrometry (MS) for identification of modified proteins (2–5). This strategy requires affinity-labeling chemistry to efficiently capture adducts and typically identifies a broad range of modified proteins. On the other hand, mechanistic studies often target a particular protein or a small family of functionally related proteins, where the analytical challenge is to selectively capture these proteins and their modified forms. This may be achieved by immunoprecipitation, but satisfactory reagents are often lacking and covalent adducts may interfere with antibody binding.

Ian R. Phillips et al. (eds.), *Cytochrome P450 Protocols*, Methods in Molecular Biology, vol. 987, DOI 10.1007/978-1-62703-321-3_15, © Springer Science+Business Media New York 2013

We have approached the problem of targeted protein adduct analysis by employing an affinity-tagged, small-molecule inhibitor to capture target proteins (6). In our studies, we focused on the heat-shock protein 90 (Hsp90) family proteins, which are regulators of diverse cellular functions and are targets of electrophiles (2, 4, 7). The inhibitor is a derivative of geldanamycin, a natural product that selectively targets Hsp90 and binds to the N-terminal ATPase domain (8). Geldanamycin induces a conformational change that releases Hsp90 client proteins and co-chaperones. Here we describe the use of the commercially available reagent geldanamycin-PEG-biotin (9) to isolate both cytosolic forms of Hsp90, for analysis of Hsp90 adducts by liquid chromatography-tandem mass spectrometry (LC-MS/MS). We characterized adduction sites on Hsp90 captured from cells and treated in vitro with HNE. We also identified additional adduction sites on both Hsp90α and Hsp90β isolated from HNE-treated RKO cells. Reaction rates for HNE adduction at several sites on both isoforms of Hsp90 both in vitro and in intact cells were characterized by combining geldanamycin-biotin capture with targeted, label-free LC-MS/MS quantification. Our results demonstrate the utility of protein-selective affinity capture for targeted analysis of electrophile adducts and their biological effects.

2 Materials

2.1 RKO Cell Culture, HNE Treatment, and Preparation of Cellular Lysate

1. McCoy's 5A medium (Invitrogen, Carlsbad, CA) supplemented with 10% fetal bovine serum (FBS, Atlas Biologicals, Fort Collins, CO).

2. Solution of trypsin (0.25%) (Gibco/BRL, Bethesda, MD).

3. The lipid electrophile 4-hydroxy-2-nonenal (HNE) (Cayman Chemical, Ann Arbor, MI) is dissolved in ethanol (64 mM), stored in aliquots at –80°C, and added to tissue-culture dishes as required.

4. NETN lysis buffer: 50 mM HEPES, pH 7.5, 150 mM NaCl, 0.5% Igepal. Can be prepared as a stock solution and be stored at 4°C.

5. Ethyl alcohol 200 proof, absolute, anhydrous (EtOH).

6. Sodium borohydride ($NaBH_4$). A 2 M stock solution is stored under the hood at room temperature. The final concentration of sodium borohydride in the experimental samples is 2 mM.

7. Ice-cold phosphate-buffered saline 1×, pH 7.2 (1× PBS).

8. Protease inhibitor cocktail: 1.0 mM phenylmethylsulfonylfluoride, 1.0 mM N-ethylmaleimide, leupeptin (10 µg/mL), aprotinin (10 µg/mL), pepstatin (10 µg/mL).

Targeted Protein Capture for Analysis of Electrophile-Protein Adducts   165

9. Phosphatase inhibitor cocktail: 1.0 mM sodium fluoride, 1.0 mM sodium molybdate, 1.0 mM sodium orthovanadate, 10.0 mM β-glycerophosphate.

10. BCA Protein Assay Kit (Pierce, Rockford, IL).

11. Distilled H_2O.

2.2 Geldanamycin-Biotin Protein Capture

1. Geldanamycin and Geldanamycin-Biotin (Enzo Life Sciences, Plymouth Meeting, PA).

2. High-capacity Neutravidin agarose resin beads (Thermo Scientific, Rockford, IL).

3. 1 M sodium chloride prepared in 1× PBS.

4. Dithiothreitol (DTT). A 1 M stock solution is stored at –20°C. The working concentration of DTT is 50 mM.

5. NuPAGE® LDS Sample Buffer (4×) (Invitrogen, Carlsbad, CA).

2.3 Western Blot Procedure

1. NuPAGE® Bis–Tris 10% sodium dodecyl sulfate-polyacrylamide gel electrophoresis (SDS-PAGE) gels (Invitrogen, Carlsbad, CA).

2. Precision Plus Protein Standard Kaleidoscope Molecular Weight Marker (Bio-Rad Laboratories, Hercules, CA).

3. 20×-MOPS running buffer: 50 mM MOPS, 50 mM Tris Base, 0.1% SDS, 1 mM EDTA, pH 7.7 (Invitrogen, Carlsbad, CA).

4. 20× NuPAGE® Transfer Buffer (Invitrogen, Carlsbad, CA).

5. Polyvinylidene difluoride (PVDF) membrane (Invitrogen, Carlsbad, CA).

6. 4× NuPAGE® LDS Sample Buffer (Invitrogen, Carlsbad, CA).

7. Methanol (Sigma-Aldrich, St. Louis, MO).

8. 1× TBS-Tween (25 mM Tris–HCl, 150 mM NaCl, 0.05% Tween 20, pH 7.5).

9. Blocking buffer for near-infrared fluorescent western blotting (Rockland, Gilbertsville, PA).

10. Anti-HNE-Michael reduced rabbit polyclonal antibody (EMD biosciences, San Diego, CA).

11. Anti-Hsp90 mouse monoclonal antibody, used for western blotting (BD Biosciences, San Jose, CA).

12. Anti-Hsp90 rabbit polyclonal antibody (Santa Cruz Biotechnology, Santa Cruz, CA) and protein G-agarose (Roche Applied Science, Indianapolis, IN), used for immunoprecipitation of Hsp90 protein.

13. AlexaFluor®680-conjugated fluorescent secondary antibodies (Molecular Probes, Eugene, OR).

14. Odyssey™ Infrared Imaging System and Odyssey software (Li-Cor, Lincoln, NE).

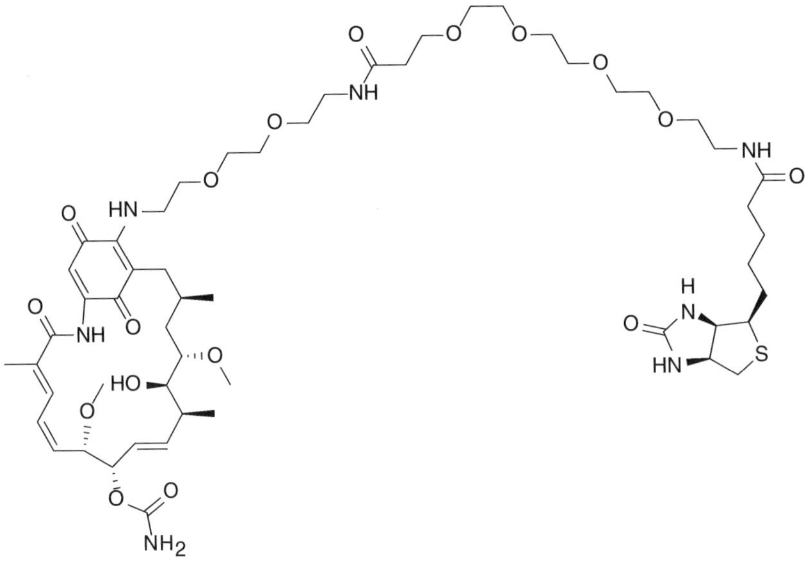

Fig. 1 Chemical structure of the Hsp90-capture reagent geldanamycin-biotin. (Reprinted from (6) with permission from the American Chemical Society)

2.4 LC-MS/MS Procedures

1. Trypsin Gold, Mass Spectrometry Grade (Promega, Madison, WI), is reconstituted in 50 mM acetic acid to a final concentration of 1 mg/mL. Aliquots are stored frozen at –20°C. Trypsin is diluted in 25 mM ammonium bicarbonate to 0.01 mg/mL and used at a ratio of 1:50 (trypsin:protein).

2. Safe Stain Blue (Invitrogen, Carlsbad, CA) is used for 1 h staining followed by distaining in water overnight.

3. Iodoacetamide (IAM); DTT; ammonium bicarbonate, acetonitrile, trifluoroacetic acid (TFA) (Sigma-Aldrich, St. Louis, MO).

4. All reagents are prepared immediately before use (see Note 1).

3 Methods

Mapping and quantitative comparison of site-specific adduction reactions on individual proteins still remains an analytical challenge. To identify sites of HNE modification of Hsp90 protein, a biotin-tagged analog of the Hsp90 inhibitor geldanamycin was used to capture endogenous protein from cellular lysates of treated and untreated cells. This compound, geldanamycin-biotin (Fig. 1) (9), binds tightly to all isoforms of Hsp90 and displaces proteins bound to Hsp90 as client proteins and co-chaperones by inducing conformational changes around the ATP-binding site (8). To generate adducts on the Hsp90 protein, we treat RKO cells with exogenous HNE *in culture*, and we use LC-MS/MS to identify and

Targeted Protein Capture for Analysis of Electrophile-Protein Adducts   167

characterize adduction sites on protein isolated from treated cells. Following capture with Neutravidin-agarose, the biotinylated affinity-tagged proteins are eluted and resolved by one-dimensional SDS-PAGE and digested with trypsin and the resulting peptides are analyzed by LC-MS/MS. The subsequent immunoblot analysis allows for a rapid screening of the Hsp90 protein-selective affinity capture and determination of whether these proteins have been adducted. Reaction rates for HNE adduction at several sites on both isoforms of Hsp90 in intact cells were characterized by combining geldanamycin-biotin capture with targeted, label-free LC-MS/MS quantification. Our results demonstrate the utility of protein-selective affinity capture for targeted analysis of electrophile adducts.

3.1 Isolation of Hsp90 from Cellular Lysates Using Affinity-Tagged Inhibitor Geldanamycin

3.1.1 RKO Cell Culture and Preparation of Cellular Lysate

1. Grow RKO human colorectal carcinoma cells to 80% confluence in McCoy's 5A medium supplemented with 10% fetal bovine serum, at 37°C in an atmosphere of 95% air/5% CO_2.

2. Wash the confluent cells plated in 150 mm culture dishes with 5 mL of cold 1× phosphate-buffered saline. Use a disposable cell scraper to harvest the cells directly in 5 mL of fresh 1× phosphate-buffered saline, and centrifuge at $100 \times g$ for 5 min. The phosphate-buffered saline, pH 7.2, is slowly aspirated off the cell pellet. Cell pellet can be stored at −80°C until further use.

3. Lyse the cell pellets from each 150 mm culture plate on ice in 2 mL of cold NETN buffer supplemented with protease inhibitor and phosphatase inhibitor cocktails. Sonicate the lysate and incubate on ice for 30 min.

4. Clear the lysate by centrifugation at $10,000 \times g$ for 10 min to remove cellular debris. Determine the total protein concentration of the supernatant using the BCA protein assay.

5. Adjust the protein concentration to 1 mg/mL for each sample using lysis buffer and use it fresh each time.

6. A 50 μL aliquot from each sample is transferred into a new pre-labeled 1.5 mL Eppendorf tube. To these sample aliquots are added 3.5 μL of 1 M DTT and 16.5 μL of NuPAGE LDS sample buffer (4×), and the samples are heated for 10 min at 95°C, and then stored at −20°C. This sample is referred to as the *input* (Fig. 2).

3.1.2 Geldanamycin-Biotin Protein Capture

1. Incubate a 250 μL aliquot of diluted cell lysate (1 mg/mL) with geldanamycin-biotin (40 μg in DMSO, 20 μg/μL stock) for at least 3 h at 4°C with gentle rotation (Fig. 1). Before application to the resin, dilute the lysate further, to a final volume of 1 mL, to give a protein concentration of 0.25 mg/mL.

2. Equilibrate Neutravidin-agarose resin by washing a 50:50 (w/v) bead slurry three times with lysis buffer. Use 375 μL of bead

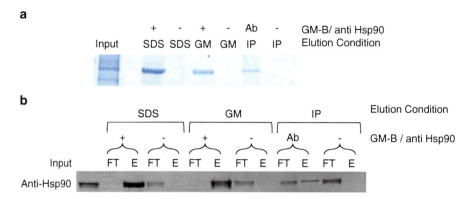

Fig. 2 Geldanamycin-biotin efficiently captures Hsp90 from cell lysates. (**a**) SDS-PAGE analysis with safe stain blue staining of Hsp90 captured with geldanamycin-biotin (GM-B) or Hsp90 antibody. The input lane contains 5 μg of cell lysate. Lanes are labeled as negative control (−), with GM-B (+), or with anti-Hsp90 (Ab). GM-B-captured protein was eluted either by boiling in LDS sample buffer or with 2 mM geldanamycin (GM). Immunoprecipitated Hsp90 was eluted by boiling in LDS sample buffer. (**b**) Western blot analysis with anti-Hsp90 antibody of the flow-through and elution from GM-B capture and immunoprecipitation of Hsp90. Lanes are labeled as flow-through (FT) or eluant (E), with elution and capture conditions noted as in part (**a**). Flow-through lanes contain 2.5 μg of total protein. (Reprinted from (6) with permission from the American Chemical Society)

slurry per sample. Each time the beads are centrifuged at 10,000×g for 2 min and the supernatant is carefully discarded.

3. Incubate diluted lysates containing the affinity capture reagent geldanamycin-biotin with Neutravidin-agarose resin previously equilibrated with lysis buffer, and rotate samples at 4°C overnight.

4. At the end of the incubation period, centrifuge samples at 10,000×g for 2 min and transfer 50 μL of the supernatant into a new pre-labeled 1.5 mL Eppendorf tube. DTT (50 mM) and NuPAGE LDS sample buffer (4×) are added to each sample. Heat samples for 10 min at 95°C and store at −20°C. This sample is referred to as the *flow through* (Fig. 2). Transfer the remainder of the supernatant into another 1.5 mL Eppendorf tube and store at −20°C.

5. Wash the bound proteins four times with 1 mL of lysis buffer. After each wash step, centrifuge the beads at 10,000×g for 1 min and discard the supernatant.

6. Elute proteins from the beads in 150 μL of either NuPAGE LDS sample buffer (4×) with DTT (50 mM) or 2 mM geldanamycin-biotin in lysis buffer. Heat samples for 10 min at 95°C. These samples are referred to as the *eluates* (Fig. 2).

7. Subject samples of the protein *input*, *flow through*, and *eluates* to gel separation by SDS-PAGE (Fig. 2a) for further LC-MS/MS analyses or immunoblot analyses with antibodies directed against Hsp90 proteins (Fig. 2b).

Targeted Protein Capture for Analysis of Electrophile-Protein Adducts   169

3.1.3 Immuno-
precipitation of Hsp90

1. Incubate a 1 mL aliquot of diluted cell lysate (1 mg/mL) with anti-Hsp90 rabbit polyclonal antibody (20 µg) for at least 3 h at 4°C with gentle rotation.

2. Equilibrate protein G-agarose resin by washing a 50:50 (w/v) bead slurry three times with lysis buffer. Use 50 µL of bead slurry per sample. Each time the beads are centrifuged at $10,000 \times g$ for 2 min and the supernatant is carefully discarded.

3. Incubate diluted lysates containing the anti-Hsp90 rabbit polyclonal antibody bound to Hsp90 proteins with protein G-agarose resin previously equilibrated with lysis buffer, and rotate samples at 4°C overnight.

4. At the end of the incubation period, centrifuge samples at $10,000 \times g$ for 2 min and transfer 50 µL of the supernatant into a new pre-labeled 1.5 mL Eppendorf tube. Add DTT (50 mM) and NuPAGE LDS sample buffer (4×) to each sample. Heat samples for 10 min at 95°C and store at –20°C. This sample is referred to as the *flow through* (Fig. 2). Transfer the remainder of the supernatant into another 1.5 mL Eppendorf tube and store at –20°C.

5. Wash the bound proteins four times with 1 mL of lysis buffer. After each wash step, centrifuge the beads at $10,000 \times g$ for 1 min and discard the supernatant.

6. Elute proteins from the beads in 100 µL of NuPAGE LDS sample buffer (4×) with DTT (50 mM). Heat samples for 10 min at 95°C. These samples are referred to as the *eluates* (Fig. 2).

3.1.4 Western Blotting

1. Subject samples of the protein *input, flow through*, and *eluates* (10 µL) from each experimental condition to immunoblot analysis with rabbit polyclonal antibody directed against Hsp90 proteins.

2. Heat the *input, flow through*, and *eluate* samples from each experimental condition for 10 min at 95°C.

3. Set up a NuPAGE Bis–Tris 10% SDS-PAGE gel and follow the manufacturer's directions to run the gel. One lane of the gel is reserved for the protein standard by adding 5 µL of the Precision Plus Protein Standard Kaleidoscope Molecular Weight Marker.

4. Load 10 µL of the samples into the gel in the following order: *input, flow through*, and *eluate*. This will allow for a more efficient comparison of bands during analysis.

5. Transfer proteins electrophoretically onto a PVDF membrane, by following the directions of the manufacturer of the transfer apparatus. Block nonspecific primary-antibody binding by placing the membrane into 5 mL of blocking buffer (1:1 1× TBS-Tween: blocking buffer) for 1 h at room temperature on a rocking platform (see Note 2).

170　Rebecca E. Connor et al.

6. Prepare dilutions of the primary antibody to proteins of interest, according to the manufacturer's directions.

7. Incubate the membrane with primary antibody overnight at 4°C while shaking on an orbital shaker.

8. Using multiple changes of 1× TBS-Tween, wash membranes for a total of 30 min before adding the secondary antibody.

9. Prepare appropriate dilutions of AlexaFluor®680-labeled secondary antibodies and incubate with the membrane for 1 h at room temperature while shaking. Wash the membrane using multiple changes of 1× TBS-Tween for a total of 30 min, before scanning.

10. Immunoreactive proteins are visualized using the Odyssey™ System and software as described by the manufacturer.

3.1.5 In-Gel Trypsin Digestion and MS Analysis

1. Resolve the protein, purified either by Neutravidin capture, using the geldanamycin-biotin affinity probe, or by immunoprecipitation, using anti-Hsp90 polyclonal antibody as described above (*eluate*), by 10% SDS-PAGE using NuPAGE Bis–Tris gels and stain with safe stain blue for 1 h, followed by detaining in water overnight.

2. Desired bands corresponding to the correct molecular weight (~90 kDa) are excised from the gel and subjected to in-gel digestion with trypsin. Carefully chop each excised band into 1 mm cubes, place in a 1.5 mL Eppendorf tube containing 100 μL of 100 mM ammonium bicarbonate, pH 8.0, and incubate at room temperature for 15 min.

3. Reduce samples with 10 μL of 45 mM DTT for 20 min at 55°C and alkylate with 10 μL of 100 mM IAM for 20 min at room temperature in the dark.

4. Discard the liquid and add 100 μL of acetonitrile:50 mM ammonium bicarbonate (50:50, v/v) to distain the samples. Incubate at room temperature for 15 min, and then discard the liquid. Repeat this step twice.

5. Dehydrate the gel pieces by incubation for 15 min at room temperature in 100 μL of acetonitrile, and discard the supernatant.

6. Digest the rehydrated gel pieces with trypsin (50 μL of Trypsin Gold (0.01 mg/mL in 25 mM ammonium bicarbonate)) overnight at 37°C.

7. Extract the peptides twice with 100 μL of 60% acetonitrile, 0.1% TFA, each for 15 min at room temperature; combine the extracts. Evaporate the liquid under vacuum and resuspend peptides in 10–20 μL of H_2O (0.1% formic acid) for LC-MS/MS analysis.

8. Perform LC-MS/MS analysis using an LTQ ion-trap mass spectrometer (Thermo Electron, San Jose, CA) equipped with

Targeted Protein Capture for Analysis of Electrophile-Protein Adducts   171

an Eksigent nanoLC (Dublin, CA) and Thermo Surveyor HPLC pump, nanospray source, and Xcalibur 1.4 instrument control.

9. Carry out the liquid chromatography at ambient temperature at a flow rate of 0.6 µL/min using a gradient mixture of 0.1% (v/v) formic acid in water (solvent A) and 0.1% (v/v) formic acid in acetonitrile (solvent B). Acquire centroid MS/MS scans using an isolation width of 2 m/z, an activation time of 30 ms, an activation Q of 0.250, and 30% normalized collision energy, using one microscan with a max ion time of 100 ms for each MS/MS scan.

10. Before analysis, tune the mass spectrometer using the synthetic peptide TpepK (AVAGKAGAR). Some parameters may vary slightly from experiment to experiment, but typically the tune parameters are as follows: spray voltage of 2 kV, a capillary temperature of 150°C, a capillary voltage of 50 V, and tube lens of 120 V.

11. Resolve the peptides on 100 µm × 11 cm fused-silica capillary column (Polymicro Technologies, LLC Phoenix, AZ) packed with 5 µm, 300 Å Jupiter C18 (Phenomenex, Torrance, CA). Introduce the peptides eluting from the capillary tip into the LTQ source with a capillary voltage of approximately 2 kV. The heated capillary is operated at 150°C and 40 V. Acquire the MS/MS spectra in the data-dependent scanning mode, consisting of a full scan obtained for eluting peptides in the range of 350–2,000 m/z, followed by four data-dependent MS/MS scans. Record MS/MS spectra using dynamic exclusion of previously analyzed precursors for 30 s with a repeat duration of 2 min.

3.1.6 Database Searching

1. The "ScanSifter" algorithm, an in-house-developed software (10), reads MS/MS spectra stored as centroid peak lists from Thermo RAW files and transcodes them to mzData v1.05 files. Spectra that contain fewer than six peaks do not result in mzData files. Only MS/MS scans are written to the mzData files; MS scans are excluded. If 90% of the intensity of a tandem mass spectrum appears at a lower m/z than the precursor ion, a single precursor charge is assumed; otherwise, the spectrum is processed under both double and triple precursor charge assumptions.

2. Tandem mass spectra are assigned to peptides from the IPI Human database version 3.56 (May 05, 2009; 153,182 proteins) by the MyriMatch algorithm (11). To estimate false discovery rates, each sequence of the database was reversed and concatenated to the database. Candidate peptides are required to feature trypsin cleavages or protein termini at both ends, though any number of missed cleavages is permitted.

All cysteines are expected to undergo carboxamidomethylation and are assigned a mass of 160 kDa. All methionines are allowed to be oxidized. Precursor ions are required to fall within 1.25 m/z of the position expected from their average masses, and fragment ions are required to fall within 0.5 m/z of their monoisotopic positions. The database searches produced raw identifications in pepXML format.

3. Peptide identification, filtering, and protein assembly are done with the IDPicker algorithm (12). Initial filtering takes place in multiple stages. First, IDPicker filters raw peptide identification to a target false-discovery rate (FDR) of 5%. The peptide filtering employs reversed-sequence database-match information to determine thresholds that yield an estimated 5% FDR for the identifications of each charge state by the formula (13) $FDR = (2R)/(R+F)$, where R is the number of passing reversed-peptide identifications and F is the number of passing forward (normal orientation)-peptide identifications. The second round of filtering removes proteins supported by less than two distinct peptide identifications in the analyses. Indistinguishable proteins are recognized and grouped. Parsimony rules are applied to generate a minimal list of proteins that explain all of the peptides that pass the entry criteria.

3.2 Analysis of Electrophile-Protein Adducts

3.2.1 Hsp90 Modification by HNE in RKO Cells

1. Grow RKO human colorectal carcinoma cells to 80% confluence in McCoy's 5A medium supplemented with 10% fetal bovine serum, at 37°C in an atmosphere of 95% air/5% CO_2.

2. Carry out the treatments with either varying concentrations of HNE dissolved in ethanol for 1 h at 37°C or 250 µM HNE for varying time intervals between 0 and 60 min. Wash confluent cells plated in 150 mm culture dishes first with 5 mL of cold phosphate-buffered saline, then incubate with 0, 50, 100 or 250 µM HNE delivered in 10 mL fresh McCoy's 5A medium without fetal bovine serum. The total concentration of ethanol per culture should be ≤0.1% of the total medium volume.

3. Expose cells to electrophile for 1 h at 37°C in an atmosphere of 95% air/5% CO_2, then, using a disposable cell scraper, scrape off cells from the culture dishes directly in the treatment medium, and centrifuge at $100 \times g$ for 5 min. Slowly aspirate off the treatment medium (supernatant) and wash the cell pellets twice with cold phosphate-buffered saline, pH 7.4, before freezing.

4. Lyse cell pellets, as described above, and reduce the Michael adducts of HNE in cell lysates with 2 mM $NaBH_4$, before capture with geldanamycin-biotin.

5. Visualize the accumulation of HNE-adducted proteins by immunoblot analysis of treated cellular lysates with an anti-HNE rabbit polyclonal antibody directed against reduced Michael adducts of HNE (Fig. 3).

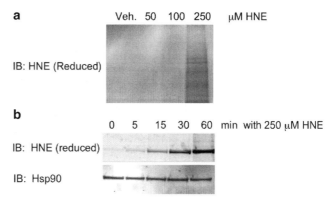

Fig. 3 Analysis of HNE adduction in RKO cells. (**a**) Immunoblot analysis of lysates from cells treated with EtOH, 50, 100, or 250 μM HNE and detected with an antibody to reduced HNE adducts. (**b**) Immunoblot analysis of Hsp90 isolated from RKO cells treated with 250 μM HNE for 0, 5, 15, 30, or 60 min and detected with an antibody to reduced HNE adducts. Loading control for total Hsp90 detected with anti-Hsp90 is shown below. (Reprinted from (6) with permission from the American Chemical Society)

6. For isolation of adducted Hsp90 protein from HNE-treated cellular lysates, reduce the incubation with geldanamycin-biotin to 1 h and the incubation with Neutravidin resin to 1.5 h at 4°C, followed by four washes with 1 mL of lysis buffer. After the last wash is removed, add 1/5 bed volume of LDS sample buffer to the resin and incubate at 95°C for 10 min, to elute Hsp90. This sample is then subjected to gel electrophoresis and the Hsp90 band is excised for digestion trypsin and analysis by LC-MS/MS.

7. To provide accurate mass characterization of HNE adducts on Hsp90, a Thermo LTQ-Orbitrap instrument with an Eksigent Nano 1D Plus pump and autosampler is used. Liquid chromatography is performed as described above. The data-dependent inclusion list screening is carried out with the following parameters, derived from previously described methods (14). An Orbitrap MS scan from m/z 300–2,000 at 60,000 resolution is followed by 10 LTQ ion trap MS/MS scans. If eight or more ions on the inclusion list are present, then the eight most intense ions are selected for tandem MS analysis. If fewer than eight ions on the inclusion list are present, then those ions plus the most intense ions in the initial scan (up to eight) are targeted. Dynamic exclusion is enabled, with a repeat count of three and a repeat duration of 10 s. The exclusion list size is 50 and the exclusion duration is 20 s. Threshold intensity for triggering peak detection is set at 100 with collision energy of 28% set for the entire list. The data are analyzed using MonsterMod, an in-house-developed algorithm (15), to identify MS/MS

spectra corresponding to Hsp90 peptides with mass shifts greater than 1 Da. Mass shifts of 158 Da correspond to the reduced Michael adducts of HNE. Spectra of the adducted precursor and fragment ions are verified manually, with a requirement of less than 10 ppm error for peptide-adduct precursor m/z measurements.

3.2.2 Kinetic Analysis of Hsp90 Adduction by HNE in Culture

1. Kinetic analysis of modification sites is performed using a label-free quantification approach we described previously (6, 16). This approach measures signals for adducted peptides by LC-MS/MS using a Thermo LTQ instrument. Each peptide adduct is monitored by targeting the m/z of the doubly or triply charged precursor for MS/MS. Two unmodified peptides from each protein are also targeted in the same manner. Specific product ions generated by MS/MS fragmentation of the targeted peptide adducts and reference peptides are extracted with Thermo Xcalibur software, and peak areas are integrated. Three product-ion signals are monitored for each peptide or peptide adduct and the peak area for each MS/MS transition is summed, to generate a peak area for each peptide. MS/MS data for both doubly and triply charged precursor ions are acquired in some cases and the product ions yielding the greatest signal are used for subsequent analysis.

2. The peak area for each peptide adduct is normalized to the average signal of the two unmodified reference peptides at each sample time point. The peak area reflecting HNE adduction after 4 h treatment is used as the endpoint for reactions with isolated Hsp90, in order to determine an observed rate of reaction.

3. Values of k_{obs} are calculated from plots of the ratio of normalized peak areas for adducted peptides at each time point to the average normalized peak area for the adducted peptide at the specified endpoint.

4. Data are fitted to a single-exponential association with GraphPad Software (Fig. 4).

4 Notes

1. All chemical reagents are purchased from commercial sources and are used without further purification. All reagents should be prepared fresh before each use.

2. Incubation with primary antibody overnight at 4°C gives a much stronger signal for western blotting than does 2-h incubation at room temperature.

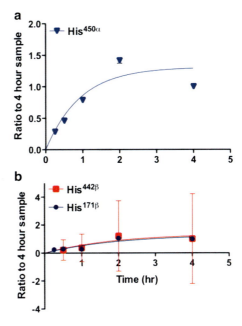

Fig. 4 Determination of the relative reaction rates of HNE-modified Hsp90 peptides after treatment of RKO cells with 250 μM HNE. (**a**) Reaction kinetics of His$^{450\alpha}$ with HNE in Hsp90α. Replicate analyses of 4–8 samples were used for each time point. Nonlinear single-exponential regression was used to fit the curve. (**b**) Reaction kinetics of His$^{442\beta}$ and His$^{171\beta}$ with HNE in Hsp90β. Replicate analyses of 4–8 samples were used for each time point. Nonlinear single-exponential regression was used to fit the curve. (Reprinted from (6) with permission from the American Chemical Society)

Acknowledgment

This work was supported by National Institutes of Health Grants ES013125 and the National Foundation for Cancer Research.

References

1. Liebler DC (2008) Protein damage by reactive electrophiles: targets and consequences. Chem Res Toxicol 21:117–128
2. Codreanu SG, Zhang B, Sobecki SM, Billheimer DD, Liebler DC (2009) Global analysis of protein damage by the lipid electrophile 4-hydroxy-2-nonenal. Mol Cell Proteomics 8:670–680
3. Kim HY, Tallman KA, Liebler DC, Porter NA (2009) An azido-biotin reagent for use in the isolation of protein adducts of lipid-derived electrophiles by streptavidin catch and photorelease. Mol Cell Proteomics 8:2080–2089
4. Vila A, Tallman KA, Jacobs AT, Liebler DC, Porter NA, Marnett LJ (2008) Identification of protein targets of 4-hydroxynonenal using click chemistry for ex vivo biotinylation of azido and alkynyl derivatives. Chem Res Toxicol 21:432–444
5. Liu J, Li Q, Yang X, van Breemen RB, Bolton JL, Thatcher GR (2005) Analysis of protein covalent modification by xenobiotics using a covert oxidatively activated tag: raloxifene proof-of-principle study. Chem Res Toxicol 18:1485–1496
6. Connor RE, Marnett LJ, Liebler DC (2011) Protein-selective capture to analyze electrophile adduction of hsp90 by 4-hydroxynonenal. Chem Res Toxicol 24:1275–1282

7. Carbone DL, Doorn JA, Kiebler Z, Ickes BR, Petersen DR (2005) Modification of heat shock protein 90 by 4-hydroxynonenal in a rat model of chronic alcoholic liver disease. J Pharmacol Exp Ther 315:8–15

8. Whitesell L, Mimnaugh EG, De Costa B, Myers CE, Neckers LM (1994) Inhibition of heat shock protein HSP90-pp 60v-src heteroprotein complex formation by benzoquinone ansamycins: essential role for stress proteins in oncogenic transformation. Proc Natl Acad Sci U S A 91:8324–8328

9. Clevenger RC, Raibel JM, Peck AM, Blagg BS (2004) Biotinylated geldanamycin. J Org Chem 69:4375–4380

10. Ma ZQ, Dasari S, Chambers MC, Litton MD, Sobecki SM, Zimmerman LJ, Halvey PJ, Schilling B, Drake PM, Gibson BW, Tabb DL (2009) IDPicker 2.0: improved protein assembly with high discrimination peptide identification filtering. J Proteome Res 8:3872–3881

11. Tabb DL, Fernando CG, Chambers MC (2007) MyriMatch: highly accurate tandem mass spectral peptide identification by multivariate hypergeometric analysis. J Proteome Res 6:654–661

12. Zhang B, Chambers MC, Tabb DL (2007) Proteomic parsimony through bipartite graph analysis improves accuracy and transparency. J Proteome Res 6:3549–3557

13. Elias JE, Gygi SP (2007) Target-decoy search strategy for increased confidence in large-scale protein identifications by mass spectrometry. Nat Methods 4:207–214

14. Jaffe JD, Keshishian H, Chang B, Addona TA, Gillette MA, Carr SA (2008) Accurate inclusion mass screening: a bridge from unbiased discovery to targeted assay development for biomarker verification. Mol Cell Proteomics 7:1952–1962

15. Hansen BT, Davey SW, Ham AJ, Liebler DC (2005) P-Mod: an algorithm and software to map modifications to peptide sequences using tandem MS data. J Proteome Res 4:358–368

16. Rachakonda G, Xiong Y, Sekhar KR, Stamer SL, Liebler DC, Freeman ML (2008) Covalent modification at Cys151 dissociates the electrophile sensor Keap1 from the ubiquitin ligase CUL3. Chem Res Toxicol 21:705–710

Chapter 16

DNA Shuffling of Cytochrome P450 Enzymes

James B.Y.H. Behrendorff, Wayne A. Johnston, and Elizabeth M.J. Gillam

Abstract

DNA family shuffling is an efficient method for creating libraries of novel enzymes, in which a high proportion of mutants exhibit correct folding and possess catalytic properties distinct from the starting material. The evolutionary arrangement of cytochromes P450 into subfamilies of enzymes with highly similar nucleotide sequences but distinct catalytic properties renders them excellent starting material for DNA family shuffling experiments. This chapter provides a general method for creating libraries of shuffled P450s from two or more related sequences and incorporates several recent improvements to previously published methods.

Key words Cytochrome P450, DNA shuffling, Directed evolution, Biocatalysts, Mutant libraries, Protein engineering

---

1 Introduction

Creating diverse variants of P450s may yield novel biocatalysts, provide important information regarding P450 folding, and deliver insights into aspects of P450 biology such as the determinants of their remarkable catalytic promiscuity. A variety of protein engineering strategies have been applied to the diversification of P450s including error-prone PCR, site-directed saturation mutagenesis (1, 2), recombination-based methods such as DNA shuffling (3), and SCHEMA-guided protein recombination (4).

DNA family shuffling is a method for recombining homologous genes to create libraries of novel mosaic sequences (5, 6). The occurrence of subfamilies of homologous cytochrome P450 genes makes DNA family shuffling a suitable method for creating libraries of novel P450 sequences. Libraries of shuffled P450s produced from members of the CYP1A, CYP2C, and CYP3A subfamilies have been published to date (3, 7–9). Interesting properties of the mutant P450s produced by DNA shuffling include enhanced catalytic rates towards known subfamily substrates (7), turnover of substrates not metabolized by the parental enzymes (8), and altered regioselectivity (9). Additionally, this method can be used to

Ian R. Phillips et al. (eds.), *Cytochrome P450 Protocols*, Methods in Molecular Biology, vol. 987,
DOI 10.1007/978-1-62703-321-3_16, © Springer Science+Business Media New York 2013

explore the sequence determinants underlying functional differences between two highly similar P450s (e.g., differences in activity towards a given substrate or expression yield in a heterologous system) (10, 11).

The DNA family shuffling method involves fragmenting multiple similar sequences and then reassembling the fragments in a PCR without primers (5, 12). Sequences suitable for shuffling may be created by random mutagenesis of a single gene, but greater functional diversity of mutants has been observed when multiple native sequences that have a high degree of sequence identity are shuffled (5). The DNA fragments in a shuffling reaction anneal to each another where there is sufficient sequence identity, permitting hybridization between fragments derived from different sources. These cross-hybridization (crossover) events result in the creation of novel mosaic sequences containing elements inherited from more than one progenitor (parental) sequence.

The protocol presented here improves on several previous applications of DNA family shuffling with P450s (3, 7–9, 13). The reassembly PCR method is based on that of Abecassis et al. (3), but restriction enzymes, rather than DNaseI, are used in the fragmentation step, as described by Kikuchi et al. (14). The use of restriction enzymes reduced the occurrence of parental sequences in the resultant libraries. Importantly, a new method for isolating DNA fragments to be used as template in the shuffling reaction is presented which improves the fidelity of the resultant library. Mutants produced using the protocol outlined below exhibit frequent recombination (>6 crossovers per mutant), and a high degree of structural and functional integrity, with up to 85% of mutants in a library expressed as folded haemoprotein detectable in whole-cell assays and up to 96% of mutants showing detectable activity against probe substrates (9).

Parental P450 sequences used in this protocol are typically inserted into the same plasmid vector (see Note 1) (15). Although this method can be applied to P450 sequences encoded on any plasmid, the process is simplified when the parental cDNAs are encoded on the same vector that is to be used for subsequent expression of the library. NADPH-P450 reductase is also included in a bicistronic format with the parental P450 sequences. Inclusion of the relevant reductase partner in the expression vector allows for the use of whole-cell methods for subsequent screening (8, 16).

2 Materials

1. The desired P450 coding sequences for DNA shuffling encoded on a common plasmid. In the current protocol the pCW' expression vector is used with NADPH-cytochrome P450 reductase coexpressed in a bicistronic format (see Note 2).

Table 1
Sequences of primers employed for cloning shuffled P450s into the pCW' expression vector in a bicistronic format with human NADPH-P450 reductase and subsequent DNA sequencing

Designation	Sequence
pCW'/P450-fwd (for cloning)	5'GGAAACAGGATCCATCGATGCTTAGGAGGTCATATG3'
pCW'/P450/hNPR-rev (for cloning and sequencing)	5'GAGGTCAATGTCTGAATTTTGGTGAACTCGGGGAC3'
pCW'/P450-seqfwd (for sequencing)	5'GAAACAGGATCCATCGATGCT3'

Standard alkaline lysis preparations of plasmid DNA (17) are of sufficient quality for DNA shuffling.

2. Ultracompetent DH5αF'IQ™ strain *Escherichia coli* (Invitrogen, Carlsbad, CA) (see Note 3).

3. Restriction enzymes and manufacturer-recommended buffers.

4. General molecular biology supplies for PCR amplifications and molecular cloning.

5. Oligonucleotide primers for amplification of shuffled products and for DNA sequencing (see Table 1).

2.1 Agarose Gel-Based Fragment Selection

1. SYBR® Safe DNA gel stain (Invitrogen, Carlsbad, CA, USA).

2. 1% (w/v) agarose in 1× TAE buffer (40 mM Tris–acetate, 1 mM EDTA, pH 8.0).

3. 10× DNA loading buffer (40 mM Tris–acetate, 10 mM EDTA, 30% (v/v) glycerol, pH 8.0).

4. Quickload 100-bp DNA ladder (New England Biolabs, Ipswich, MA, USA) or other equivalent molecular size marker.

5. Blue light source (see Note 4) and amber light filter (see Note 5).

6. QIAquick Gel Extraction kit (QIAGEN, Doncaster, VIC, Australia).

2.2 Selection of Fragments with PCR Filter Units

1. Montage PCR filter units (Millipore, Billerica, MA, USA).

3 Methods

The general workflow for creating a library of shuffled P450s involves the fragmentation of homologous P450 coding sequences using two different sets of restriction enzymes, separation of the desired fragments from undigested material, reassembling the fragments into

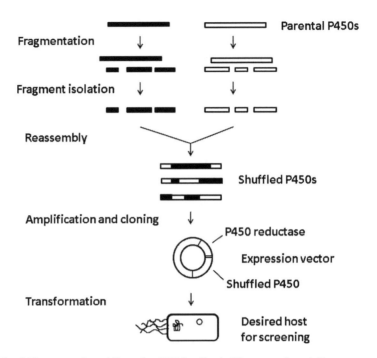

Fig. 1 The general workflow of a DNA family shuffling experiment. Two or more P450 sequences sharing greater than 70% nucleotide sequence identity are selected and fragmented. Fragments are isolated on the basis of size then reassembled in a primerless PCR mix. The shuffled products are then amplified with flanking primers, cloned into an appropriate expression vector and used to transform the desired host for subsequent screening. Inclusion of the appropriate P450 reductase in the expression vector facilitates high-throughput screening for catalytic activities

mosaic sequences using PCR without primers, amplifying the shuffled sequences via a conventional PCR, and cloning the novel sequences into a suitable vector for expression in the desired host organism (see Fig. 1).

3.1 Fragmentation of Parental cDNAs

Two or more P450 coding sequences for shuffling should be selected and these parental sequences should have >70% nucleotide sequence identity to facilitate efficient recombination during the shuffling reaction (see Note 6). Each plasmid carrying a different P450 coding sequence should be fragmented with two different combinations of restriction enzymes, in order to facilitate hybridization of overlapping fragments.

1. Identify two different sets of restriction enzymes that can be used to fragment the parental P450 coding sequences into segments that are generally between 100 and 300 bp in length (see Notes 7 and 8). Useful free tools that can aid in this planning step are RE Cut (developed in this laboratory by Mr. Weiliang Huang and available for download at http://www.

DNA Shuffling of Cytochrome P450 Enzymes   181

scmb.uq.edu.au/staff/elizabeth-gillam or http://dl.dropbox.com/u/50010004/RECUT.zip) or NEB Cutter v2.0 (available at http://tools.neb.com/NEBcutter2).

2. In duplicate sterile 0.6-mL tubes (see Note 9), combine parental plasmids in equal quantities so that the total mass of DNA per tube is 50 μg (e.g., 25 μg of each plasmid when using two parental sequences, or 12.5 μg of each plasmid when using four parents). One duplicate will be used for each combination of restriction enzymes.

3. To each tube add the desired restriction enzymes plus buffers to achieve reaction conditions in accordance with the enzyme manufacturer's recommendations, in a total reaction volume of 50 μL.

4. Mix gently and incubate at the recommended temperature for 4 h (see Note 10).

3.2 Selection of Fragments

Following fragmentation with restriction enzymes, large DNA fragments, including undigested plasmid and any full-length parental P450 coding sequences, must be removed to avoid contamination of the shuffled library with native sequences. Two principal methods have been used to accomplish this: the use of PCR filter units (3) (see Subheading 3.2.2) or physical separation of DNA fragments by an agarose gel-based technique outlined below (see Subheading 3.2.1, Fig. 2). The latter method is recommended, based on a greater ability to customize the fragment range selected and better resultant library quality (see Note 11).

3.2.1 Agarose Gel-Based Fragment Selection

In a modification of the method by Kadokami and Lewis (18), agarose gel-based fragment selection can be used for the isolation and recovery of DNA fragments within a user-defined size range from a complex mixture.

1. Prepare a 1% agarose gel in TAE buffer with the comb positioned ~3 cm from the top end of the gel (see Note 12).

2. Before loading samples, carefully precut the gel as shown in Fig. 2a. Cuts are made precisely with a ruler and scalpel so that gel sections can be moved without tearing (see Fig. 2c). Make the transverse cut above the wells at least 1 cm from the edge of the gel to avoid the gel becoming fragile at the top.

3. To each 50-μL mixture of digested DNA fragments, add 5 μL of 10× DNA loading buffer and 5 μL SYBR Safe dye diluted to 1,000× in sterile milliQ water (diluted from the manufacturer's 10,000× stock solution in DMSO). Also mix 10 μL of a standard DNA ladder (e.g., Quickload 100-bp DNA ladder) with 1 μL SYBR Safe dye (1,000×). Incubate samples for 15 min at room temperature. Alternatively, the SYBR Safe dye can be

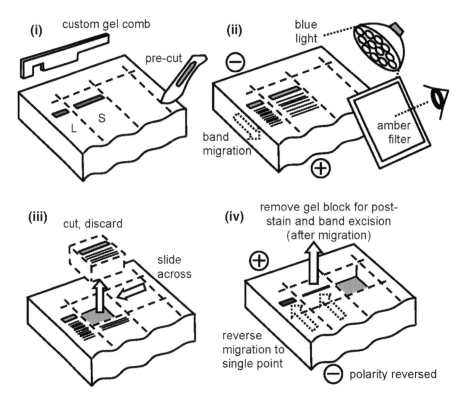

Fig. 2 Schematic representation of DNA fragment exclusion via gel electrophoresis in four stages: (**i**) preparation of gel; (**ii**) band migration visualized under blue light/amber filter; (**iii**) cutting at desired gel size based on ladder migration, removal of gel block containing large fragment sizes and substitution with fresh gel block; (**iv**) reversal of polarity during second electrophoresis to reconcentrate small DNA fragments at the origin. *S* denotes sample lanes and *L* denotes the molecular weight size marker (*ladder*)

added at a concentration of 1× directly to the molten agarose gel immediately before pouring (see Note 13).

4. Load the DNA marker ladder and each fragmentation mixture, and begin electrophoresis. During electrophoresis the DNA fragments can be visualized using a blue light and amber filter (see Fig. 2b).

5. When bands in the DNA ladder are sufficiently resolved, turn off the current and cut the gel at the appropriate point to separate and discard DNA fragments with sizes greater than 1,000 bp or at any other desired molecular weight cutoff. Also discard the sample-loading wells (see Fig. 2c). Replace the excised section with agarose gel containing no DNA.

6. Reverse the direction of the electric field and rerun the samples for the same length of time as in the initial electrophoresis, such that the bands containing DNA fragments <1,000 bp converge to form a single band (see Fig. 2d and Note 14).

DNA Shuffling of Cytochrome P450 Enzymes   183

7. Excise the bands containing the two different sets of DNA fragments and extract the DNA using the QIAquick Gel Extraction kit in accordance with the manufacturer's instructions.

8. Quantify DNA by measuring absorbance at 260 nm.

3.2.2 Fragment Selection with PCR Filter Units

Filter units designed for cleaning up PCR reactions can be used to recover fragments for DNA shuffling experiments using a modification to the manufacturer's instructions (3).

1. Place PCR filter unit sample reservoirs into collection tubes.

2. Dilute samples to a total volume of 400 µL in sterile milliQ water. Load samples into sample reservoirs and centrifuge at $1,000 \times g$ for 15 min. Discard sample reservoirs.

3. Precipitate DNA using 0.1 volumes of 3 M sodium acetate and 2.5 volumes of 100% ethanol for 1 h at −20°C. Pellet the DNA by centrifugation, wash in 70% ethanol, and evaporate to dryness. Resuspend the DNA pellet in sterile milliQ water.

4. Quantify DNA by measuring absorbance at 260 nm.

3.3 The DNA Family Shuffling (Reassembly) Reaction

1. Combine equal quantities of DNA fragments from each fragmentation reaction in a 50-µL PCR mix without primers and using routine buffering and reagent conditions (50 mM KCl, 0.2 mM dNTPs, 1.5 mM $MgCl_2$, and 0.5 U/µL *Taq* polymerase in 20 mM Tris-Cl (pH 8.4)). Cycling conditions are adapted from Abecassis et al. (3) and are as follows: an initial hot start at 94°C while *Taq* polymerase is added, followed by 35 cycles of denaturation at 94°C for 30 s, a 13.5-min hybridization step where the temperature is decreased from 65 to 41°C, and an extension step at 72°C for 90 s. This is followed by a final extension at 72°C for 7 min to ensure complete extension of any remaining incomplete amplifications. The hybridization step can also be performed in twenty-five 1-min steps rather than as a 13.5-min smooth gradient.

2. Analyze reassembly products by agarose gel electrophoresis. Reassembly products should appear as a smear, indicating the successful assembly of polynucleotides of various lengths (see Note 15).

3.4 Amplification of Shuffled Products and Cloning into E. coli

The products of the reassembly reaction are used as template in a conventional PCR with flanking primers to permit efficient isolation of products that are the same length as the full-length parental P450s. The primers used anneal to vector sequence upstream and downstream of the P450 open reading frame (see Note 16).

1. Prepare a 50-µL PCR using 5 µL of reassembly products as template DNA, and 2.5 µM each of primers pCW'/P450-fwd and pCW'/P450/hNPR-rev (see Table 1). Buffer

184   James B.Y.H. Behrendorff et al.

conditions are as recommended by the manufacturer of the *Taq* polymerase used.

2. Amplify the shuffled products using the following cycling conditions: an initial hot start at 94°C while *Taq* polymerase is added, followed by ten cycles of denaturation at 94°C for 30 s, hybridization at 60°C for 30 s, and an extension step at 72°C for 90 s. This is followed by a further 25 cycles of the same conditions with the difference that 10 s are added to the extension-step time in each successive cycle. A final extension at 72°C for 7 min is used to facilitate completion of all partially extended products.

3. Clean up reactions using a QIAquick PCR cleanup kit in accordance with the manufacturer's instructions.

4. Digest PCR products with the appropriate restriction enzymes for cloning into the expression vector and prepare the vector similarly.

5. Ligate PCR products of the correct length into the double-digested vector source and transform ultracompetent *E. coli* DH5αF'IQ containing the pGro7 plasmid (or other desired host) with the ligation mixture.

6. Pick colonies of successful transformants to inoculate 100 μL of LB medium containing appropriate selection antibiotics in 96-well microtiter plates. Incubate overnight at 37°C with shaking. Make glycerol stocks by adding to each well 100 μL of double-concentration (2×) LB medium containing 30% (v/v) glycerol. Seal the plate with an X-Pierce sealing membrane (see Note 17) and mix thoroughly before storing at –80°C.

3.5 Verification of Shuffled Products

Plasmid DNA from a random selection of mutants should be prepared and sequenced, in order to confirm that shuffling of the parental genes has occurred. The primers pCW'/P450-seqfwd and pCW'/P450/hNPR-rev (see Table 1 and Note 18) can be used to sequence shuffled P450s cloned in the pCW'/P450/hNPR bicistronic format. Crossover events and spontaneous mutations can be identified by comparing shuffled mutants with the parental sequences in a multiple-sequence alignment.

Archived glycerol stocks of shuffled mutants can be used to inoculate small-scale expression cultures for measurement of P450 and screening assays ((7, 8, 16) and chapter 17 of this book).

4   Notes

1. The pCW' plasmid (15) has been used previously for the expression of many P450s in *E. coli*.

2. The parental coding sequences should be available in the same expression vector format as that which will be used for expression

of the shuffled library. Using the same expression vector as is intended for expression of the shuffled library simplifies library cloning. Furthermore, using parental cDNAs with the same vector format prevents mishybridization between vector-containing fragments at the 5′ and 3′ ends of the P450 coding sequence, and limits the possibility of mishybridization between P450 and vector fragments, as perfectly matching vector fragments will preferentially anneal to one another. Use of a vector containing the NADPH-cytochrome P450 coding sequence permits reconstitution of a functional assay system (19) in whole *E. coli* cells, enabling the use of whole-cell library screening assays.

3. The use of the cloning strain *E. coli* DH5αF'IQ™ aids in stable maintenance of shuffled-library sequences. Competent *E. coli* DH5αF'IQ™ transformed with the pGro7 plasmid (Takara Bio Inc., Shiga, Japan) may also be used, as co-expression of the GroEL/ES chaperone system may improve folding of some shuffled P450s. Although it is important to maximize transformation efficiencies to ensure that as much of the library diversity can be sampled as possible, cells of a sufficient competency can be prepared by the method of Inoue et al. (20).

4. A commercially available blue-light transilluminator may be used for excitation of the dye (Invitrogen SafeImager, Invitrogen, Carlsbad, CA, USA), but in this laboratory gels are successfully illuminated with a simple blue multi-LED light (20 LED MR-16 lamp, available from most retail electronics suppliers). Such blue LED-based lights show broad emission spectra around 500 nm, suitable for exciting SYBR dyes (SYBR DNA stains are maximally excited at 494 nm (blue) and emit at 521 nm (green), with secondary excitation at 284 and 382 nm).

5. Clear visualization of the SYBR Safe-stained sample under blue light requires an appropriate light filter. For ease of use during gel manipulations, Safe Imager™ viewing glasses or equivalent are recommended. Safe Imager viewing glasses can be purchased separately from the manufacturer (Invitrogen, Carlsbad, CA, USA).

6. The most diverse P450s used in DNA shuffling experiments reported to date are human P450s CYP1A1 and CYP1A2 (74% nucleotide sequence identity) (8). Unpublished experiments from this laboratory have shown shuffling to be inefficient when using P450 coding sequences with less than 70% sequence identity, with many mutants containing only a single crossover event and reassembly of unshuffled parents being favored instead.

7. It is not critical that all fragments lie within the 100- to 600-bp size range. Data from this laboratory have demonstrated that the mean length of fragments used in DNA shuffling influences the recombination frequency (10). As expected, longer fragments resulted in fewer crossover events, whereas shorter fragments resulted in more frequent crossovers.

8. Care should be taken to select combinations of enzymes that have compatible buffer conditions.

9. Use of microtubes smaller than 1.5 mL reduces changes in reaction conditions due to evaporation.

10. Where the selected restriction enzymes have different optimal reaction temperatures, the reactions can be incubated for 2 h at the lower temperature followed by 2 h at the higher temperature. Similarly, buffer compositions may be adjusted part way through an incubation if, for example, the two conditions differ only in the concentration of salt or some other simple parameter. Complete digestion at all restriction sites is unnecessary, and incomplete digestion may result in a greater diversity of fragments.

11. Early publications in this area cite the use of PCR filter units to remove large undigested fragments (3, 7, 8). However, recent work in this laboratory has found that this approach, although technically simple, produces libraries of inferior quality, since it strongly favors the recovery of DNA fragments <100 bp in length. Larger quantities of fragments are required to obtain full-length sequences, due to the bias in fragment size and because the frequency of mishybridization events and spontaneous mutations (generated in the reassembly PCR) is greater when there is a larger proportion of shorter fragments.

12. The placement or the design of the gel comb should be adjusted such that wells can accommodate 50 μL of sample.

13. Pre-binding the dye at high concentration in the samples, instead of including SYBR Safe dye throughout the gel, serves to reduce costs.

14. Concentrating the desired DNA fragments into a single band facilitates more efficient extraction of the DNA from the agarose (18).

15. The mean length of reassembly products typically increases with the quantity of DNA fragments added to the reassembly reaction. If the reassembly smear does not contain products of length equal to a full-length parental coding sequence, a greater quantity of DNA fragments should be used in the reassembly reaction.

16. The primer pCW'/P450-fwd anneals to the start codon and the preceding 11 upstream codons in the pCW' sequence, and

facilitates cloning of shuffled P450s into the pCW' expression vector at the *Nde*I restriction site. The reverse primer pCW'/P450/hNPR-rev anneals to the nucleotides 184–150 of the human NADPH-P450 reductase sequence, 191 nucleotides downstream of the P450 stop codon. Any suitable restriction site downstream of the stop codon in the original parental constructs can be used for cloning. Use of the pCW'/P450/hNPR-rev primer is appropriate only when the parental P450s used in shuffling originated from the bicistronic pCW'/P450/hNPR vector. For vectors containing the parental sequences that do not include the hNPR cDNA in bicistronic format, a different reverse primer must be designed.

17. X-Pierce sealing membranes feature a precut crossover each well to facilitate rapid sampling from 96-well plates.

18. The pCW'/P450-seqfwd primer anneals to the ribosome-binding site of the pCW' vector and is sufficiently upstream from the P450 open reading frame to allow sequencing from the start codon. The pCW'/P450/hNPR-rev primer used for cloning can also be used for sequencing in the reverse direction.

References

1. Wong TS, Arnold FH, Schwaneberg U (2004) Laboratory evolution of cytochrome P450 BM-3 monooxygenase for organic cosolvents. Biotechnol Bioeng 85:351–358

2. Li HM, Mei LH, Urlacher VB, Schmid RD (2008) Cytochrome P450 BM-3 evolved by random and saturation mutagenesis as an effective indole-hydroxylating catalyst. Appl Biochem Biotechnol 144:27–36

3. Abecassis V, Pompon D, Truan G (2000) High efficiency family shuffling based on multi-step PCR and in vivo DNA recombination in yeast: statistical and functional analysis of a combinatorial library between human cytochrome P450 1A1 and 1A2. Nucleic Acids Res 28:E88

4. Voigt CA, Martinez C, Wang ZG, Mayo SL, Arnold FH (2002) Protein building blocks preserved by recombination. Nat Struct Biol 9:553–558

5. Crameri A, Raillard SA, Bermudez E, Stemmer WP (1998) DNA shuffling of a family of genes from diverse species accelerates directed evolution. Nature 391:288–291

6. Stemmer WP (1994) DNA shuffling by random fragmentation and reassembly: in vitro recombination for molecular evolution. Proc Natl Acad Sci U S A 91:10747–10751

7. Huang W, Johnston WA, Hayes MA, De Voss JJ, Gillam EM (2007) A shuffled CYP2C library with a high degree of structural integrity and functional versatility. Arch Biochem Biophys 467:193–205

8. Johnston WA, Huang W, De Voss JJ, Hayes MA, Gillam EM (2007) A shuffled CYP1A library shows both structural integrity and functional diversity. Drug Metab Dispos 35:2177–2185

9. Hunter DJ, Behrendorff JB, Johnston WA, Hayes PY, Huang W, Bonn B, Hayes MA, De Voss JJ, Gillam EM (2011) Facile production of minor metabolites for drug development using a CYP3A shuffled library. Metab Eng 13:682–693

10. Behrendorff JBYH (2011) Investigations of cytochromes P450 using the DNA family shuffling method. Ph.D. Thesis, The University of Queensland

11. Behrendorff JBYH, Moore CD, Kim K-H, Kim D-H, Smith CA, Johnston WA, Yun C-H, Yost GS, Gillam EMJ (2012) Directed evolution reveals requisite sequence elements in the functional expression of P450 2F1 in Escherichia coli. Chem Res Toxicol 25: 1964–1974.

12. Stemmer WP (1994) Rapid evolution of a protein in vitro by DNA shuffling. Nature 370:389–391

13. Rosic NN, Huang W, Johnston WA, DeVoss JJ, Gillam EM (2007) Extending the diversity of

cytochrome P450 enzymes by DNA family shuffling. Gene 395:40–48

14. Kikuchi M, Ohnishi K, Harayama S (1999) Novel family shuffling methods for the in vitro evolution of enzymes. Gene 236: 159–167

15. Barnes HJ, Arlotto MP, Waterman MR (1991) Expression and enzymatic activity of recombinant cytochrome P450 17 alpha-hydroxylase in *Escherichia coli*. Proc Natl Acad Sci U S A 88: 5597–5601

16. Johnston WA, Huang W, De Voss JJ, Hayes MA, Gillam EM (2008) Quantitative whole-cell cytochrome P450 measurement suitable for high-throughput application. J Biomol Screen 13:135–141

17. Sambrook J, Russell DW (2006) Using plasmid vectors in molecular cloning. In: Samuel D, Russell DW (eds) The condensed protocols from molecular cloning: a laboratory manual. Cold Spring Harbor Laboratory Press, Cold Spring Harbor, NY, pp 2–8

18. Kadokami Y, Lewis RV (1995) Reverse electrophoresis to concentrate DNA fractions. Anal Biochem 226:193–195

19. Parikh A, Gillam EM, Guengerich FP (1997) Drug metabolism by Escherichia coli expressing human cytochromes P450. Nat Biotechnol 15:784–788

20. Inoue H, Nojima H, Okayama H (1990) High efficiency transformation of Escherichia coli with plasmids. Gene 96:23–28

Chapter 17

Measurement of P450 Difference Spectra Using Intact Cells

Wayne A. Johnston and Elizabeth M.J. Gillam

Abstract

Whole-cell assays provide a rapid means of determining expression and substrate binding for cytochrome P450 enzymes expressed heterologously in *Escherichia coli* and, potentially, other host cells. Such assays are particularly useful for screening large libraries of mutant P450s, where rapid, high-throughput assays are needed for first-tier screens that can, firstly, quantify any P450 form independent of P450 subfamily and, secondly, suggest possible ligands before more labor-intensive direct measurement of substrate turnover. Whole-cell spectral techniques are derived from methods that have been used for a long time to study P450s in microsomal or other subcellular fractions (Omura T and Sato R, J Biol Chem 239:2370–2378, 1964; Schenkman JB et al., Biochemistry 11:4243–4251, 1972), but recent studies have detailed important modifications which allow quantitative results to be obtained in whole cells (Otey CR, Methods in Molecular Biology, vol. 230, Humana, Totowa, NJ, pp. 137–139, 2003; Johnston WA et al., J Biomol Screen 13:135–141, 2008). A general method is presented here for the measurement of difference spectra on recombinant P450 cultures that can be applied to both carbon monoxide and any number of alternative ligands that alter the characteristic spectral signature of P450s.

Key words Cytochrome P450, Directed evolution, *Escherichia coli*, Difference spectroscopy, Fe(II)·CO versus Fe(II) difference spectra, High-throughput screening, P450 libraries, DNA shuffling, Whole cells

---

1 Introduction

As heme-thiolate proteins, cytochrome P450 enzymes possess a screening advantage held by few other enzymes of biotechnological interest: the enzyme has a spectral signature in the visible region which indicates whether the enzyme is correctly folded and whether the thiolate ligand to the heme, which provides P450s with their exceptional properties, is intact. The catalytically functional, thiolate-liganded form contains a heme prosthetic group that absorbs at 450 nm in the carbon monoxide (CO)-bound ferrous form. The CO-Fe(II) versus Fe(II) difference spectrum provides a sensitive and specific means by which to quantify P450s in crude mixtures. Although such a screen does not directly indicate

Ian R. Phillips et al. (eds.), *Cytochrome P450 Protocols*, Methods in Molecular Biology, vol. 987, DOI 10.1007/978-1-62703-321-3_17, © Springer Science+Business Media New York 2013

that the P450s measured can turn over a substrate, it provides a "next-best" alternative independent of the nature of the individual P450, based on the assumption that, to be catalytically active, P450s need to have an intact thiolate ligand to the heme. Additionally, other ligands binding close to the heme frequently perturb the spin state of the iron and thereby influence the visible absorption spectra of the P450, resulting in characteristic "binding spectra" indicative of different types of ligands. Type I/II and reverse type I ligand-binding spectra allow detection of potential substrates or inhibitors that bind tightly to the heme. Type I ligands displace the aqua ligand to the ferric form, stabilizing the high-spin form of the enzyme. Type II spectra typically result from nitrogenous compounds (usually amines or azoles) which provide an alternative ligand to the heme and stabilize the low-spin form, whereas reverse type I ligands appear to provide an alternative hydroxyl ligand, which again stabilizes the low-spin form.

In recent years P450s have attracted attention as potential biocatalysts, leading to numerous efforts to create libraries of P450s by random mutagenesis and other artificial evolution strategies. Such libraries can be screened to find novel, industrially useful forms (1–10). High-throughput measurement of P450 expression can allow culling of mutants that are misfolded or poorly expressed. Screening for substrate binding, while not a direct indication of activity toward a particular substrate, can still allow prioritization of mutants for more detailed reaction profiling (i.e., measurement of turnover of substrates via LC–MS or similar analytical techniques).

Measurements on intact cells allow P450 libraries to be rapidly screened within the same *Escherichia coli* cells in which they were produced (11), without subcellular fractionation or partial purification of the enzyme, thus reducing the time and effort required for screening, which is important for efficient directed-evolution strategies. This procedure involves (1) growth, induction, and concentration of cultures; (2) reduction of heme by sodium dithionite; (3) measurement of the CO-Fe(II) versus Fe(II) difference spectrum; and finally, (4) data processing.

Several differences between P450 measurements on subcellular fractions and whole cells exist that must be addressed for quantitative measurements. A primary difference results from the increase in turbidity of suspensions of whole cells. In typical measurements of culture density by optical methods, light is scattered by intact cells out of the spectrophotometric beam path and is not detected. This gives an optical density (OD) reading comprised primarily of loss of intensity due to light scattering within the turbid suspension, but which also includes absorbance due to chromophores in the sample. Light scattering is highly nonlinear with increasing cell density, with typical spectrophotometers showing nonlinearity at approximately $OD > 0.5$; however, the absolute readings and linear range are dependent on the geometry of the light path for the particular spectrophotometer used (12). Typically, OD readings taken to measure culture densities above this are made by diluting cultures before measurement, with correction for the dilution factor during calculation of the final OD. The nonlinearity is due to light initially scattered by cells being redirected back toward the detector by further scattering, and is dependent on the distance from the sample and photomultiplier, along with the slit width (12).

The accurate measurement of P450 spectra in whole cells relies on this nonlinear nature of light scattering by whole cells. Light scattered within cell suspensions contains the spectral signature of any absorbing species within the cells, specifically the ligand-bound P450 heme in this case. In the linear range for OD (i.e., very dilute cell resuspensions), concentrating cellular suspensions linearly scales both the background OD and P450 absorbance simultaneously, leading to no improvement in signal-to-noise (S/N) ratio with concentration. However, once the linear OD range is exceeded, further concentration of cells leads to a linear increase in the P450 absorbance signal, with little further increase in background OD. This increases S/N and, hence, measurement precision. Although eventually the signal due to the P450 absorbance will also become nonlinear (i.e., Beer's law will no longer be valid; absorbance >1.0 for typical spectrophotometers), it is practically impossible to achieve sufficiently high expression of P450s in cultures of *E. coli* for this to be a problem. Hence, it is possible to concentrate cells to improve the sensitivity of measuring the P450 signal but not significantly increase the interference due to light scattering (the noise) (13). In practical terms, the useful limit of concentrating cells for whole-cell P450-spectral measurements is determined by individual spectrophotometer characteristics and the volume of culture available for each P450 before concentration.

The second factor that facilitates accurate quantification of P450s within intact cells is the vertical light-beam geometry of typical microplate readers, as suspensions of cells settle over the course of measurements. In cuvette spectrophotometers (typical horizontal light-beam geometry), this means the density of cells through which the light passes changes over the course of the measurement (unless samples are constantly stirred). Settling occurs at right angles to the incident light. However, in microplate readers, measurements are typically taken with the light path oriented vertically, normal to the plane of the plate, and aligned exactly with the direction of the settling. This means that the number of cells in the optical window does not alter with settling, minimizing artefactual distortions in the spectra.

192   Wayne A. Johnston and Elizabeth M.J. Gillam

2 Materials

2.1 Reagents

1. *E. coli* transformed with vector coding for P450(s) of interest, for example, a library of altered P450s (see Note 1), stored as glycerol stocks at −80°C.

2. LB medium (14) containing antibiotics at concentrations that select for cells transformed with the appropriate plasmids.

3. Modified Terrific Broth (final medium concentrations provided): Bactotryptone (12 g/l), yeast extract (24 g/l), bactopeptone (2 g/l), 1 mM NaCl, 4% (v/v) glycerol with phosphate buffer (KH_2PO_4(2.31 g/l) K_2HPO_4(12.5 g/l)) autoclaved separately from other medium components and then combined.

   In addition, the following additives should be filter sterilized and added to media once cooled:

   (a) Trace element solution (250 µl/l) (2.7 g/100 ml $FeCl_3$·$6H_2O$; 0.2 g/100 ml $ZnCl_2$·$4H_2O$; 0.2 g/100 ml $CoCl_2$·$6H_2O$; 0.2 g/100 ml Na_2MoO_4·$2H_2O$; 0.1 g/100 ml $CaCl_2$·$2H_2O$; 0.1 g/100 ml $CuCl_2$; 0.05 g/100 ml H_3BO_3); powdered stocks should be dissolved in 10 ml of conc. HCl then made up to 100 ml with milli-Q water and the solution autoclaved (15).

   (b) 1 mM thiamine.

   (c) Ampicillin (0.1 mg/ml).

   (d) Chloramphenicol (0.2 mg/ml) (for maintenance of pGro7 chaperone vector, if appropriate).

4. Isopropyl-β-D-thiogalactopyranoside (IPTG).

5. D-arabinose (for induction of chaperone expression from the pGro7 vector, if appropriate).

6. δ-aminolevulinic acid.

7. Sodium dithionite (sodium hydrosulfite).

8. Whole-cell assay buffer (WCAB: 100 mM potassium phosphate, 6 mM Mg acetate, 10 mM (+)glucose, pH 7.4).

2.2 Equipment and Consumables

1. 24-well microplates.

2. 96-well microplates.

3. Shaking microplate incubator.

4. Microplate reader (e.g., Spectramax M2, Molecular Devices, Sunnyvale CA). A cuvette spectrophotometer is generally not suitable (see Note 2).

5. Centrifuge capable of accommodating microplates and capable of speeds sufficient to sediment *E. coli* cells, i.e., ~2,000×*g*.

6. BreatheEasy polyurethane membranes (Diversified Biotech, Boston, MA).

Measurement of P450 Difference Spectra Using Intact Cells   193

3 Methods

3.1 Growth of Culture and Concentration of Cell Suspensions

A suitable format for screening cultures is in 1-ml Modified Terrific Broth (TB) cultures grown in 24-well microplates, in a shaking microplate incubator (gyratory radius ~3–5 mm). Micro-aerobic growth is generally best for screening, and is achieved by covering cultures with membranes that are partially permeable to oxygen (BreathEasy™, Diversified Biotech, Boston, MA); this also prevents cross contamination of cultures. However, measurements are feasible on cultures grown both aerobically and microaerobically (see Note 3). Details are provided below for *E. coli* expressing mammalian P450s using the pCW' expression plasmid, with GroES and GroEL chaperones coexpressed from the pGro7 plasmid (see Note 1):

1. Inoculate suitable P450-expressing recombinant *E. coli* into LB medium containing appropriate antibiotics for maintenance of plasmids. Incubate for 16 h (or overnight) at 37°C, shaking at 400 rpm in a microplate shaker.

2. Inoculate 1 ml of TB medium in 24-well plates with 50 µl of overnight culture (5% v/v). Seal plates with BreathEasy membranes for microaerobic conditions (microplate lid omitted). Alternatively, cover cultures with the microplate lid alone for aerobic growth.

3. Incubate cells at 25°C in shaking incubator (see Note 4).

4. Induce cultures at $t=5$ h (i.e., 5 h post inoculation of TB) with 1 mM IPTG and D-arabinose (4 g/l), with 0.5 mM δ-aminolevulinic acid added to supplement heme synthesis.

5. Harvest cells at $t=72$ h (72 h after inoculation into TB).

6. Centrifuge cells at $2,000 \times g$ in a microplate centrifuge for 10 min. Resuspend 1 ml cultures in 0.5 ml of ice-cold WCAB (see Note 5). For increased precision, resuspend in 0.25 ml of WCAB (see Note 6).

7. Place resuspensions at 4°C before recording absorption spectra. Cells should be stored for a maximum of 2 h.

3.2 Reduction of P450 and Measurement of Spectra

In order to quantify the total amount of P450 in cells, the P450 heme must be exclusively in the ferrous state. In lysed cells or subcellular fractions such as microsomes or in purified protein the P450 heme is generally in the oxidized (ferric) form, although some ferrous P450s have been detected in cell lysates (DeVoss JJ and Stok JE, *personal communication*). For the isolated ferric form, brief contact with sodium dithionite is sufficient to render all P450 heme ferrous and able to bind CO. For whole-cell P450 measurements, a longer incubation time with sodium dithionite is necessary in order to allow a plateau in dithionite response. A typical incubation time is 30–60 min with 10 mg/ml dithionite, added

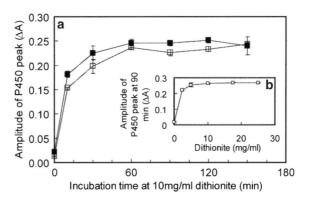

Fig. 1 Response of CYP176A1 P450 peak size in Fe(II) CO versus Fe(II) difference spectra to changes in dithionite incubation conditions. (**a**) Variation in the amplitude of P450 spectra with increasing incubation times in 10 mg/ml dithionite. *Filled squares*, unstirred incubation in air; *open squares*, unstirred incubation in 1 atm argon. *Inset* (**b**) P450 peak size after 90 min incubation with increasing dithionite concentrations. Data represent the mean ± SD of triplicate determinations for all measurements

from a freshly made 100 mg/ml dithionite solution (see Notes 7 and 8, and Fig. 1). Importantly, the measurement of P450 relies on an assumption of complete reduction of the P450 pool by dithionite, i.e., the dithionite-reducible P450 pool represents the total P450 (Fig. 1a).

A recent study has shown that P450s in whole cells are generally largely reduced (16). Thus, the amounts of P450 detected are not always increased by dithionite incubation. Amongst several human drug-metabolizing P450s examined (16), only CYP2A6 and CYP3A4 required the addition of dithionite to reduce ferric P450 to ferrous, and in these cases only when their cognate electron-donor partners were not co-expressed in bicistronic expression format. Where the reduction state (ratio of endogenous ferrous P450 to total dithionite-reducible P450) is to be determined, spectra should be recorded in both the presence and absence of dithionite. A typical 1-ml culture concentrated to 0.5 ml after resuspension in WCAB provides enough material for two 0.2-ml spectral determinations (in the presence and absence of dithionite).

As well as ensuring sufficient time is allowed for dithionite to reduce the intracellular P450 before measurement, it is also necessary to ensure CO can enter and bind to the ferrous heme between the measurement of the baseline and difference spectra. Although CO is a small molecule, studies of the time dependence of the spectral response show that approximately 10 min contact time at 1 atm CO is necessary and sufficient for ingress of CO in typical whole-cell suspensions (Fig. 2).

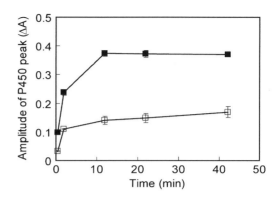

Fig. 2 Variation in the CYP176A1 P450 peak size in Fe(II)·CO versus Fe(II) difference spectra with increasing CO contact time. *Filled squares*, cell suspension incubated for 60 min with 10 mg/ml sodium dithionite prior to read. *Open squares*, minus dithionite. Data represent the mean ± SD of triplicate determinations for all measurements

1. Transfer cell resuspensions in 200-μl aliquots into individual wells of a 96-well plate (see Note 9).

2. To one set of wells, add 20 μl of freshly prepared sodium dithionite (100 mg/ml) (plus dithionite; total P450). Add 20 μl of milli-Q water to the remaining wells. Briefly mix (see Note 10).

3. Incubate for 90 min at ambient temperature covered with the microplate lid, without stirring or agitation.

4. Mix briefly (see Note 10).

5. Transfer the plate to the microplate reader and record a spectrum between 400 and 500 nm in 2-nm increments. Store this (Fe(II) only) spectrum so that it can be used for background subtraction (see Note 11).

6. Transfer plate to the CO chamber (see Note 12) and incubate for 10 min at 1 atm CO.

7. Transfer plate to microplate reader and record the spectrum of the Fe(II)·CO form of the P450 between 400 and 500 nm in 2-nm increments.

8. From the difference spectrum (Fe(II)·CO minus Fe(II) only), calculate the P450 concentration for each sample (well) according to Subheading 3.3 below.

3.3 Post-measurement Data Processing

1. For relatively concentrated cells and/or high P450 amounts, P450 can be quantified by the well-established method of quantifying the P450 peak in the difference spectrum as $Abs_{450nm} - Abs_{490nm}$, subtracting any $Abs_{450nm} - Abs_{490nm}$ seen in the baseline spectrum, and using the extinction coefficient ($\varepsilon = 91$ mM^{-1} cm^{-1}). It is important to consider the specific path length relevant to the fill volume of the microplate wells in this calculation (17, 18) (see Note 13).

2. In cultures in which P450 expression is low (e.g., recombinant expression of mammalian rather than bacterial P450 forms), the background spectrum due to bacterial hemoproteins can impinge on measurement. In addition, slow settling of cultures in the plate reader over the course of measurement leads to a steady decrease in light scattering, which manifests as a gentle slope in spectra from one end of the spectrum to the other; this is more pronounced when many samples are measured concurrently, due to the longer overall read times for plates with more samples. In such cases, precision can be improved by estimating the effect of such baseline distortion by drawing a line between suitable points on each side of the 450-nm peak, and using this adjusted baseline to calculate the amplitude of the difference spectrum. Based on previous work, suitable values are 436 nm and 470 nm (see Note 14). Effectively, this correction reduces variation seen between replicate measurements on the same sample by modeling the baseline distortion as linear in the spectral region directly under the P450 difference peak. For rapid, high-throughput measurement, it is possible to make spectral reads only at 436, 450, and 470 nm. The formulae for calculating the P450 peak size from the single difference spectrum (or isolated reads) are as follows. Firstly, in expanded form:

$$\text{Amplitude of P450 peak} = \text{Abs}_{450} - \text{Abs}_{436} - ((450 - 436)/(470 - 436)) \times (\text{Abs}_{470} - \text{Abs}_{436})$$

where Abs_x is Absorbance at $x$nm on the Fe (II)·CO versus Fe(II) difference spectrum.

If different wavelength values for calculating the baseline correction appear more appropriate (e.g., due to alterations in the endogenous spectrum, see Note 14), they can be substituted in the formula above (by substituting the Abs_{436} and/or Abs_{470} value and corresponding 436 and/or 470 numerical value). If not, the formula can be simplified to:

$$\text{Amplitude of P450 peak} = \text{Abs}_{450} - \text{Abs}_{436} - 0.412 \times (\text{Abs}_{470} - \text{Abs}_{436})$$

The extinction coefficient ($\varepsilon = 91$ mM^{-1} cm^{-1}) is the same as before.

For high-throughput data processing, the software supplied with plate readers generally allows output of data in a spreadsheet-compatible format (e.g., .csv). It is straightforward to develop a standard spreadsheet template to which the actual data can be copied in the original format (including the appropriate absorbance readings at 436, 450, and 470 nm), and which can be used to calculate the size of the P450 peaks using the formulae above.

3.4 Substrate-Binding Spectra and Screening

The approach outlined above can also be applied to other ligands that bind in the vicinity of the P450 heme group, such as potential substrates and inhibitors that produce types I and II or reverse type I ligand-binding spectra. Type I interactions are characterized by a spectral peak at ~390 nm and trough at ~420 nm, whereas type II and reverse type I spectra show troughs at ~392 nm and peaks at ~425–435 nm (type II) and ~420 nm (reverse type I), respectively ((19) and references therein). In such cases, the results obtained are purely qualitative, since it is effectively impossible to quantify the concentration of ligand that reaches the P450 active site. Moreover, since some compounds are unable to enter the cell, high-throughput screens based on ligand-binding spectra recorded in whole cells are characterized by a high rate of false negatives. In addition, not all ligands of P450s elicit binding spectra. Nevertheless, such qualitative data are still useful as an initial screen to prioritize P450 forms or mutants that may interact with chemicals of interest, before detailed investigation of the P450-ligand interaction by other, more resource-intensive, means. Example spectra from CYP2C9 plus ibuprofen (type I), diclofenac (type I), and imidazole (type II) are provided in Fig. 3.

1. Transfer cell resuspensions in 200-µl aliquots into individual wells of a 96-well plate (see Note 9).

2. If relevant, add 20 µl of freshly prepared sodium dithionite (100 mg/ml) and agitate vigorously for 10 min in air (see Note 15).

3. Transfer the plate to the microplate reader and record a spectrum between 350 and 550 nm in 2-nm increments. Store this spectrum so that it can be used for background subtraction (see Notes 11 and 16).

4. Add 10–20 µl of a suitable concentration of the ligand of interest at high concentration (e.g., 100 µM, see Note 17) and incubate for 10 min. If ligands are present in a vehicle (e.g., dimethyl sulfoxide, DMSO), a vehicle control should be included.

5. Transfer plate to microplate reader and record the spectrum of the ligand-bound P450 between 350 and 550 nm in 2-nm increments (see Note 16).

6. Interpretations regarding the presence and nature of P450-ligand interactions can be made by observation of the existence and shape of any resultant difference spectrum (ligand bound minus baseline).

4 Notes

1. In the authors' laboratory, mammalian P450s were encoded as native or modified cDNA ORFs in bicistronic pCW'/P450/hCPR, where hCPR is human NADPH cytochrome P450 reductase (cognate electron donor). The *E. coli* used in expression was DH5αF'IQ (Invitrogen, Mulgrave Australia). A vector (pGro7) encoding folding chaperones GroES and

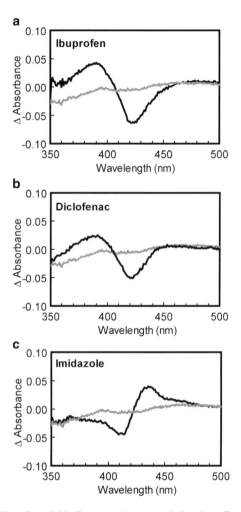

Fig. 3 Qualitative ligand-binding spectra recorded using *E. coli* expressing CYP2C9 incubated with (**a**) 100 μM ibuprofen; (**b**) 100 μM diclofenac; and (**c**) 100 μM imidazole. The vehicle control (10% DMSO) is plotted for all (*grey line*)

GroEl was co-expressed. pCW' and pGro7 required isopropyl-β-D-thiogalactopyranoside (IPTG) and arabinose for promoter induction, respectively, while maintenance of plasmids required ampicillin (100 μg/ml) and chloramphenicol (20 μg/ml), respectively. Further details on P450 growth and maintenance are provided in (9).

2. Measurement of difference spectra in whole cells is better done using a microplate reader rather than conventional cuvette spectrophotometer, due to the different orientation of the optical path. Intact cells settle rapidly under the influence of gravity in the concentrated cell suspensions used in the assay. In plate reader spectrophotometers the light path is vertical, while in cuvette-type spectrophotometers the light beam is horizontal. The vertical light path minimizes spectral artifacts resulting

from cells settling out of the light path and markedly improves the quality of data that can be obtained from whole-cell P450 spectra. However, even with the vertical light beam, variations in mixing samples should be minimized, via the use of thorough and consistent mixing protocols. A conventional horizontal-beam spectrophotometer may be appropriate if vigorous in-cuvette mixing during reads is possible, but has not been attempted by the authors.

3. Microaerobic growth occurs more slowly than aerobic, with a generally greater specific content of P450. The maximal P450 concentrations are typically achieved at approximately 72–96 h post inoculation. Under aerobic growth conditions, the maximal P450 concentrations are achieved at approximately 48–72 h post inoculation, and total P450 yields are generally greater than with microaerobic conditions. However, bacterial hemoproteins that absorb at ~417 and ~435 nm are more pronounced in spectra from aerobic cultures, and can interfere with the interpretation of spectra where P450 expression is poor. Moreover, under aerobic conditions P450 content can decrease rapidly after maximal expression is achieved, so it is harder to select a single optimal harvest time for screening (unpublished results).

4. It is important to ensure sufficient agitation to resuspend cells that have settled.

5. Individual well supernatants can be carefully removed by multichannel pipette. Decanting via careful inversion of the plate into a suitable receptacle and blotting with paper towel is also possible, although this carries some risk of cross contamination between wells.

6. The extent to which cells should be concentrated before assay depends on the expected P450 concentration and the precision required. For typical microaerobic cultures of a selection of human, drug-metabolizing P450s (CYP1A2, CYP2A6, CYP2A13, CYP2C9, CYP2C19, CYP3A4), culture density at harvest was OD600 ~8.0 and cells were concentrated two- to fourfold in WCAB. For typical cultures with low P450 concentrations (equivalent to ~100 nM in original cultures), the coefficient of variation was less than 15% (unpublished results). The sensitivity limit was considered to be ~50 nM in harvested cultures using these procedures. In general, the sensitivity limit can be decreased by further concentration of cells, and the degree to which cells are concentrated should be determined experimentally, based on the final cell concentration reached and the sensitivity of the particular plate reader used (i.e., to a sufficient extent that an acceptable degree of precision can be obtained between multiple determinations).

7. Care should be taken to minimize oxygenation of cell suspensions after dithionite addition for measurement of P450 concentrations in whole cells. This is because agitation and/or aeration of the cell resuspension after dithionite addition leads to oxidation of the P450 heme, the opposite effect to that intended, resulting in an underestimation of the total P450 pool (16). However, this counterintuitive effect can be used to increase the magnitude of substrate-binding spectra for screening enzyme-ligand interactions, which require the P450 heme to be in ferric form (see Subheading 3.4). Incubating cell resuspensions for 90 min after dithionite addition in stationary, multiwell plates with lids (after an initial mix to distribute dithionite) is sufficient to allow complete reduction of the P450 heme. Resuspensions can be incubated under argon to exclude oxygen, but this is generally not necessary, as the P450 contents measured with and without this additional step are not significantly different (Fig. 1a). It is important to use only freshly prepared sodium dithionite solution, as once made it rapidly reacts with oxygen in air.

8. Sodium dithionite undergoes successive reactions with oxygen to yield a cascade of reactions and products (20, 21). It is unknown which if any intermediates in the cascade are able to cross the cell membrane and reduce the P450s in whole cells; however, the ability of dithionite to reduce P450s within whole cells can be demonstrated by kinetic studies showing augmentation of the CO-$Fe(II)$ versus $Fe(II)$ difference spectrum observed following incubation of cells with dithionite for increasing lengths of time. Typically, the difference spectrum achieves a maximum within 30–90 min after dithionite addition to cell suspensions (Fig. 1a). We recommend 30 min as a typical dithionite incubation time, as the amplitude of the CO-$Fe(II)$ versus $Fe(II)$ difference spectrum decreases with longer incubation times with some P450 forms (unpublished data).

9. It is critical that a new plastic microplate is used for each incubation with CO. CO adsorbs to the plastic and will leach into any samples subsequently placed into any reused plates, leading to artefactually low readings.

10. Adequate mixing of the cell resuspensions, particularly immediately before recording of spectra, is critical for good-quality data. Unfortunately, the shaking function of microplate readers may not be sufficient to resuspend cells that have been allowed to settle. In the authors' experience, either individual mixing of samples in each well by repeated pipetting up and down, or vigorous mixing (800 rpm momentarily) in a shaker with a narrow gyratory radius (~3 mm; e.g., VorTemp™ 56 microplate shaker) can be used to resuspend settled cultures effectively.

Measurement of P450 Difference Spectra Using Intact Cells   201

11. The technique relies on sequential measurement of wavelength spectra before and after CO incubation, allowing calculation of a difference spectrum between Fe(II)- and Fe(II)·CO-bound forms of the P450. In traditional measurements of such difference spectra, the P450 is first reduced and the absorbance spectrum set to zero on this sample. Then, a baseline spectrum is recorded before incubation with CO, to check for any deviations from zero. This is critical for accurate quantification, as the P450 content of the sample is determined from the difference between the absorbance at 450 nm and 490 nm (an isosbestic wavelength) corrected for any difference between the same wavelengths in the baseline (Fe(II) only) spectrum; any deviations from zero in the baseline (e.g., due to instrument issues) will otherwise affect the accuracy of the measurement. An equivalent step (i.e., verification of the flatness or otherwise of the baseline) can be included in the current protocol by recording the baseline spectrum before the incubation with CO. This additional step may reveal a failure to satisfactorily resuspend the settled cells or other causes of spectral artifacts. If necessary the instrument can be set to zero again before incubation with CO and measurement of the difference spectrum. When large number of samples are to be measured, this additional step adds substantially to the overall time taken for the assay and the time the cells are exposed to dithionite. In the authors' experience, however, baselines are generally straight as long as cells are resuspended well, so this additional check can generally be omitted in the interests of high-throughput measurement of spectra. The slight slope that can be seen, due to settling of cells in experiments involving the analysis of many samples concurrently, is accounted for in the calculation presented above.

12. Extreme care must be taken in gassing plates with carbon monoxide (CO), and all manipulations should take place in a chemical fume hood that is regularly checked for correct operation. It is not appropriate to use high-pressure chambers for plate gassing, even within a working fume hood. A good alternative is to modify a typical air-tight "lunch box"-style container (of an appropriate size to fit several microplates) (11) with a connector and tubing to a three-way valve, with the valve further connected to the chamber, vacuum, and CO (Fig. 4). When closed, the lid of the chamber can be secured with two thick rubber bands. This forms a gas-tight vessel able to hold slight sequential underpressure (evacuation of chamber when switched to vacuum) and overpressure (filling of chamber switched to CO). The final stage outlet pressure of the CO cylinder should be set to just above atmospheric pressure, and it is good practice to also incorporate a needle flow valve in

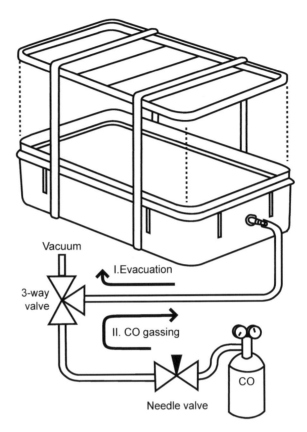

Fig. 4 Schematic view of the adapted plate-gassing system used for Fe(II)·CO versus Fe(II) difference spectra. The vessel itself is a lunch-box-style container with a hole bored in one sidewall into which a barbed tubing connector has been inserted and affixed using epoxy glue. The barbed connection is plumbed via silicon or similar tubing to a small three-way valve (medical drip bag type or similar). This allows sequential chamber evacuation via underpressure, followed by slight CO overpressure (for plate incubation, controlled carefully using a needle valve). It is critical that the device be used only in a working fume hood, as leaks of small amounts of CO are likely. The cylinder CO pressure should be set to the minimum value at which the regulator allows CO flow (<5 psi is recommended). In case of inadvertent overpressurization of the CO chamber, it is useful to constrain the lid to avoid the container opening suddenly at relatively low pressures. A suitable method is to place several elastic bands around the chamber and lid. Importantly, such a system will still not allow high overpressurization and possible sudden CO release into the fume hood, as the chamber lid will open slightly and vent before this occurs

series (to restrict the maximum CO flow to a low level). Such a system does not allow high overpressure and possible sudden CO release into the fume hood, as the chamber lid will open slightly and vent before this occurs. The pressurization with CO can be confirmed visually from the slight outward deformation of the gassing chamber lid.

13. For plate readers, it is necessary to derive the path length (cm) for quantification of P450 concentration via the extinction coefficient (ε=91 mM^{-1} cm^{-1}). For total volume 220 µl in a 96-well microplate, the path length is ~0.5 cm, but will vary according to microplate geometry. Automatic path-length correction, such as that the PathCheck option provided in Spectramax (Molecular Devices, Sunnyvale, CA) instruments, is not recommended, as such readings are affected by turbidity in concentrated whole-cell suspensions (22). An alternative is to derive the microplate path length by measuring the absorbance of a solution of an absorbing dye in WCAB (at an appropriate wavelength for the dye) in the plate reader/microplate to be used (with desired fill volume) alongside a conventional 1-cm path-length cuvette spectrophotometer. The ratio of absorbances between microplate and cuvette can be used to derive the microplate path length. Some minor path-length error, due to increased viscosity of the concentrated cell suspensions (altering surface meniscus shape in microplate), can be expected.

14. Originally an appropriate set of wavelengths for the baseline correction was proposed to be 438 nm/470 nm (13); however, subsequently 436 nm/470 nm was used in the authors' laboratory and these values are recommended. If desired, the baseline-correction wavelengths can be refined empirically for a particular set of duplicate test spectra (n=5 or greater), as outlined previously (13).

15. For most ligands of interest, the difference spectrum is more significant for the Fe(III) form of the P450. However, as noted above (in Subheading 3.2) the P450 is typically predominantly found in the reduced form within the intact cell environment (16). Although a ligand-binding spectrum may still be seen, it may be advantageous to convert the Fe(II) form to Fe(III) to improve the sensitivity. This can be done by vigorously aerating the P450 in the presence of dithionite (16) (see also Note 7). Optimization experiments have indicated that with 900 rpm agitation in air with sodium dithionite (10 mg/ml), the amplitude of the type I spectrum reaches a plateau at approximately 10 min (unpublished data). Conversely, for ligands that bind predominantly to the reduced form of the P450, it may be useful to add dithionite without the agitation step.

16. The wavelength range should be adjusted according to the type of spectrum expected.

17. At the time of writing, whole ligand-binding spectra have been observed for human CYP2C9 (ibuprofen, diclofenac, hexobarbital, and imidazole), CYP1A2 (phenacetin and imidazole), CYP3A4 (midazolam), as well as bacterial CYP101A1 (camphor) and CYP176A1 (cineole) ((16), unpublished data).

References

1. Abécassis V, Pompon D, Truan G (2000) High efficiency family shuffling based on multi-step PCR and in vivo DNA recombination in yeast: statistical and functional analysis of a combinatorial library between human cytochrome P450 1A1 and 1A2. Nucleic Acids Res 28:e88
2. Landwehr M, Carbone M, Otey CR, Li Y, Arnold FH (2007) Diversification of catalytic function in a synthetic family of chimeric cytochrome P450s. Chem Biol 14:237–238
3. Meinhold P, Peters MW, Hartwick A, Hernandez AR, Arnold FH (2006) Engineering cytochrome P450BM3 for terminal alkane hydroxylation. Adv Synth Catal 348:763–772
4. Sawayama AM, Chen MMY, Kulanthaivel P, Kuo MS, Hemmerle H, Arnold FH (2009) A panel of cytochrome P450 BM3 variants to produce drug metabolites and diversify lead compounds. Chemistry 15:11723–11729
5. Glieder A, Farinas ET, Arnold FH (2002) Laboratory evolution of a soluble, self-sufficient, highly active alkane hydroxylase. Nat Biotechnol 20:1135–1139
6. Parikh A, Josephy PD, Guengerich FP (1999) Selection and characterization of human cytochrome P450 1A2 mutants with altered catalytic properties. Biochemistry 38:5283–5289
7. Kim D, Guengerich FP (2004) Enhancement of 7-methoxyresorufin O-demethylation activity of human cytochrome P450 1A2 by molecular breeding. Arch Biochem Biophys 432:102–108
8. Johnston WA, Huang W, De Voss JJ, Hayes MA, Gillam EMJ (2007) A shuffled CYP1A library shows both structural integrity and functional diversity. Drug Metab Dispos 35:2177–2185
9. Huang W, Johnston WA, Hayes MA, De Voss JJ, Gillam EMJ (2007) A shuffled CYP2C library with a high degree of structural integrity and functional versatility. Arch Biochem Biophys 467:193–205
10. Hunter DJB et al (2011) Facile production of minor metabolites for drug development using a CYP3A shuffled library. Metab Eng 13:682–693
11. Otey CR (2003) High-throughput carbon monoxide binding assay for cytochromes P450. In: Arnold FH, Georgiou G (eds) Methods in molecular biology, vol 230. Humana, Totowa, NJ, pp 137–139
12. Lawrence JV, Maier S (1977) Correction for the inherent error in optical density readings. Appl Environ Microbiol 33:482–484
13. Johnston WA, Huang W, Hayes MA, De Voss JJ, Gillam EMJ (2008) Quantitative whole cell cytochrome P450 measurement suitable for high throughput application. J Biomol Screen 13:135–141
14. Sambrook J, Russell DW (2001) Molecular cloning: a laboratory manual. Cold Spring Harbor Laboratory Press, Cold Spring Harbor, NY
15. Bauer S, Shiloach J (1974) Maximal exponential growth rate and yield of E. coli obtainable in a bench-scale fermentor. Biotechnol Bioeng 16:933–941
16. Johnston WA et al (2011) Cytochrome P450 is present in both ferrous and ferric forms in the resting state within intact Escherichia coli and hepatocytes. J Biol Chem 286:40750–40759
17. Omura T, Sato R (1964) The carbon monoxide-binding pigment of liver microsomes. I. Evidence for its hemoprotein nature. J Biol Chem 239:2370–2378
18. Guengerich FP (1994) Analysis and characterization of enzymes. In: Hayes AW (ed) Principles and methods of toxicology. Raven, New York, pp 1259–1313
19. Schenkman JB, Cinti DL, Orrenius S, Moldeus P, Kraschnitz R (1972) The nature of the reverse type I (modified type II) spectral change in liver microsomes. Biochemistry 11:4243–4251
20. Lambeth DO, Palmer G (1973) The kinetics and mechanism of reduction of electron transfer proteins and other compounds of biological interest by dithionite. J Biol Chem 248:6095–6103
21. Dixon M (1971) Acceptor specificity of flavins and flavoproteins.1. Techniques for anaerobic spectrophotometry. Biochim Biophys Acta 226:241–258
22. SpectraMax® M2/M2e user guide, Molecular Devices, Sunnyvale, CA, USA

Chapter 18

DNA Shuffling of Cytochromes P450 for Indigoid Pigment Production

Nedeljka N. Rosic

Abstract

DNA family shuffling is a powerful method of directed evolution applied for the generation of novel enzymes with the aim of improving their existing features or creating completely new enzyme properties. This method of evolution in vitro requires parental sequences containing a high level of sequence similarity, such as is found in family members of cytochrome P450 enzymes. Cytochromes P450 (P450s or CYPs) are capable of catalyzing a variety of chemical reactions and generating a wide range of products including dye production (e.g., pigments indigo and indirubin). Application of the method of DNA family shuffling described here has enabled us to create novel P450 enzymes and to further extend the capacity of P450 to oxidize indole, leading to pigment formation.

Key words Directed evolution, DNA family shuffling, CYP, Molecular breeding, Indole metabolism, Indigo

1 Introduction

Cytochromes P450 are hemoproteins described as monooxygenase enzymes as they typically catalyze the inclusion of a single atom of oxygen into a substrate. Because of their broad substrate specificity, P450s are recognized as an ideal starting material for directed evolution and the development of novel enzymes with improved catalytic properties (for more information see recent reviews (1–4)). Directed evolution methods mimic the natural evolutionary processes of mutagenesis, recombination, and selection in vitro. For the success of the procedure there are two basic requirements: first, a successful method for the generation in vitro of molecular diversity and second, an efficient method for the identification of the improved variants. As some mammalian P450s are able to oxidize indole and produce indigoid pigments when expressed within *Escherichia coli* together with their redox partner, reduced nicotinamide adenine dinucleotide phosphate (NADPH)-cytochrome P450 reductase (5), we aimed to further extend the capacity of

Ian R. Phillips et al. (eds.), *Cytochrome P450 Protocols*, Methods in Molecular Biology, vol. 987,
DOI 10.1007/978-1-62703-321-3_18, © Springer Science+Business Media New York 2013

P450s for pigment formation by directed evolution applying a modified DNA family shuffling procedure (6, 7). Previously, random mutagenesis was used on CYP2A6 by Guengerich and coworkers (8), confirming the capacity of P450 enzymes in the development of novel dyes. The DNA family shuffling method follows the basic principles of Stemmer's original DNA shuffling method (9, 10) and the established DNA family shuffling method for recombination between homologous genes (11, 12). The procedure used for the construction of the CYP shuffled library here was adapted from a method described by Abecassis (13). We applied a modified DNA family shuffling methodology to mammalian P450 enzymes from the 2C subfamily that share more than 80% amino acid sequence identity (1, 14) (see Note 35). This P450 subfamily has a broad pharmacological importance as one of the most complex found in humans (15) and, consequently, holds a great potential for future biotechnological applications. The modification of traditional DNA family shuffling protocols that was employed here involved the application of a nonrandom digestion using different sets of the restriction enzymes (REs) (16) followed by isolation of DNA fragments under 300 bp, by size-selective filtration, to avoid the recovery of non-shuffled parental sequences (6). Three mammalian P450 genes (*CYP2C9*, *CYP2C11*, and *CYP2C19*) were initially fragmented by a set of restriction enzymes, then reassembled in a primerless polymerase chain reaction (PCR), because highly similar fragments acted as primers for each other, and, finally, the shuffled P450s were amplified by a PCR using flanking primers (Fig. 1). The modified family shuffling technique applied to parental P450 2C sequences resulted in an improved level of chimeragenesis in the areas of low sequence similarity, such as six substrate-recognition sites (SRSs). Using this approach, a biased recombination towards more homologous regions when using random fragmentation (17) was overcome by the application of nonrandom digestion and restriction enzymes (14). The resultant shuffled P450 library was characterized with a high proportion of functional mutants and lack of parental "contamination" within the shuffled progeny (6, 14). Application of this modified DNA shuffling procedure, originally created and successfully tested on three CYP2C family members (6, 7, 14), was later applied for the creation of the additional CYP2C (18) and CYP1A (19) shuffled libraries, confirming its utility and efficiency for the generation of novel P450s.

2 Materials

Prepare all buffers and solutions using water purified by reverse osmosis, via the MQ Ultrapure Water System ("Milli-Q water"). Perform sterilization by filtration or autoclaving at 120°C, with 1.36 atm (20 psi), for 30 min.

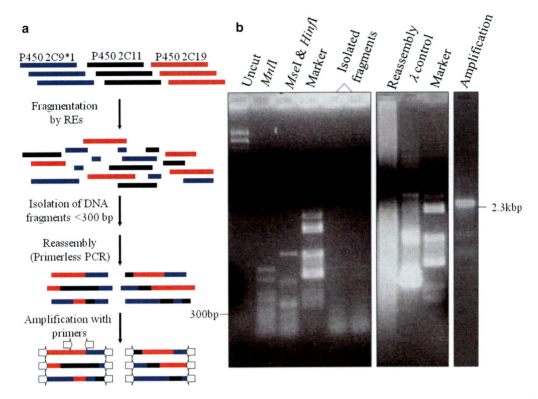

Fig. 1 A DNA family shuffling method used for the generation of novel P450s. (**a**) The schematic overview of the modified DNA family shuffling procedure applied here. P450 genes are initially fragmented by a set of restriction enzymes; then reassembled by a primerless PCR, in which homologous DNA fragments act as primers for each other; and finally amplified in a PCR using flanking primers. Electrophoresis through a 1% (w/v) agarose gel of DNA shuffling products. (**b**) Non-digested mixture of P450s 2C9, 2C11, and 2C19 plasmid DNA (bicistronic expression constructs—pCW'/P450/hNPR) is in *lane 1*, P450 2C digested with two sets of REs are in *lanes 2* (the *Mnl*I digest) and *3* (the *Mse*I and *Hinf*I digest), the molecular weight marker is in *lane 4*, and fragments under 300 bp isolated with Microcon PCR filter units is in *lane 5*. Primerless PCR (PCR without primers) was performed with DNA fragments (10.5 μg of DNA fragments from both digests) smaller than 300 bp. The product of a primerless PCR is in *lane 6*; lambda positive control is in *lane 7*, with the expected 500-bp band; and DNA marker is in *lane 8*. Finally, the PCR-amplified CYP2C shuffled library using 1/100 of primerless PCR product as a template is in *lane 9*

2.1 Bacterial Strains and Plasmids

1. Bacterial strains *E. coli* DH5αF'IQ™ and JM110 can be purchased from GIBCO-BRL and Invitrogen, respectively.

2. The original expression vector, pCW, was obtained from Prof. F. W. Dalquist (University of Oregon, Eugene, OR, USA). The pCW' vector was created by removal of the *CheW* gene from the pCW vector and used for preparation of all other expression plasmids used in this study.

3. Bicistronic constructs for the concurrent expression of P450s and their redox partner human NADPH-cytochrome P450 reductase (hNPR) were labelled as pCW'/P450 form/hNPR, e.g., the bicistronic construct for *CYP2C9* is pCW'/2C9/hNPR.

208   Nedeljka N. Rosic

They were prepared in the Gillam or Guengerich laboratories as previously reported in the listed studies: P450 3A4 (20), P450 2A6 (5), P450 2C9 (wild type, P450 2C9*1), P450 2C11, and P450 2C19 (21). The monocistronic expression vector pCW'/hNPR was generated in the laboratory of Prof. F. P. Guengerich (Vanderbilt University, Nashville, TN, USA).

2.2 Chemicals and Reagents for Bacterial Expression

1. *LBglu medium*: Bactotryptone (10 g/L), yeast extract (5 g/L), 0.2 N NaCl, adjust to pH 7 with NaOH and autoclave before addition of 0.2% (w/v) glucose (see Note 1).

2. *SOB medium*: Bactotryptone (20 g/L), yeast extract (5 g/L), NaCl (0.5 g/L), 2.5 mM KCl, at pH 6.7–7.0, and add 10 mM $MgCl_2$ and $MgSO_4$ before use. Store at room temperature.

3. *SOC medium*: Bactotryptone (20 g/L), yeast extract (5 g/L), NaCl (0.5 g/L), 2.5 mM KCl, 20 mM glucose, pH 6.7–7.0, and add 10 mM $MgCl_2$ and $MgSO_4$ before use (see Note 2).

4. *TB* (Terrific Broth) *medium* for P450 expression in *E. coli*: Bactotryptone (12 g/L), yeast extract (24 g/L), bactopeptone (2 g/L), and glycerol (4 mL/L) in 900 mL of Milli-Q water and autoclave. Supplement TB medium just before use with 100 mL of potassium phosphate buffer (KPi) for TB (see Note 3), ampicillin (100 µg/mL), 1.0 mM IPTG, 1.0 mM thiamine, 0.5 mM δ-aminolevulenic acid (δ-ALA), and trace elements (250 µL/L). (Trace elements contain $FeCl_3$ (27 g/L), $ZnCl_2$ (2 g/L), $CoCl_2$ (2 g/L), $NaMoO_4$ (2 g/L), $CoCl_2$ (1 g/L), $CuCl_2$ (1 g/L), H_3BO_3 (0.5 g/L), and concentrated HCl (100 mL/L) (22).) See Notes 4 and 7–11.

5. *TB solution*: 10 mM Pipes, 15 mM $CaCl_2$, 250 mM KCl, adjust to pH 6.7 with 5 N KOH before adding $MnCl_2$ to a final concentration of 55 mM. Store at 4°C.

6. *Protease inhibitors*: 1.0 mM phenylmethylsulfonyl fluoride (PMSF, see Note 5), 2 µM leupeptin, 0.04 U/mL aprotinin, and 1.0 µM bestatin (see Notes 6 and 13).

7. *Screening on solid medium* for pigment formation: Apply the transformation mixture (23) to LB agar medium containing ampicillin (100 µg/mL), chloramphenicol (20 µg/mL), L-(+)-arabinose (4 mg/mL), 0.5 mM δ-ALA, 1.0 mM IPTG, trace elements, and 1.0 mM thiamine (see Notes 7–11).

8. *Sonication buffer* for harvesting of bacterial cells: 100 mM potassium phosphate buffer (pH 7.6), 6 mM magnesium acetate, 20% (v/v) glycerol, and 0.10 mM DL-dithiothreitol (DTT, see Note 12). Store at 4°C.

9. *TES (1×)* for membrane fractions: 50 mM Tris–acetate buffer (pH 7.6), containing 250 mM sucrose and 0.25 mM EDTA. Store at 4°C (see Note 13).

10. *TES (2×)* for harvesting of bacterial cells: 100 mM Tris–acetate buffer (pH 7.6), containing 500 mM sucrose and 0.5 mM EDTA. Store at 4°C (see Note 13).

11. *NADPH-generating system*: 250 µM β-nicotinamide adenine dinucleotide phosphate (NADP$^+$), 10 mM glucose-6-phosphate (G-6-P), and glucose-6-phosphate dehydrogenase (G-6-P DH) (0.5 U/mL) (see Note 14).

12. *Spectral assay buffer* for P450 determination: 100 mM potassium phosphate (pH 7.4), containing 20% (v/v) glycerol and 0.20% (w/v) Emulgen 913. Store at 4°C (see Note 15).

13. *Phosphate-buffered saline* (PBS): 10 mM potassium phosphate buffer, pH 7.4, supplemented with 0.9% (w/v) NaCl. Store at 4°C.

14. Chemicals: δ-aminolevulinic acid (δ-ALA); L-(+)-arabinose; bestatin; bovine serum albumin (fraction V, BSA) (see Note 16) (DTT); glucose-6-phosphate (G-6-P); horse heart cytochrome *c*; leupeptin; β-nicotinamide adenine dinucleotide phosphate (NADP$^+$) (see Note 11); reduced β-nicotinamide adenine dinucleotide phosphate (NADPH) (see Note 17); PMSF (see Note 5); and thiamine can be purchased from Sigma.

    Ampicillin, chloramphenicol, and isopropyl β-D-thiogalactopyranoside (IPTG) can be purchased from Sigma-Aldrich.

15. *BCA protein determination kit* can be obtained from Pierce Chemical Company.

16. *Restriction enzymes*, T$_4$ DNA ligase, and other enzymes can be obtained from New England Biolabs and used as recommended by manufacturer. Store at −20°C.

17. Primary polyclonal *antibodies* raised in rabbit against P450 2C19 were previously prepared in the Gillam lab. Secondary goat anti-rabbit IgG (H + L) antibody can be obtained from Amersham Biosciences Corp.

18. *Agarose gel-loading buffer*: 5% (w/v) sucrose, 0.2% (w/v) SDS, 5 mM Tris–HCl, pH 8.0, and 2 mM EDTA with 0.05% (w/v) bromophenol blue, and/or 0.05% (w/v) xylene cyanol (see Note 18).

19. *Chemicals for Pigment Production*: Indole, 5-aminoindole (5-AI), 5-bromoindole (5-BI), 5-fluoroindole (5-FI), 5-cyanoindole (5-CI), 5-nitroindole (5-NI), 5-methylindole (5-MI), 5-methoxyindole (5-MXI), 6-chloroindole (6-CHI), 5-hydroxyindole (5-HI), 5-hydroxyindole-3-acetic acid (5-HIAA), and 2-methylindole (2-MI). Molecular and structural formulae for indole and indole derivatives are presented in Fig. 2. Prepare indole stocks and other indole derivatives at concentrations of 10 or 33 mM (see also Note 19).

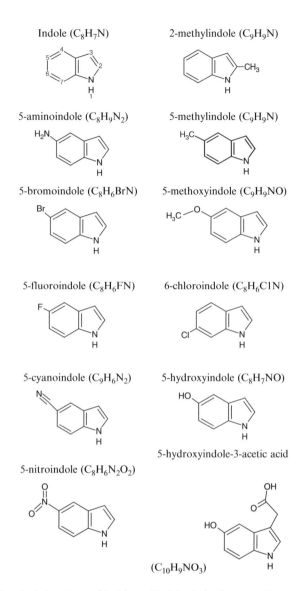

Fig. 2 Chemical structures of indole and indole derivatives used here

2.3 Oligonucleotide Primers

1. Lambda phage DNA and lambda primers for PCR amplification correspond to the lambda positive control for PCR as recommended by Perkin Elmer.

2. The oligonucleotide primers used for this work are presented below (Table 1).

3. Mutants from the CYP2C shuffled library (named 2Cs20, 2Cs48, 2Cs53, 2Cs1B, and 2Cs2B) are obtained after the first round of shuffling.

Shuffled CYPs for Dye Production   211

Table 1
Oligonucleotide primers used in this project

Primer name	Sequence
Flanking/forward	5′ GGAAACAGGATCCATCGATGCTTAGGAGGT<u>CATATG</u> 3′
Flanking/reverse	5′ GCCGCATCTATGTCGGTGTGGACCACAAGC 3′
2Cseq#1	5′ ATGGAAGGAGATCCGGCGTTTC 3′
2Cseq#2 (antisense)	5′ GAAACGCCGGATCTCCTTCCAT 3′
2Cseq#3	5′ CTGGGCTGTGCTCCCTGCAATGT 3′
2Cseq#4 (antisense)	5′ ACATTGCAGGGAGCACAGCCCAG 3′
2Cseq#5	5′ GTCACAGCTAAAGTCCAGGAAGAGAT 3′
2Cseq#6 (antisense)	5′ ATCTCTTCCTGGACTTTAGCTGTGAC 3′
2Cseq#7	5′ TACTTYATGCCWTTCTCAGCAGGAAA 3′
2Cseq#8 (antisense)	5′ TTTCCTGCTGAGAAWGGCATRAAGTA 3′
2Cseq#9	5′ ATTTTRCAGAAYTTTAACCTGAARTCT 3′
2Cseq#10 (antisense)	5′ AGAYTTCAGGTTAAARTTCTGYAAAAT 3′
Lambda sense	5′ GATGAGTTCGTGTCCGTACAACTGG 3′
Lambda antisense	5′ GGTTATCGAAATCAGCCACAGCGCC 3′

The *Nde*I restriction site is underlined

3 Methods

3.1 Construction of a CYP2C Shuffled Library

1. Perform the DNA shuffling procedure (see Fig. 1) using cDNAs of *CYP2C9*, *CYP2C11*, and *CYP2C19* in the format of bicistronic plasmids (pCW'/P450/hNPR). Digest equal amounts of parental DNA (100 μg of total DNA) with *Mnl*I in one mixture and *Mse*I–*Hinf*I in a second mixture, for 2 h at 37°C. The fragments produced by *Mnl*I or *Mse*I and *Hinf*I digestion of parental P450 cDNAs are listed in Table 2, and the positions of *Mnl*I or *Mse*I and *Hinf*I restriction sites within the P450 sequences are presented in Fig. 3. Isolate DNA fragments of less than ~300 bp using Microcon PCR filter units (Millipore), then precipitate with 0.1 volumes of 3 M sodium acetate and 2.5 volumes of 100% ethanol and resuspend in sterile Milli-Q water (100 μL).

2. In the second step, do a reassembly of full-length coding sequences by PCR without primers (primerless PCR) in a reaction mixture (100 μL) consisted of 10.5 μg of DNA fragments from each restriction digest, 0.2 mM dNTPs, 2.2 mM MgCl$_2$, 2.5 U recombinant *Taq* polymerase (Invitrogen), and buffer (20 mM Tris–HCl, pH 8.4, 50 mM KCl). Use the following

Table 2
List of DNA fragments (bp) produced with *Mnl*I or *Mse*I and *Hinf*I digestion of parental P450 cDNAs (*CYP2C9*, *CYP2C11*, and *CYP2C19*) used in the format of bicistronic expression vector (pCW'/P450/hNPR)

pCW'/P450/hNPR	*Mnl*I	*Mse*I and *Hinf*I
CYP2C9 (1,417 bp)	3, 3, 3, 3, 3, 6, 8, 9, **10**, **15**, **15**, **16**, **16**, 18, 20, **21**, 24, 27, **28**, 29, 30, **30**, **33**, **39**, 39, 43, 45, 49, 56, 68, 70, 72, 74, **75**, **81**, 81, 81, 103, **108**, 117, 121, **124**, 143, 147, **148**, **154**, **157**, **182**, 182, 202, 206, 209, 213, 250, 264, 266, 267, **296**, 302, 324, 400, 415, 483, 662, 798	5, 5, **5**, 6, 9, 11, 11, 12, 13, 14, 17, 18, **20**, 25, 28, 36, 39, **41**, 52, 53, **56**, 57, 59, 59, 67, 75, 77, 80, 81, 88, 89, **94**, 95, 111, **111**, **112**, 122, 122, 123, 127, 153, **164**, **173**, 175, 176, 194, 211, 246, 258, 268, 281, 334, 355, **364**, 365, 396, **424**, 631, 1093
CYP2C11 (1,447 bp)	**3**, 3, 3, 3, 3, 3, 6, **8**, 8, 9, 10, **15**, 15, **16**, 18, 18, 20, **21**, 24, 27, **28**, 29, 30, **30**, 39, 43, 45, **49**, 49, 56, 68, 70, 72, 74, 81, 81, 103, **103**, 117, 121, **123**, **131**, 143, 147, **154**, **154**, **159**, 182, **198**, 202, 206, 209, 213, 250, 264, 266, 267, 302, 324, **327**, 400, 415, 483, 662, 798	5, 5, 6, 9, 11, 11, 12, 13, 14, **15**, 17, 18, **18**, 25, 28, **32**, 36, 39, 52, 53, **56**, 57, 59, 59, 65, 67, 75, 77, 80, 81, 88, 89, 95, **99**, 111, 111, **113**, 122, 122, 123, 127, 153, **169**, 175, 176, 194, 211, **220**, 246, 258, 268, 281, **330**, 334, **350**, 355, 365, 396, 631, 1093
CYP2C19 (1,417 bp)	3, 3, 3, 3, 3, 6, 8, 9, **10**, **15**, **15**, **16**, **16**, 18, 20, **21**, 24, 27, **28**, 29, 30, 30, **33**, **33**, 37, 39, 43, 45, 49, **55**, 56, **57**, 68, 70, 70, 72, 74, **75**, 81, 81, 103, **108**, 117, 121, 143, 147, **154**, **157**, 182, 202, 206, 209, 213, 250, 264, 266, 267, **272**, 302, 324, **335**, 400, 415, 483, 662, 798	5, 5, 6, 9, 11, 11, 12, 13, 14, 17, 18, **20**, 25, 28, 36, 39, **41**, 52, 53, **56**, 57, 59, 59, 67, 75, 77, 80, 81, 88, 89, **94**, 95, **111**, 111, **117**, 122, 122, 123, 127, 153, 153, **172**, **173**, 175, 176, 194, 211, 246, **252**, 258, 268, 281, 334, 355, **364**, 365, 396, 631, 1093

In the table are listed fragments obtained, where fragments within the P450 sequence are in bold letters

conditions for PCR amplification: 1 min of denaturation at 94°C, then 35 cycles of 30 s at 94°C, annealing for 13 min 30 s, during which time gradually decrease the temperature from 65 to 41°C, followed by elongation for 1 min 30 s at 72°C, and finally a 7-min step at 72°C.

3. In the final stage, perform a PCR with flanking primers (Table 1) to multiply the established P450 2C library, with an expected amplification product size of 2.2 kbp (Fig. 1b). For a 100-μL reaction mixture use 1/100 product from the primerless PCR (i.e., 1 μL of the 100 μL reaction), 0.25 μM forward and reverse primers (Table 1), 0.2 mM dNTPs, 2.2 mM $MgCl_2$, 2.5 U *Taq* polymerase (recombinant, Invitrogen), and buffer (20 mM Tris–HCl, pH 8.4, 50 mM KCl). Use the following PCR program: 1 min of denaturation at 94°C, 10 cycles

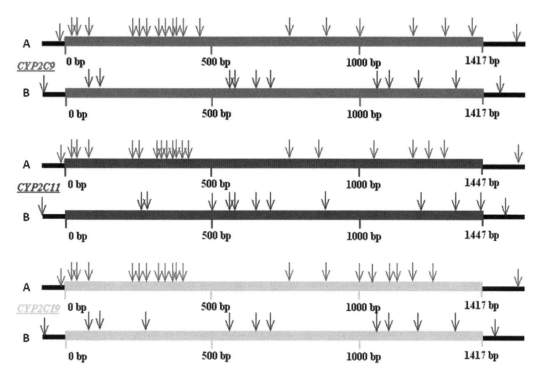

Fig. 3 Scheme illustrating the *Mnl*I (*A*) and *Mse*I and *Hinf*I (*B*) restriction sites within the P450 sequences (*CYP2C9*, *CYP2C11*, and *CYP2C19*) used for performing the fragmentation step of the DNA shuffling. The restriction sites bordering P450-coding sequence are also presented

of denaturation at 94°C for 30 s, annealing at 55°C for 30 s, and elongation at 72°C for 1 min 30 s, followed by 25 cycles of 30 s at 94°C, 30 s at 55°C, and 1 min 30 s at 72°C, with a 5-s extension per cycle, and finally a 7-min step at 72°C.

3.2 Subcloning and Archiving of Amplified Chimeric genes

1. Digest the PCR product of the amplified 2C shuffled library (Fig. 1b) with *Nde*I and *Bsm*I, extract with phenol–chloroform and chloroform as explained under second phenol–chloroform extraction (see ref. 24 and Notes 20 and 21), then precipitate with 0.5 volumes of 7.5 M ammonium acetate and 2.5 volumes of 100% ethanol, and resuspend in Milli-Q water.

2. Digest the pCW'/3A4/hNPR expression vector with *Nde*I and *Bsm*I; separate products on a 1% (w/v) agarose gel (see Note 18) and purify the desired DNA fragment (~6.2 kb) using the Geneclean procedure (Geneclean® III kit) or QIAquick gel extraction kit (Qiagen). Sub-clone the digested PCR product into the cognate sites of the pCW'/3A4/hNPR expression vector (20) by replacing the *Nde*I–*Bsm*I fragment of the P450 3A4 cDNA. Perform the ligation using T4 DNA Ligase according to the manufacturer's recommendations (New England Biolabs), with addition of 1 mM ATP to the reaction (see Note 22).

In parallel, prepare two negative controls containing only insert or vector fragments.

3. Use the CYP2C library ligation mixture to transform competent *E. coli* DH5α F'IQ following the method of Inoue et al. (*23*), and then apply transformation mixtures to LB agar containing ampicillin (100 μg/mL) followed by incubation overnight at 37°C (see Notes 23 and 24). Randomly select around 50 or more bacterial colonies for further characterization of the resultant shuffled library. Isolate the plasmid DNA using the alkaline lysis method or prepared kits (see Notes 18, 25, and 34).

4. Assess the quality of isolated plasmid DNA on 1% (w/v) agarose gel (see Note 18) and additionally by diagnostic digestion following the recommendations of the manufacturer of the restriction enzymes (New England Biolabs). Measure the DNA quality and DNA quantity from the ratio of absorbance at 260 nm to that at 280 nm (see Note 26).

5. Two approaches can be used for archiving the newly produced P450 forms. First, use the CYP2C library ligation mixture to transform competent *E. coli* DH5α F'IQ cells, pre-transformed with the chaperone coexpression plasmid pGro7 (see ref. 25). Apply the transformation mixture to LBglu agar plates supplemented with ampicillin (100 μg/mL) and chloramphenicol (20 μg/mL) and incubate overnight at 37°C (see Note 27). In the second method, use individual colonies to inoculate 2× LBamp/cam medium in a 1-mL 96-well microtiter plate (see Note 28).

3.3 Analysis of the CYP2C Shuffled Library

1. Perform restriction analyses according to the enzyme manufacturer's recommendations (New England Biolabs). Digest the shuffled mutants (e.g., we digested 54 from ~200 bacterial colonies) and their parents (P450s 2C9, 2C11, and 2C19) using *Apo*I, *Ear*I, and *Bst*YI (Fig. 1b). Perform an additional digestion using *Nde*I and *Bsm*I (i.e., the cloning sites). The observed differences in the restriction fragment length patterns should suggest that the library is free from parental contamination and highly diverse.

2. *Bacterial expression and analysis of recombinant proteins.* Express the selected mutants from the shuffled library in *E. coli* (21, 26) with and without chaperone co-expression (GroES and GroEL chaperones, in plasmid pGro7), as described below (see Notes 4 and 29). Also express parental P450s (2C9*1, 2C11, and 2C19) and P450 2A6, using the bicistronic expression vector pCW'/P450/hNPR, whereas for expression of a negative control hNPR use a monocistronic expression vector (pCW'/hNPR). Harvest cells after 48 h incubation, then prepare S10,000 (from 10,000 ×g centrifugation) and membrane

Shuffled CYPs for Dye Production 215

fractions following previously described procedure (see ref. 27 and Note 13). Estimate the total amount of protein in membrane fractions, using the BCA assay (Pierce; see Note 16).

3. Cytochrome P450 determination: Assess P450 hemoprotein expression by Fe(II) CO versus Fe(II) difference spectroscopy. Measure P450 content of S10,000 fractions and of membrane fractions. Use reduced CO (Fe^{2+}-CO) versus reduced (Fe^{2+}) difference spectra according to the method of Omura and Sato (1964) for quantification of P450 content in subcellular fractions of *E. coli* (S10,000 and/or membrane fractions) diluted in spectral assay buffer, at room temperature (see Note 30).

4. Estimate apoprotein expression by immunoblotting different cellular fractions isolated from 3 mg wet weight of the bacterial cells (27). Separate proteins by electrophoresis on SDS 8.3% (w/v) polyacrylamide gels and transfer to nitrocellulose. Incubate the blots with a 1/1,000 dilution of primary antibodies raised in rabbit against α CYP 2C19, followed by incubation with the same dilution of goat anti-rabbit IgG (H+L) (see Note 31). Develop the blots using 4-chloro-1-naphthol as a chromogen as previously described (see ref. 28 and Note 32).

3.4 Screening of the CYP2C Library for Indigoid Pigment Formation

1. *Screening for Pigment Formation in Liquid Cultures*

Estimate indigoid pigment production by visible spectroscopy, in the range of 400–700 nm, using ethyl acetate extracts from bacterial cultures (29). Incubate bacterial cultures under conditions described above and sample after a minimum of 48 h incubation (26). Extract 1 mL of bacterial culture twice with a total of 1 mL of ethyl acetate to isolate indigoid pigments (Fig. 4). Estimate the concentrations of indigo and indirubin with reference to a standard curve of synthetic indigo (λ_{max} = 598 nm) and indirubin (λ_{max} = 522 nm) dissolved in ethyl acetate in the concentration range of 0–10 mg/L (see Notes 33 and 34).

2. *Screening for Pigment Formation on Solid Medium*

Perform screening for mutant colonies with enhanced indole metabolism on plates using a modified protocol previously established by Nakamura et al. (8). Introduce library plasmid DNA into competent *E. coli* DH5α F'IQ, co-transformed with the chaperone expression vector pGro7 (26), following the method of Inoue et al. (23). Apply the transformation mixture to LB agar medium containing ampicillin (100 μg/mL), chloramphenicol (20 μg/mL), arabinose (4 mg/mL), 0.5 mM δ-aminolevulinic acid, 1.0 mM IPTG, trace elements, and 1.0 mM thiamine. Incubate plates for 5 days at 26°C. Select the colonies that develop a blue color for further evaluation of hemoprotein expression and indigo production.

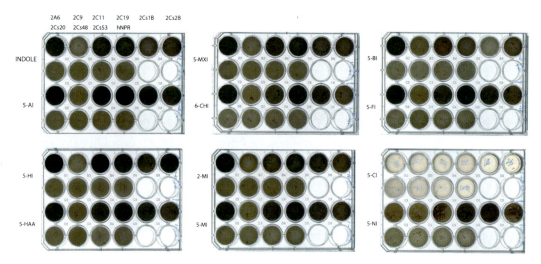

Fig. 4 Bacterial cultures of CYP2A6, CYP2C9, CYP2C11, CYP2C19, and the CYP2C mutants, using bicistronic expression vectors (pCW'/P450/hNPR), and hNPR, using monocistronic expression vector (pCW'/hNPR) with chaperone co-expression, were used for the generation of dyes in *Escherichia coli*. The P450 expression was done as outlined above (see main text and Notes 4 and 19) in the presence of indole derivatives at a concentration of 1.0 mM. The sample order was the same in all 12 systems, as summarized at the top of the picture. Indole derivatives used here were 5-aminoindole (5-AI), 5-bromoindole (5-BI), 5-fluoroindole (5-FI), 5-cyanoindole (5-CI), 5-nitroindole (5-NI), 2-methylindole (2-MI), 5-methylindole (5-MI), 5-methoxyindole (5-MXI), 6-chloroindole (6-CHI), 5-hydroxyindole (5-HI), and 5-hydroxyindole-3-acetic acid (5-HAA). See also Notes 33 and 34

In addition, perform the screening on the LB agar medium containing additives for promoting P450 expression, by inoculating plates with suitably diluted glycerol stock in LB medium, to obtain separate colonies (around ~250 colonies per plate). Estimate the cell densities of glycerol stocks by measuring the absorbance at 600 nm (OD_{600}), where 8×10^8 cells/mL have an absorbance of 1. Prepare glycerol stocks of P450s 2A6, 2C9, 2C11, and 2C19 (containing plasmid DNA in the format of pCW'/P450/hNPR) and hNPR (pCW'/hNPR) and use these in parallel for verification of P450 expression and development of color.

3. *Indole Oxidation Assay*

For the estimation of indigo production in vitro use indole oxidation assay as previously described (30). Incubate bacterial membranes containing 0.2 μM recombinant P450 enzymes and hNPR at 37°C with 3.3 mM indole in 100 mM Tris–HCl (pH 7.4) supplemented with an NADPH-generating system (250 μM NADP⁺, 10 mM glucose-6-phosphate (G-6-P), and glucose-6-phosphate dehydrogenase (G-6-P DH) (0.5 U/mL)). For the negative controls perform the reactions in the absence of substrate, the NADPH-generating system, or P450. Use bacterial membranes from cells transformed with the empty pCW' vector or expressing hNPR alone as P450-deficient controls. Stop the reactions at different time points (0, 15, 30, 45, and 60 min) by adding an equal volume of 1% (w/v) sodium dodecyl sulfate (SDS). Finally, transfer 200 μL of reaction mixture to a microtiter plate and measure the absorbance at 685 nm using a TECAN Rainbow Xread Plus microtiter plate reader (TECAN). Estimate indigo production by measuring absorbance at 685 nm, that is, the λ_{max} for indigo dissolved in 0.5% (w/v) SDS. In addition, before each experiment test the activity of the NADPH-generating system. Briefly, mix 2.5 mM G-6-P and 1.0 mM NADP⁺ in 100 mM Tris–HCl (pH 7.4). Record a baseline at 340 nm, and then add G-6-P DH (0.5 U/mL). Measure the absorbance versus time for 3 min and, on average, a plateau should be reached at 1.4 absorbance units.

Perform the indole oxidation assay also using different indole derivatives (5AI, 5BI, 6CHI, and 5HI) as substrates at a concentration of 3.3 mM. Conduct the experiments at two time points (0 min and 60 min). Analyze pigment formation in membrane fractions from three separate expression trials of P450s 2A6, 2C9, 2C11, 2C19, and the CYP2C mutants (2Cs28, 2Cs34, 2Cs1B, and 2Cs2B) co-expressed with hNPR, as well as using membranes containing only hNPR.

3.5 Determination of Reductase Activity

Assess NADPH-cytochrome P450 reductase indirectly by measuring its NADPH-cytochrome *c* reductase activity (28). Briefly, dilute bacterial membranes in 0.3 M potassium phosphate buffer (pH 7.7) and incubate with 0.04 mM horse heart cytochrome *c* at 30°C. Record a baseline, then add NADPH (0.1 mM, see Note 17), and measure absorbance at 550 nm over 3 min. Rates can be measured provided that a linear increase in absorbance versus time is obtained over the period of the assay (with rate <0.2 per min). Calculate the reductase activity using the following equation:

$$\Delta A_{550} \ \mathrm{min}^{-1}/\Delta\varepsilon_{550} = \text{nmol cytochrome } c \text{ reduced/min}$$

where $\Delta\varepsilon_{550} = 0.021 \ \mu M^{-1} \ cm^{-1}$ represents absorption coefficient for reduced horse heart cytochrome *c*. Use a specific activity of 3,200 nmol reduced cytochrome *c*/min/nmol reductase, based on a purified rabbit NADPH-P450 preparation (31), for estimation of NADPH-P450 reductase yield.

4   Notes

1. LBglu medium: Store medium at room temperature and add filter-sterilized glucose to medium before use.

2. SOC medium: This medium is the same as SOB medium with the addition of 20 mM glucose. Filter sterilize before use.

3. Potassium phosphate buffer (KPi) for TB: Dissolve 23.1 g KH_2PO_4 and 125.4 g K_2HPO_4 in up to 1 L with Milli-Q water and autoclave.

4. Cytochrome P450 expression in *E. coli* with chaperone co-expression. The expression of P450s with GroEL/GroES co-expression can be carried out as described above with a slight modification (26). Use single colonies selected on LBglu agar containing ampicillin (100 μg/mL) and chloramphenicol (20 μg/mL) for inoculation of LBamp/cam medium and grown overnight, with a shaking speed of 200 rpm, at 37°C. Dilute the bacterial cultures 1/100 in TB medium supplemented with ampicillin (100 μg/mL), chloramphenicol (20 μg/mL), 1.0 mM thiamine, and trace elements, as described above. After incubation for 5 h, with shaking at 160–180 rpm, at 25°C, start the induction by the addition of arabinose (4 mg/mL), 1.0 mM IPTG, and 0.5 mM δ-ALA. Incubate cultures for an additional 43 h, using the same conditions, before harvesting.

5. Make a PMSF working stock of 100 mM in isopropanol and keep at –20°C.

6. Protease inhibitors: keep a 4 U/mL stock of aprotinin in the fridge, whereas 1 mg/mL stocks of bestatin and leupeptin should be stored at –20°C and freshly diluted in Milli-Q water before use.

7. Make a working stock of 500 mM δ-ALA (84 mg/mL Milli-Q water) and filter sterilize.

8. For 1 M IPTG (isopropylthiogalactoside): Dissolve 714.6 mg of IPTG in 3 mL of sterile Milli-Q water. Make up immediately before use and filter to sterilize (0.45-μm filter units).

9. For 1 M thiamine: Take 1.012 g of thiamine and dissolve in 3 mL of sterile Milli-Q water. Make up immediately before use and filter to sterilize (0.45-μm filter units).

10. Trace element stock: 2.7 g $FeCl_3 \cdot 6H_2O$; 0.2 g $ZnCl_2 \cdot 4H_2O$; 0.2 g $CoCl_2 \cdot 6H_2O$; 0.2 g $Na_2MoO_4 \cdot 2H_2O$; 0.1 g $CaCl_2 \cdot 2H_2O$; 0.1 g $CuCl_2$, 0.05 g H_3BO_3; and 10 mL HCl, and make up to 100 mL with Milli-Q water and then autoclave.

11. Dissolve antibiotics ampicillin (in Milli-Q water) and chloramphenicol (in 100% ethanol) separately at concentrations of 100 mg/mL and 20 mg/mL, respectively. Store aliquots at –20°C. Avoid refreezing.

12. DTT: Sonication buffer is supplemented freshly with DTT before adding to spheroplasts. Freshly add 15.4 μL of a 100 mg/mL solution of DTT to 100 mL of buffer.

13. Keep cultures chilled on ice and then centrifuge at $5,000 \times g$, for 10 min at 4°C. Measure the mass of the pellet and resuspend in 2× TES (100 mM Tris–acetate buffer (pH 7.6), containing

500 mM sucrose and 0.5 mM EDTA) at a ratio of 15 mL per gram of cells wet weight. Add lysozyme (300 µg/g cells) (see also Note 20) with gentle agitation and then dilute the suspension twofold with chilled H_2O before incubating on ice for 30 min. Pellet the resulting spheroplasts by centrifugation at $10,000 \times g$ (4°C, 10 min), collect the pellets (but do not resuspend completely) in sonication buffer and store frozen at –70°C before fractionation. Subsequently, thaw spheroplast preparations at room temperature and supplement with protease inhibitors: 1.0 mM PMSF, 2 µM leupeptin, aprotinin (0.04 U/mL), and 1.0 µM bestatin (see Note 6). Cool suspensions in a salt bath during sonication with three 20-s bursts using a Vibra-cell sonicator (Sonic Materials, Danbury, CT, USA) and then centrifuge sonicates at $10,000 \times g$ (4°C, 20 min). Carefully remove supernatants (S10,000 $\times g$ fraction) and recentrifuge at $180,000 \times g$ (4°C, 65 min). Freeze aliquots (0.5 mL) of both the $10,000 \times g$ pellet and supernatant and store at –70°C. Resuspend membrane fractions in 1× TES using a small Potter-Elvehjem Teflon/glass homogenizer. Freeze sub-aliquots of membrane preparations with liquid nitrogen and store at –70°C until use.

14. *NADPH-generating system*: Make fresh before use 250 µM β-nicotinamide adenine dinucleotide phosphate ($NADP^+$), 10 mM glucose-6-phosphate (G-6-P), and 0.5 U/mL glucose-6-phosphate dehydrogenase. Use a 100 mM G-6-P stock stored at –20°C, defrost an aliquot of suitable size, and add $NADP^+$ and then G6PDH freshly.

15. *Spectral assay buffer*: 100 mL 1 M KPi, pH 7.4, 200 g glycerol, and 10 mL 20% (w/v) Emulgen 913; make up to 1 L with Milli-Q water.

16. Estimate the total amount of protein in membrane fractions using the *BCA assay* (Pierce). Dilute samples in TES mixed with BCA reagent, followed by incubation for 30 min, at 37°C. Measure absorbance at 562 nm and calculate protein concentrations with reference to a standard curve of BSA in TES (0–1 mg/mL). Subsequently, thaw bacterial spheroplast preparations at room temperature and supplement with protease inhibitors.

17. For *reductase assay* make a fresh stock of reduced β-nicotinamide adenine dinucleotide phosphate (NADPH), 10 mg/mL in Milli-Q water, and keep on ice.

18. Prepare *agarose gel*: Usually use a final concentration of 1% (w/v) agar in TAE buffer (0.4 M Tris–acetate, 1 mM EDTA, pH 7.6). Add gel-loading buffer to DNA samples and perform gel electrophoresis at 100 V for 1.5–2 h, using a powerPac300 (Bio-Rad Ltd.). Stain gels in TAE buffer containing ethidium bromide (0.5 µg/mL) for at least 1 h. DNA fragments can be

220   Nedeljka N. Rosic

detected under UV light using an ultraviolet transilluminator and analyzed with the GelDoc Multi-Analyst Version 1.1 software (Bio-Rad Labs Inc.).

19. Prepare *indole stocks* in Milli-Q water with gentle warming at 37°C. Indole derivatives 5-AI, 5-NI, and 2-MI can be dissolved in dimethyl sulfoxide (DMSO), whereas stock solutions of all other indole derivatives can be prepared in ethanol.

20. Isolate *plasmid DNA* from bacterial cells using the alkaline lysis method (24). In brief, inoculate 3–15 mL of TB medium, supplemented with ampicillin (100 µg/mL), with a single bacterial colony, and incubate overnight at 37°C with shaking. The next day, centrifuge bacterial cultures ($750 \times g$, 10 min, 4°C) using Beckman J2-MC centrifuge (Beckman Coulter), then resuspend pellet in "TEN" (10 mM Tris–HCl, 1 mM EDTA, 100 mM NaCl, pH 8.0) and recentrifuge. Then, resuspend the pellet in 500 µL of lysis buffer (50 mM glucose, 10 mM EDTA, and 25 mM Tris–HCl, pH 8.0), containing lysozyme (5 mg/mL), and keep for 5 min at room temperature with occasional mixing. Next, add 1 mL of fresh alkaline solution (0.2 N NaOH, 1% (w/v) SDS) and incubate on ice for 5 min. Addition of 750 µL of 7.5 M potassium acetate solution will allow the precipitation of genomic DNA after a minimum of 5 min incubation on ice. Obtain the pellet after centrifugation (bench centrifuge) at $1,000 \times g$ for 20 min at 4°C. Extract plasmid DNA by adding an equal volume of phenol/chloroform mixture, where the chloroform component contained 1 part of isoamyl alcohol and 24 parts of chloroform, and then precipitate by adding two volumes of 100% ethanol. Wash precipitated DNA with 70% (v/v) ethanol, then remove RNA from the pellet by treatment with RNAse (0.2–0.4 mg/mL) prepared in TE buffer (10 mM Tris–HCl, 0.1 mM EDTA, pH 8.0). Perform a second phenol/chloroform DNA extraction, as described above, with an additional extraction step with an equal volume of chloroform, followed by a precipitation step with 0.5 volumes of 7.5 M ammonium acetate (pH 5–6) and 2.5 volumes of 100% ethanol. Wash pellet with 70% (v/v) ethanol and resuspend plasmid DNA in 100 µL sterile Milli-Q water.

21. For *lysozyme solution*, 50 mg/mL in Milli-Q water should be prepared immediately before use.

22. *Different molar ratios* of insert and vector DNA fragments can be used, though most often the ligation reaction mixture contains 0.15 pmol of insert and 0.05 pmol of vector. Incubate the reaction mixture overnight at 16°C.

23. Prepare the *ultracompetent E. coli* strain DH5αF'IQ™ cells according to Inoue et al. (23). Inoculate a single colony of *E. coli* into LBglu medium and grow for 24 h, at 37°C, with

orbital shaking at 250 rpm. Then, dilute the bacterial culture 1/100 in SOB medium and grow at 18°C, with orbital shaking at 240 rpm. Harvesting should be done after the OD_{600} (optical density) of the culture has reached 0.2–0.4 (~30 h of incubation). Wash cells in ice-cold TB solution, then centrifuge (2,500×g) and finally resuspend in TB solution. Make cells competent by the addition of 7% (v/v) dimethyl sulfoxide (DMSO) and freeze aliquots in liquid nitrogen. Store at –70°C.

24. For the *transformation procedure*, thaw an aliquot (100 μL) of ultracompetent DH5α cells and add 180–200 ng of plasmid DNA (pCW'/P450/hNPR, pCW'/hNPR, or pCW'), accompanied by 180–200 ng of pGro7 in the case of chaperone GroEL/GroES co-expression (25, 32, 33). Incubate cells on ice for 30 min and subject to heat shock at 42°C for 30 s. Return the transformation mixture to ice and add 400 μL of freshly filter-sterilized SOC medium. Incubate for 1 h at 37°C. Apply aliquots of 100–200 μL to LBamp or LBamp/cam plates (LBglu medium supplemented with bactoagar (15 g/L), ampicillin (100 μg/mL), and chloramphenicol (20 μg/mL)) and incubate overnight at 37°C. Ampicillin is used for selection of transformants containing pCW-derived plasmids and chloramphenicol in the case of chaperone GroEL/GroES co-expression. Evaluate the size of DNA samples after electrophoresis, using a DNA ladder prepared from plasmid pCW'/NF14 digested with *Nci*I and *Nde*I. The sizes of DNA fragments generally produced are 1,640, 1,010, 910, 810, 701, 696, 440, 350, and 35 bp. In addition, most of the DNA markers prepared in this manner had an additional band of 2,047 bp, due to incomplete digestion of pCW'/NF14 with *Nde*I.

25. For *sequencing* purposes, use a higher purity of plasmid DNA, obtained using the QIAGEN DNA Mini Prep Kit. DNA extraction from agarose gels can be performed using the geneclean procedure (Geneclean® III kit) or QIAGEN DNA gel extraction kit, according to the manufacturers' instructions.

26. The *DNA concentration* can be estimated by measuring the absorbance at 260 and 280 nm wavelengths, where 50 μg/mL of double-stranded DNA has an absorbance of 1, at 260 nm.

27. Collect *bacterial colonies* using a bent glass rod and transfer to 15% (w/v) glycerol LB medium. Gently mix colonies in LB medium to disperse clumps and to obtain an even distribution of cells. Then, take 0.5-mL aliquots, freeze in liquid nitrogen and store at –70°C, until use.

28. Grow *cultures* overnight by shaking at 240 rpm at 37°C in 2xLBamp/cam. After incubation supplement cultures with sterile glycerol (15% glycerol in the final concentration), and then archive plates at –70°C.

29. Perform *expression* of cytochrome P450 enzymes in *E. coli* DH5αF'IQ™ following the conditions previously established (27). Select single colonies from LBamp plates and grow overnight in LBamp medium, with shaking at 200 rpm, at 37°C. Dilute the cultures 1/100 in TB medium supplemented with ampicillin (100 µg/mL), 1.0 mM IPTG, 1.0 mM thiamine, 0.5 mM δ-ALA, and trace elements (see ref. 22 and also Notes 4 and 7–11). Before harvesting, grow cells at a shaking speed of 180–200 rpm, at 30°C, for a period of 24 h.

30. Dilute subcellular fractions of *E. coli* (S10,000 and/or membrane fractions) in spectral-assay buffer, at room temperature. Record the differences in absorbance between Fe^{2+}-CO and Fe^{2+} in the wavelength range of 400–500 nm, using a Varian Cary 100 spectrophotometer. Calculate the P450 content using the following equation:

$$\left[(A_{450-490})_{observed} - (A_{450-490})_{baseline}\right]\Big/ \varepsilon = \text{nmol P450} /$$

where $\varepsilon = 0.091$ μM^{-1} cm^{-1} represents the micromolar $\Delta E_{450-490}$ absorbance coefficient for Fe^{2+}-CO versus Fe^{2+}-difference spectra for P450s (34, 35).

31. Optionally, redevelop the blots using a higher concentration of the same antibodies (1/500 dilution).

32. Briefly, dissolve 4-chloro-1-naphthol in methanol and dilute with PBS. Before use add H_2O_2 and pour the solution over the nitrocellulose membrane. Then wash the membrane three times with PBS and two times with water and dry between filter papers.

33. Synthesize indirubin as previously described (36). In brief, add 6.25 mL of methanol (under argon) to 250 mg of indoxyl acetate and stir. Also under argon, add 211 mg of isatin and 325 mg of sodium carbonate to the resulting solution and stir for 30 min at room temperature, then leave to stand for 24 h. Then filter the residue and wash with methanol and cold Milli-Q water until neutral reaction (pH 7) of the filtrate is achieved. Dry the dark-violet product over potassium hydroxide in a desiccator.

34. Perform the UV-visible spectroscopy of the newly synthesized indirubin and authentic indigo (Aldrich Chemical) in the wavelength range of 300–800 nm (30). Dissolve indirubin and indigo in chloroform at a concentration of 1 mg/mL, then dilute further by 1/10 (100 µg/mL) and 1/100 (10 µg/mL) for the recording of spectra.

35. *DNA sequencing*. Sequence the mutants using specific primers (Table 1), which can be designed by comparison of CYP2C sequences to determine regions of maximum sequence identity.

Data analyses can be performed using web-based BLAST services, whereas the ClustalW algorithm in Vector NTI, InforMax, can be used for the multiple-sequence alignment. The information about the sequences origin can be obtained using the BLAST search option. Also, alignments of mutant sequences with parental cDNAs can be used for the determination of sequence diversity in the analyzed mutants (14).

Acknowledgments

An International Postgraduate Research Scholarship and University of Queensland Graduate School Award (to N.N.R.) supported this research. This work was also supported by a UQ Postdoctoral Fellowship for Women and by a Discovery Early Career Researcher Award (DECRA) to N.N.R.

References

1. Rosic NN (2009) Versatile capacity of shuffled cytochrome P450s for dye production. Appl Microbiol Biotechnol 82:203–210
2. Grogan G (2011) Cytochromes P450: exploiting diversity and enabling application as biocatalysts. Curr Opin Chem Biol 15:241–248
3. Kumar S (2010) Engineering cytochrome P450 biocatalysts for biotechnology, medicine and bioremediation. Expert Opin Drug Metab Toxicol 6:115–131
4. Pompon D, Truan G, Urban P (2008) Cytochrome P450 engineering. Ingénierie des cytochromes Biofutur 288:34–38
5. Gillam EM et al (1999) Formation of indigo by recombinant mammalian cytochrome P450. Biochem Biophys Res Commun 265:469–472
6. Rosic N (2005) Molecular breeding of cytochrome P450s for indigoid pigment production. PhD, University of Queensland, Brisbane
7. Rosic N et al (2003) Directed evolution of mammalian cytochrome P450 enzymes involved in xenobiotic metabolism. Drug Metab Rev 91:46–46
8. Nakamura K, Martin MV, Guengerich PF (2001) Random mutagenesis of human cytochrome P450 2A6 and screening with indole oxidation. Arch Biochem Biophys 395(1):25–31
9. Stemmer WP (1994) Rapid evolution of a protein in vitro by DNA shuffling. Nature 370:389–391
10. Stemmer WP (1994) DNA shuffling by random fragmentation and reassembly: in vitro recombination for molecular evolution. Proc Natl Acad Sci U S A 91:10747–10751
11. Schmidt-Dannert C (2001) Directed evolution of single proteins, metabolic pathways, and viruses. Biochemistry 40:13125–13136
12. Kurtzman AL et al (2001) Advances in directed protein evolution by recursive genetic recombination: applications to therapeutic proteins. Curr Opin Biotechnol 12:361–370
13. Abecassis V, Pompon D, Truan G (2000) High efficiency family shuffling based on multi-step PCR and in vivo DNA recombination in yeast: statistical and functional analysis of a combinatorial library between human cytochrome P450 1A1 and 1A2. Nucleic Acids Res 28:E88
14. Rosic NN et al (2007) Extending the diversity of cytochrome P450 enzymes by DNA family shuffling. Gene 395:40–48
15. Guengerich FP (1995) Human cytochrome P450 enzymes. In: de Ortiz Montellano PR (ed) Cytochrome P450: structure, mechanism and biochemistry, vol 2. Plenum Press, New York, pp 473–535
16. Kikuchi M, Ohnishi K, Harayama S (1999) Novel family shuffling methods for the in vitro evolution of enzymes. Gene 236:159–167
17. Joern JM, Meinhold P, Arnold FH (2002) Analysis of shuffled gene libraries. J Mol Biol 316:643–656
18. Huang W et al (2007) A shuffled CYP2C library with a high degree of structural integrity and functional versatility. Arch Biochem Biophys 467:193–205
19. Johnston WA et al (2007) A shuffled CYP1A library shows both structural integrity and functional diversity. Drug Metab Dispos 35:2177–2185
20. Parikh A, Gillam EM, Guengerich FP (1997) Drug metabolism by Escherichia coli expressing human cytochromes P450. Nat Biotechnol 15:784–788

21. Cuttle L et al (2000) Phenytoin metabolism by human cytochrome P450: involvement of P450 3A and 2C forms in secondary metabolism and drug-protein adduct formation. Drug Metab Dispos 28:945–950

22. Fisher CW et al (1992) High-level expression in Escherichia coli of enzymatically active fusion proteins containing the domains of mammalian cytochromes P450 and NADPH-P450 reductase flavoprotein. Proc Natl Acad Sci U S A 89: 10817–10821

23. Inoue H, Nojima H, Okayama H (1990) High efficiency transformation of Escherichia coli with plasmids. Gene 96:23–28

24. Sambrook J, Fritsch EF, Maniatis T (1989) Small-scale preparations of plasmid DNA molecular cloning, a laboratory manual, 2nd edn. Cold Spring Harbor Laboratory Press, New York, pp. 1.25–21.28

25. Nishihara K et al (1998) Chaperone coexpression plasmids: differential and synergistic roles of DnaK-DnaJ-GrpE and GroEL-GroES in assisting folding of an allergen of Japanese cedar pollen, Cryj2, in Escherichia coli. Appl Environ Microbiol 64:1694–1699

26. Notley LM et al (2002) Bioactivation of tamoxifen by recombinant human cytochrome p450 enzymes. Chem Res Toxicol 15:614–622

27. Gillam EM et al (1993) Expression of modified human cytochrome P450 3A4 in Escherichia coli and purification and reconstitution of the enzyme. Arch Biochem Biophys 305:123–131

28. Guengerich FP (1994) Principles and methods of toxicology, 3rd Edition: Analysis and characterization of enzymes. In: Hayes AW (ed) Principles and methods of toxicology, edn. Raven Press, New York, pp. 1259–1313

29. Ensley BD et al (1983) Expression of naphthalene oxidation genes in Escherichia coli results in the biosynthesis of indigo. Science 222:167–169

30. Gillam EM et al (2000) Oxidation of indole by cytochrome P450 enzymes. Biochemistry 39:13817–13824

31. Yasukochi Y, Masters BS (1976) Some properties of a detergent-solubilized NADPH-cytochrome c(cytochrome P-450) reductase purified by biospecific affinity chromatography. J Biol Chem 251:5337–5344

32. Georgiou G, Valax P (1996) Expression of correctly folded proteins in Escherichia coli. Curr Opin Biotechnol 7:190–197

33. Thomas JG, Ayling A, Baneyx F (1997) Molecular chaperones, folding catalysts, and the recovery of active recombinant proteins from E. coli. To fold or to refold. Appl Biochem Biotechnol 66:197–238

34. Ryan D et al (1975) Multiple forms of cytochrome P-450 in phenobarbital- and 3-methylcholanthrene-treated rats. Separation and spectral properties. J Biol Chem 250:2157–2163

35. Haugen DA, Coon MJ (1976) Properties of electrophoretically homogeneous phenobarbital-inducible and beta-naphthoflavone-inducible forms of liver microsomal cytochrome P-450. J Biol Chem 251:7929–7939

36. Hoessel R et al (1999) Indirubin, the active constituent of a Chinese antileukaemia medecine, inhibits cyclin-dependent kinases. Nat Cell Biol 1:60–67

Chapter 19

P450 Oxidoreductase: Genotyping, Expression, Purification of Recombinant Protein, and Activity Assessments of Wild-Type and Mutant Protein

Vishal Agrawal and Walter L. Miller

Abstract

P450 oxidoreductase (POR) is the flavoprotein that transfers electrons from NADPH to microsomal cytochrome P450 enzymes and to some other proteins. Protocols for genotyping human POR for common polymorphisms are described. Expression in *E. coli* of recombinant human POR, its purification, and different methods of assessing the effect of amino-acid sequence variants of POR on the activity of various cytochromes P450 are also described.

Key words P450 reductase, Genotyping, Mutant, Heterologous expression, Purification, Enzyme activity

1 Introduction

P450 oxidoreductase (POR) is a 78-kDa, di-flavin, microsomal protein that transfers electrons from NADPH to all 50 human microsomal cytochromes P450, including those involved in the biosynthesis of cholesterol, sterols, and steroid hormones and those that metabolize more than 80% of clinically used drugs, in addition to countless endobiotics and xenobiotics. Human POR deficiency causes a complex disorder of steroidogenesis, and, in severe cases, a complex skeletal disorder termed Antley–Bixler syndrome (1). The *POR* gene is highly polymorphic: sequencing the *POR* gene in 842 healthy individuals from four different ethnic groups identified 140 single-nucleotide polymorphisms (SNPs) (2). These include the amino-acid sequence variant A503V found on 19–37% of alleles, depending on ethnicity, and three promoter polymorphisms found on 1–13% of alleles. These POR polymorphisms can affect drug metabolism and steroid biosynthesis by altering the activities of type II cytochrome P450 enzymes (3–7). Thus, POR may contribute to pharmacogenetic variation in drug response and variations in steroid synthesis.

Ian R. Phillips et al. (eds.), *Cytochrome P450 Protocols*, Methods in Molecular Biology, vol. 987,
DOI 10.1007/978-1-62703-321-3_19, © Springer Science+Business Media New York 2013

2 Materials

2.1 POR Genotyping

2.1.1 Preparation of DNA from Human Blood and Polymerase Chain Reaction

1. Human blood: Collect 1–5 ml blood using ethylenediaminetetraacetic acid (EDTA) as anticoagulant. Heparin is not preferred, as it tends to stick to DNA, interfering with subsequent manipulations. Blood should be processed as soon as possible to isolate genomic DNA; DNA in blood cells is stable for a few days at 4°C, but yields decrease and degradation increases with storage.

2. TE buffer: 10 mM Tris–HCl, pH 7.4, 1 mM EDTA.

3. Nucleotide mix: A nucleotide mix containing 2 mM of each of the four dNTPs can be purchased from commercial suppliers or can be purchased individually and mixed to make a final stock of 2 mM. The stock is stored in small aliquots at –20°C.

4. Thermostable DNA polymerase: DNA polymerase is available from a variety of commercial suppliers (see Note 1).

5. 10× polymerase chain reaction (PCR) reaction buffer: Most suppliers of DNA polymerase will include 10× PCR buffer along with the polymerase; we recommend using the supplier's buffer. If a buffer is not supplied, you may make the following 10× buffer: 670 mM Tris–HCl (pH 9.0), 166 mM ammonium sulfate, and 0.17 mM bovine serum albumin; the buffer is filter sterilized. We find that addition to the PCR mix of dimethyl sulfoxide (DMSO) to a final concentration of 2–4% gives better results.

6. $MgCl_2$: Usually a 50 mM stock is provided by the commercial suppliers of DNA polymerase.

7. DMSO.

8. Thermocycler.

9. PCR primers: We obtain satisfactory results from standard desalted oligonucleotides, which can be purchased from many suppliers; we use the sequences shown in Table 1.

10. 10× Tris–borate–EDTA (TBE) buffer: 0.9 M Tris–borate, 20 mM EDTA.

11. Agarose: prepare 1.5% (w/v) agarose gel in 1× TBE buffer.

12. Molecular weight markers: 100-bp DNA ladder.

13. UV transilluminator.

14. ExoSAP-IT: degrades DNA of size less than 100 bp and over 20 kb. Used to clean PCR products from primers, nucleotides, and genomic DNA.

15. Access to DNA sequencing facility.

Table 1
Primers for PCR and sequencing various exons of POR

Exon	Primer	Sequence of primer	Size of product amplified (bp)
1U	1U_F	5′-ACAGCCACAGTTCTGCAGTG-3′	588
	1U_R	5′-GAGATCAGCTCTAGGGGAAGG-3′	
1	1_F	5′-GTCTGTTATGTCAGCCCCAGTC-3′	549
	1_R	5′-GAGCTCAAGCACAATCTAGCAA-3	
2	2_F	5′-TTACTGTAGGGGAAATGGGAAG-3′	562
	2_R	5′-GGGAGAAGCTTCGTGAGTTAGA-3′	
3	3_F	5′-GCAAGTCCCAGAGGAACTTAGA-3′	568
	3_R	5′-GTTTGGTTTGGGAGATGTGG-3′	
4	4_F	5′-CCCTCCGTGTTGTTACTTCTCT-3′	550
	4_R	5′-CCTTTCTTGCCTTTAGTCTCCA-3′	
5	5_F	5′-AGGTCAACCAGATGAAGCCTCT-3′	489
	5_R	5′-CAAGCCGAAAAGCAAAACTG-3′	
6	6_F	5′-CTTCCTGATGCTCTGGGTTTAT-3	496
	6_R	5′-CAAAGTTGAACCTAGCCACAGA-3′	
7	7_F	5′-GCTTCCTTACCTTCTCCCAGAT-3′	583
	7_R	5′-TGCAGAGTAAGGTGGCTAAGTG-3′	
8 and 9	8_F	5′-GTAACCGGTGAGATTTCCTCAT-3′	692
	9_R	5′-ACTATGACAGTGACGGGGTAGG-3′	
10 and 11	10_F	5′-AGGGAGGCATCAGAGAGCATAG-3′	781
	11_R	5′-GGCTGGACAGATGCTGAGAA-3′	
12 and 13	12_F	5′-GAGGGGGCCTCTGAGGTTTG-3′	764
	13_R	5′-ACAGGTGCTCTCGGTCTTGCTT-3′	
14 and 15	14_F	5′-GGGAGACGCTGCTGTACTACG-3′	686
	15_R	5′-GCCCAGAGGAGTCTTTGTCACT-3′	

The primer sequences have been taken from Huang et al. (2)

2.2 Expression of Recombinant POR in E. coli and Isolation of E. coli Membranes Containing POR

2.2.1 Expression of POR in E. coli

1. *E. coli* strain. Commonly used *E. coli* strains for expression, such as BL21(DE3) or JM109, can be used. We have used *E. coli* strain C41(DE3) pLysS, which is derived from BL21(DE3). The C41(DE3) strain has at least one uncharacterized mutation. This strain allows expression of toxic and membrane proteins.

2. Luria-Bertani (LB) medium: Dissolve 10 g of tryptone, 5 g of yeast extract, and 10 g of sodium chloride in 1 l of water. Mix and autoclave to sterilize.

3. Carbenicillin stock (100 mg/ml): Dissolve carbenicillin in water (deionized and double distilled or equivalent) at a concentration of 100 mg/ml and filter sterilize by passing through a 0.2-μm filter. The stock can be stored at –20°C in small aliquots.

228 Vishal Agrawal and Walter L. Miller

4. Chloramphenicol stock (32 mg/ml): Dissolve chloramphenicol in ethanol at a concentration of 32 mg/ml. Chloramphenicol solution can be stored at −20°C.

5. Terrific broth: Dissolve 12 g of tryptone, 24 g of yeast extract, and 8 ml of glycerol in water. Make the volume to 900 ml with water and autoclave to sterilize. Allow to cool to room temperature and add 100 ml of a filter-sterilized solution of 0.17 M KH_2PO_4, 0.72 M K_2HPO_4.

6. Trace elements stock: For 100 ml of 4,000× stock dissolve 2.7 g of $FeCl_3 \cdot 6H_2O$, 0.2 g of $ZnCl_2 \cdot 4H_2O$, 0.2 g of $CoCl_2 \cdot 6H_2O$, 0.2 g of $Na_2MoO_4 \cdot 2H_2O$, 0.1 g of $CaCl_2 \cdot 2H_2O$, 0.1 g of $CuCl_2$, and 0.05 g of H_3BO_3 in 1.2 M HCl. Store the stock at 4°C.

7. Riboflavin.

8. Isopropyl-β-D-thiogalactopyranoside (IPTG): Prepare 1 M stock by dissolving in water. IPTG stock can be stored at −20°C.

9. Centrifuge bottles.

2.2.2 Isolation of E. coli Membranes Containing POR

1. TES buffer: 100 mM Tris-acetate (pH 7.6), 500 mM sucrose, 1 mM EDTA. Store at 4°C.

2. Lysozyme.

3. KMDG buffer: 100 mM potassium-phosphate buffer (pH 7.6), 6 mM magnesium acetate, 0.1 mM DTT, 20% (v/v) glycerol.

4. PMSF: Make a 200 mM stock in isopropanol or DMSO. Store at −20°C.

5. DNase I: Make a stock of 20 mg/ml in 50 mM Tris–HCl (pH 7.6), 20% (v/v) glycerol.

6. Tissue homogenizer: The type of homogenizer and the clearance of the pestle are not important.

7. Probe sonicator.

8. Ultracentrifuge.

2.3 Purification of POR from Bacterial Membranes

1. 10% (w/v) solution of sodium cholate.

2. 10% (v/v) solution of Triton X-100.

3. ADP-sepharose beads. ADP-sepharose beads can be purchased from commercial suppliers such as Sigma or GE Healthcare.

4. Adenosine.

5. 2′-Adenosine monophosphate.

6. Dialysis buffer: 50 mM Tris–HCl (pH 7.6), 20% (v/v) glycerol.

7. Ni-agarose/sepharose beads.

P450 Oxidoreductase: Genotyping, Expression, Purification...   229

8. Sodium chloride.

9. Equilibration buffer: 100 mM Tris–HCl (pH 7.6), 20% glycerol, 0.2% (v/v) Triton X-100, 0.2% (w/v) sodium cholate.

10. Purification buffer: 100 mM Tris–HCl (pH 7.6), 20% glycerol, 0.2% sodium cholate.

11. Histidine.

12. Electrophoresis apparatus and power supply to run SDS-PAGE gels.

13. Dialysis membranes or cassettes.

2.4 Assays with P450 Oxidoreductase

1. NADPH. Reduced nicotinamide dinucleotide phosphate can be purchased from commercial suppliers. Always prepare a fresh solution of NADPH before starting a reaction.

2. Glucose-6-phosphate. Prepare 100 mM solution of glucose-6-phosphate. The stock can be stored at 4°C.

3. Glucose-6-phosphate dehydrogenase.

4. L-1,2-Dilauroyl-*sn*-glycerol-3-phosphocholine (DLPC). A 2 mg/ml stock of DLPC is prepared in 20 mM phosphate buffer (pH 7.6), 10% (v/v) glycerol.

5. Progesterone.

6. [^{14}C]Progesterone.

7. 17-Hydroxypregnenolone.

8. [^{3}H]17-Hydroxypregnenolone.

9. Cytochrome b$_5$.

10. Thin-layer chromatography (TLC) plates: 250-μm silica-coated TLC plates can be purchased from commercial suppliers such as Whatman or GE Healthcare.

11. Iso-octane.

12. Dichloromethane.

13. Ethyl acetate.

14. Spectrophotometer.

15. Phosphorimager.

3 Methods

3.1 POR Genotyping

3.1.1 Preparation of DNA from Human Blood

1. Genomic DNA can be isolated from human blood using commercially available kits, or by published laboratory techniques (8).

2. Genomic DNA can be re-suspended in TE buffer. DNA can be stored at −20°C for long periods.

3. DNA concentration and its purity are assessed spectroscopically by measuring its absorbance at 260 and 280 nm (see Note 2).

3.1.2 Polymerase Chain Reaction

1. Prepare PCR master mix containing PCR buffer, dNTPs, DNA, and polymerase enzyme. Mix the contents of the tube by inverting a few times and spin briefly for 10 s at $5,000 \times g$ in a microcentrifuge.

2. Aliquot either 25 or 50 μl into a 200-μl thin-walled PCR tube. Add specific forward and reverse primers. Mix gently by inverting the tube and centrifuging briefly.

3. Place the PCR tubes in a thermocycler. PCR amplification is carried out under touchdown cycling conditions (2), starting at 95°C for 5 min, followed by 14 touchdown cycles of 94°C for 20 s, 65°C to 58.5°C for 20 s (the annealing temperature for each subsequent cycle is decreased by 0.5°C), and 72°C for 1 min. The touchdown PCR is followed by 35 cycles of PCR amplification at 94°C for 20 s, 58°C for 20 s, and 72°C for 1 min. The final extension is done at 72°C for 10 min, followed by a hold at 4°C (see Note 3).

3.1.3 Analysis of PCR Product and Sequencing

Upon completion of the PCR, it is important to run a gel to determine whether the amplification was successful and to monitor for the presence or absence of additional bands that represent incorrect amplification. It is essential that a DNA molecular weight standard is run on all gels; a 100-bp DNA ladder works well for this purpose.

1. Mix 4 μl of sample with 1 μl of 5× DNA-loading dye and load on a 1.5% (w/v) agarose gel containing ethidium bromide or any other DNA-staining dye.

2. Electrophoresis is done at 70–100 V for 30–45 min.

3. Examine the agarose gel on a UV transilluminator.

4. If a single PCR amplification product is seen, take 5 μl of sample, mix with 2 μl of ExoSAP-IT, and incubate at 37°C for 15 min, followed by a 15-min incubation at 80°C, to inactivate ExoSAP-IT.

5. If multiple bands are seen in the PCR product, then separate the rest of the sample by agarose-gel electrophoresis and purify the PCR band of right size, using a commercially available gel-extraction kit.

6. The PCR products treated with ExoSAP or the gel-purified DNA is used for DNA sequencing using specific oligonucleotide primers.

7. The results from DNA sequencing are analyzed against the wild-type sequence of human POR found on the NCBI database at http://www.ncbi.nlm.nih.gov/nuccore/94721355? from=83&to=2125. More than 40 different POR alleles have been described and at least 140 single-nucleotide polymorphisms (SNPs) have been described (2) where the haplotype has not been assigned (for details see homepage for P450 allele nomenclature http://www.cypalleles.ki.se/por.htm).

P450 Oxidoreductase: Genotyping, Expression, Purification...

231

3.2 Expression of Active Human P450 Oxidoreductase in E. coli

3.2.1 Expression of P450 Oxidoreductase

Human POR contains 680 amino-acid residues (rodent POR has 677). It is anchored to the membranes of the endoplasmic reticulum by approximately 54 N-terminal amino-acid residues. Our studies have routinely used N-27 POR, lacking the 27 N-terminal residues. N-27 POR becomes inserted into the membrane and is catalytically active with P450 enzymes. This is in contrast to the N-66 POR used for crystallographic studies, which is soluble. Although N-66 POR is active with cytochrome c, it is not membrane anchored and cannot donate electrons to cytochrome P450 enzymes.

1. POR expression vector is transformed into *E. coli* strain C41(DE3) pLysS and selected on LB-agar plates containing carbenicillin (100 μg/ml) and chloramphenicol (32 μg/ml) (see Note 4).

2. Pick a single colony from the LB-agar plate and inoculate into 5 ml of LB-medium containing carbenicillin (100 μg/ml) and chloramphenicol (32 μg/ml). Grow culture overnight at 37°C with shaking at 200 rpm.

3. Inoculate 5 ml of overnight-grown culture into 500 ml of phosphate-buffered Terrific Broth containing trace elements, carbenicillin (100 μg/ml), and chloramphenicol (32 μg/ml). Incubate at 28°C with shaking at 150 rpm for 4–5 h, until the OD at 600 nm reaches 0.4.

4. Add riboflavin to the culture to a final concentration of 50 mg/l.

5. 30 min after adding the riboflavin, induce POR expression by addition of IPTG to a final concentration of 0.4 mM. Continue to grow the culture with continuous shaking at 28°C for an additional 36 h.

6. Chill the flask on ice. Pellet the cells by centrifugation at $5,000 \times g$ for 10 min at 4°C and determine the wet cell mass. The pelleted cells can be stored at –80°C or can be used directly to prepare spheroplasts.

3.2.2 Isolation of E. coli Membranes Containing POR

1. Thoroughly resuspend the cells in cold TES buffer. Dilute with additional cold TES buffer to 10 ml/g of wet cell mass.

2. Add lysozyme to a concentration of 0.1 mg/ml and slowly add an equal volume of cold 0.1 mM EDTA, while gently stirring.

3. Continue to stir gently at 4°C for 20–30 min.

4. Pellet spheroplasts by centrifugation at $5,000 \times g$ for 15 min at 4°C. The pelleted spheroplasts can be stored at –80°C or one can proceed to prepare membranes.

5. Using a Teflon homogenizer, resuspend the spheroplasts in 2 ml of cold KMDG buffer per gram of spheroplasts. Add PMSF to a final concentration of 0.5 mM and DNase I to a final concentration of 20 μg/ml.

232   Vishal Agrawal and Walter L. Miller

6. Transfer to 50-ml polypropylene tubes in a salt/ice bath and sonicate using a probe sonicator, with 550 W and 20 kHZ output, for 5 min with 20-s bursts at 30–40% power, each followed by a 30-s cooling period. Care should be taken not to sonicate at high power as this may denature the protein.

7. Centrifuge at $12,000 \times g$ for 15 min at 4°C to remove cellular debris (see Note 5).

8. Collect supernatant. Transfer the supernatant to ultracentrifuge tubes and centrifuge at $100,000 \times g$ for 90 min at 4°C, to collect membranes containing active POR.

9. Discard supernatant. Resuspend pelleted membranes in 50 mM Tris–HCl (pH 7.6), 20% (v/v) glycerol. Pass the resuspended membranes through a 27-gauge needle to make a homogeneous re-suspension. Aliquot them into small fractions in Eppendorf tubes and store at –80°C. The membranes can be stored at –80°C for 1–2 years.

3.3 Purification of POR from Bacterial Membrane

The bacterial membranes containing expressed human POR can be used directly for various activity measurements. If experiments require purified proteins, they can be purified using an ADP-sepharose column. Recently, we constructed a new expression vector for human N-27 POR that has a C-terminal tail consisting of three glycines followed by six histidines (G3H6) (9). The G3H6 tail facilitates rapid one-step purification of POR to apparent homogeneity using a nickel column. Both methods of purification are described below.

3.3.1 Solubilization of E. coli Membranes Containing POR

1. To the re-suspended *E. coli* membranes, add sodium cholate and Triton X-100, each to a final concentration of 0.2%.

2. Gently swirl the tubes at 4°C for 2–3 h.

3. Centrifuge at $100,000 \times g$ for 60 min at 4°C. Collect supernatant and proceed to purification either using ADP-sepharose column or Ni-column (if expressing POR with a 6X-histidine tail).

3.3.2 Purification Using ADP-Sepharose Column

1. Soak 1–2 ml of ADP-sepharose beads in water for 15–20 min. Swirl gently, allow the beads to settle and pipet out water from the top.

2. Add three bed volumes of equilibration buffer. Swirl gently; allow the beads to settle and pipet out buffer from the top. Repeat this step once again to equilibrate the beads.

3. Add supernatant, from the solubilization of *E. coli* membranes, to the equilibrated ADP-sepharose beads. Allow the POR to bind to the beads by incubating at 4°C with gentle rotation for 2 h.

4. Pack the mix of ADP-sepharose beads and supernatant, by gently pouring into a column. Let the solubilized supernatant

P450 Oxidoreductase: Genotyping, Expression, Purification...   233

flow through. Collect the flow-through to run it on SDS-PAGE gel for analysis.

5. Wash the column with 10 bed volumes of purification buffer containing 2 mM adenosine.

6. Elute the bound POR with three bed volumes of purification buffer containing 5 mM 2′-adenosine monophosphate.

7. Dialyze the eluted POR extensively against 50 mM Tris–HCl (pH 7.6), 20% (v/v) glycerol. If required, the protein can be concentrated using a protein concentrator such as Centricon.

8. Store the protein at –80°C in small aliquots.

3.3.3 Purification Using a Nickel Column

1. Aliquot 3–4 ml of Ni beads. Wash thoroughly with water. Equilibrate with equilibration buffer (used in Subheading 3.3.2).

2. Add solubilized membrane (from Subheading 3.3.2) to the nickel beads and incubate at 4°C for 2–4 h with gentle rotation to allow the POR to bind.

3. Pack the beads into a column and let the supernatant flow-through. Collect the flow-through for analysis by SDS-PAGE.

4. Wash the nickel column with 10 bed volumes of purification buffer containing 20 mM histidine and 0.5 M sodium chloride.

5. Elute the proteins with purification buffer containing 0.5 M sodium chloride and 200 mM histidine.

6. Dialyze the eluted POR extensively against 50 mM Tris–HCl (pH 7.6), 20% (v/v) glycerol. If required, the protein can be concentrated using a protein concentrator (such as Centricon) with a molecular cutoff of 30 kDa or less.

7. Store the protein at –80°C in small aliquots.

3.4 Assays of P450 Oxidoreductase

The activity of purified or membrane-bound POR can be assessed by performing a cytochrome c reduction assay. This assay is a simple and straightforward assay of protein activity, but is not physiologic, as cytochrome c is found in mitochondria and POR is in the endoplasmic reticulum (10).

3.4.1 POR Assay Based on Cytochrome c

1. The assay buffer system contains 0.1 M Tris–HCl (pH 7.6) and NADPH-regeneration system (2 mM glucose-6-phosphate, 3 U of glucose-6-phosphate dehydrogenase, and 5 μM NADPH) in 1 ml volume.

2. Variable amounts of cytochrome c, 0.5, 1, 2, 3, 5, and 10 μM, are added to each tube.

3. The reaction is started by adding 2 pmol of POR and monitored continuously against time by measuring absorbance at 550 nm in a spectrophotometer.

234 Vishal Agrawal and Walter L. Miller

4. The amount of reduced cytochrome c formed is deduced from its molar extinction coefficient of 21.1 mM^{-1} at 550 nm (11) and velocity is expressed as (nmol of reduced cytochrome c)$\times$(pmol of POR)$^{-1}\times$(min)$^{-1}$.

5. The kinetic parameters Michaelis constant (K_m) and maximal velocity (V_{max}) are calculated by plotting the substrate concentration ((S)) versus the velocity (v) (see Note 6). The ratio of V_{max}/K_m is used to measure the catalytic efficiency.

6. To assess the loss or gain of catalytic efficiency of a POR mutant, assays can be done using mutant POR and its activity can be compared with that of wild-type POR.

3.4.2 POR Assay Based on Human P450c17

All 50 type II human microsomal cytochrome P450 enzymes depend on electron donation from POR for activity. P450c17 is an important type II enzyme involved in the biosynthesis of steroid hormones. P450c17 has two principal activities, 17α-hydroxylase and 17,20 lyase. 17α-Hydroxylase activity converts pregnenolone (Preg) to 17α-hydroxy pregnenolone (17OH-Preg) and progesterone to 17α-hydroxy progesterone (17OHP). 17,20 Lyase activity converts 17OH-Preg to dehydroepiandrosterone (DHEA). The P450c17 of rodents and ungulates also converts 17OHP to androstenedione, but human P450c17 catalyzes this reaction with only 2% of the activity to convert 17OH-Preg to DHEA (12). Therefore, conversion of progesterone to 17OHP assays only 17-hydroxylase activity and conversion of 17OH-Preg to DHEA assays 17,20 lyase activity. We describe how to assess the effect of different POR mutants on the two activities of P450c17.

1. In a total reaction volume of 200 μl, add 10 pmol of purified (or membrane-bound) human P450c17 (for expression and single-step purification of P450c17 see: (13)), 20 pmol of POR, 20 μg of phosphatidyl choline in a 50 mM potassium-phosphate buffer (pH 7.4) containing 20% (v/v) glycerol, 6 mM potassium acetate, 10 mM magnesium chloride, 1 mM reduced glutathione, 3 U of glucose-6-phosphate dehydrogenase, and 2 mM glucose-6-phosphate.

2. For 17α-hydroxylase activity, add varying concentrations of unlabeled substrate progesterone mixed with 50,000 cpm of [^{14}C]-labeled progesterone.

3. For 17,20 lyase activity, cytochrome b$_5$ is added along with P450c17 and POR. Cytochrome b$_5$ acts as an allosteric facilitator of 17,20 lyase activity (12). Varied concentrations of unlabeled substrate 17α-hydroxypregnenolone mixed with 50,000 cpm of ^{3}H-labeled 7α-hydroxypregnenolone are added to the reaction mix.

P450 Oxidoreductase: Genotyping, Expression, Purification... 235

4. Incubate the reaction mix at 37°C for 5 min. Start the reaction by adding 1 mM NADPH. Incubate at 37°C for 2 h.

5. Terminate the reaction by adding 600 μl of a 1:1 mixture of ethyl acetate and iso-octane.

6. Extract the steroids by vigorously mixing by vortexing for 30 s. Centrifuge at $10,000 \times g$ for 10 min.

7. Transfer the upper, organic phase into a different Eppendorf tube and dry the samples by evaporation under a gentle stream of nitrogen.

8. Dissolve the dried steroids in 20 μl of dichloromethane and spot on thin-layer chromatography (TLC) plates coated with silica gel. The TLC plates are developed with a 3:1 mixture of chloroform and ethyl acetate. Unlabeled steroids (17OHP and DHEA) should be spotted on the TLC plate as standards for steroid migration. The unlabeled steroid can be seen with UV light and the position can be marked with a pencil. This facilitates identifying the autoradiographic spot corresponding to the product.

9. Radiolabeled steroids are detected by exposing the TLC plates to a phosphorimager screen. The spots corresponding to unconverted substrate and product are quantified with Image Quant software.

10. The amount of product formed is determined by calculating the percent conversion of radiolabeled substrate into product. Velocity is expressed as (nmol of product)$\times$(pmol of P450)$^{-1}\times$(min)$^{-1}$.

11. The K_m and V_{max} are calculated by plotting (S) versus v, as noted in Subheading 3.4.1 (see Note 6). V_{max}/K_m is used to indicate catalytic efficiency.

12. To assess the loss or gain of catalytic efficiency of a POR mutant, assays can be done using mutant POR and their activity can be compared with that of wild-type POR.

3.4.3 POR Assays Based on Other Microsomal Cytochromes P450

As POR donates electrons to all 50 microsomal cytochrome P450 enzymes, it is useful to determine how various POR mutants will affect their activities. The ability of a POR variant to affect activity varies with the cytochrome P450 assayed (2–5, 10, 14–16) and with the substrate metabolized (4, 17, 18). Thus, the effect of a POR variant on the activity of one P450 cannot be extrapolated to other P450 enzymes and it is necessary to determine the effect of each POR variant with each P450 of interest, and often with the particular substrate being metabolized. Procedures for assays with other important cytochrome P450 enzymes are described in detail in the various chapters of the second edition of Cytochrome P450 Protocols (19).

4 Notes

1. A 20:1 mix of *Taq* DNA polymerase and *Pfu* polymerase gives good results. We have used *Taq* DNA polymerase from a variety of commercial suppliers and all have worked well when mixed with *Pfu* polymerase.

2. An absorbance of 1 at 260 nm corresponds to a DNA concentration of 50 µg/ml. A ratio of A_{260}/A_{280} is used as a measure of DNA purity. A ratio of less than 1.6 indicates protein contamination and may interfere with PCR amplification.

3. After PCR amplification, the samples can be processed immediately or can be stored at −20°C for extended periods.

4. Ampicillin can be used in place of carbenicillin in LB-agar plates to select transformants or for growing overnight cultures. For expression purposes, as the culture is grown for an extended period, we find it is best to use carbenicillin.

5. The cell debris can again be re-suspended in KMDG buffer and sonicated to increase the yield of membranes.

6. To calculate kinetic parameters, (S) versus v can be plotted using commercially available software such as Prism from GraphPad, Leonara.

References

1. Miller WL, Agrawal V, Sandee D, Tee MK, Huang N, Choi JH, Morrissey K, Giacomini KM (2011) Consequences of POR mutations and polymorphisms. Mol Cell Endocrinol 336:174–179

2. Huang N, Agrawal V, Giacomini KM, Miller WL (2008) Genetics of P450 oxidoreductase: sequence variation in 842 individuals of four ethnicities and activities of 15 missense mutations. Proc Natl Acad Sci U S A 105: 1733–1738

3. Sandee D, Morrissey K, Agrawal V, Tam HK, Kramer MA, Tracy TS, Giacomini KM, Miller WL (2010) Effects of genetic variants of human P450 oxidoreductase on catalysis by CYP2D6 in vitro. Pharmacogenet Genomics 20:677–686

4. Agrawal V, Choi JH, Giacomini KM, Miller WL (2010) Substrate-specific modulation of CYP3A4 activity by genetic variants of cytochrome P450 oxidoreductase. Pharmacogenet Genomics 20:611–618

5. Agrawal V, Huang N, Miller WL (2008) Pharmacogenetics of P450 oxidoreductase: effect of sequence variants on activities of CYP1A2 and CYP2C19. Pharmacogenet Genomics 18:569–576

6. de Jonge H, Metalidis C, Naesens M, Lambrechts D, Kuypers DR (2011) The P450 oxidoreductase*28 SNP is associated with low initial tacrolimus exposure and increased dose requirements in CYP3A5-expressing renal recipients. Pharmacogenomics 12:1281–1291

7. Oneda B, Crettol S, Jaquenoud Sirot E, Bochud M, Ansermot N, Eap CB (2009) The P450 oxidoreductase genotype is associated with CYP3A activity in vivo as measured by the midazolam phenotyping test. Pharmacogenet Genomics 19:877–883

8. Daly AK, King BP, Leathart JB (2006) Genotyping for cytochrome P450 polymorphisms. Methods Mol Biol 320:193–207

9. Sandee D, Miller WL (2011) High-yield expression of a catalytically active membrane-bound protein: human P450 oxidoreductase. Endocrinology 152:2904–2908

10. Huang N, Pandey AV, Agrawal V, Reardon W, Lapunzina PD, Mowat D, Jabs EW, Van Vliet G, Sack J, Fluck CE, Miller WL (2005) Diversity and function of mutations in P450 oxidoreductase in patients with Antley-Bixler syndrome and disordered steroidogenesis. Am J Hum Genet 76:729–749

11. van GB, Slater CE (1962) The extinction coefficient of cytochrome c. Biochim Biophys Acta 58:593–595

12. Auchus RJ, Lee TC, Miller WL (1998) Cytochrome b5 augments the 17,20-lyase activity of human P450c17 without direct electron transfer. J Biol Chem 273:3158–3165

13. Wang YH, Tee MK, Miller WL (2010) Human cytochrome P450c17: single step purification and phosphorylation of serine 258 by protein kinase a. Endocrinology 151:1677–1684

14. Gomes LG, Huang N, Agrawal V, Mendonca BB, Bachega TA, Miller WL (2008) The common P450 oxidoreductase variant A503V is not a modifier gene for 21-hydroxylase deficiency. J Clin Endocrinol Metab 93:2913–2916

15. Fluck CE, Mullis PE, Pandey AV (2010) Reduction in hepatic drug metabolizing CYP3A4 activities caused by P450 oxidoreductase mutations identified in patients with disordered steroid metabolism. Biochem Biophys Res Commun 401:149–153

16. Pandey AV, Kempna P, Hofer G, Mullis PE, Fluck CE (2007) Modulation of human CYP19A1 activity by mutant NADPH P450 oxidoreductase. Mol Endocrinol 21:2579–2595

17. Dong P-P, Fang Z-Z, Zhang Y-Y, Ge G-B, Mao Y-X, Zhu L-L, Qu Y-Q, Li W, Wang L-M, Liu C-X, Yang L (2011) Substrate-dependent modulation of the catalytic activity of CYP3A by erlotinib. Acta Pharmacol Sin 32:399–407

18. Kaspera R, Naraharisetti SB, Evangelista EA, Marciante KD, Psaty BM, Totah RA (2011) Drug metabolism by CYP2C8.3 is determined by substrate dependent interactions with cytochrome P450 reductase and cytochrome b5. Biochem Pharmacol 82:681–691

19. Phillips IR, Shephard EA (eds) (2006) Cytochrome P450 protocols, 2nd edn, Methods in molecular biology. Humana, Totowa, NJ

Chapter 20

LICRED: A Versatile Drop-In Vector for Rapid Generation of Redox-Self-Sufficient Cytochromes P450

Federico Sabbadin, Gideon Grogan, and Neil C. Bruce

Abstract

Cytochromes P450 (P450s) are a family of heme-containing oxidases with considerable potential as tools for industrial biocatalysis. Organismal genomes are revealing thousands of gene sequences that encode P450s of as yet unknown function, the exploitation of which will require high-throughput tools for their isolation and characterization. Here, we describe a new ligation-independent cloning vector (LICRED) that enables the high-throughput generation of libraries of redox-self-sufficient P450s, by fusing a range of P450 heme domains to the reductase of P450RhF (RhF-Red) in a robust and generically applicable way.

Key words LICRED, Cytochromes P450, Biocatalyst, Ligation-independent cloning, P450 fusion proteins

1 Introduction

Most prokaryotic P450s can be expressed easily as soluble proteins in *Escherichia coli* and are considered good candidates as industrial biocatalysts for the biosynthesis of fine chemicals. Unfortunately, excess amounts of redox partners are required in order to reach maximum turnover rates, and uncoupling often makes it necessary to increase the amount of pyridine cofactor in the reaction or to incorporate a NAD(P)H recycling system. Several bacterial P450s have recently emerged, where two or more components of the electron-transfer chain are naturally fused through short peptide linkers. The best example of such a multicomponent P450 is P450BM3 from *B. megaterium* (CYP102A1) (1). In this soluble, self-sufficient enzyme, the heme domain is C-terminally fused to a FAD/FMN reductase and can catalyze the hydroxylation of medium- and long-chain fatty acids with impressive turnover rates (17,000/min for arachidonic acid) and coupling efficiency (>95%) (2). Such impressive kinetics are mainly due to the proximity and orientation of the domains in the "fused" protein. Another P450 enzyme featuring a unique structural organization is P450RhF

Ian R. Phillips et al. (eds.), *Cytochrome P450 Protocols*, Methods in Molecular Biology, vol. 987, DOI 10.1007/978-1-62703-321-3_20, © Springer Science+Business Media New York 2013

(CYP116B2) isolated from the soil bacterium *Rhodococcus* sp. NCIMB 9784. Here, the heme domain at the N terminus is fused to a reductase domain containing an FMN-binding motif and an Fe$_2$S$_2$ sub-domain. P450RhF can catalyze the dealkylation of 7-ethoxycoumarin and several alkyl aryl ethers at low turnover rate, although the physiological role and the natural substrate still remain elusive (3–5). Besides naturally occurring fusion P450 systems, artificial fusion constructs have also been engineered either by genetically linking the components of the polypeptide chain or by creating chemical cross-links between individually expressed proteins. Such fusion strategies can be used for the possible characterization of large numbers of "orphan" P450s, whose function is not known, and for the generation of self-sufficient P450 biocatalysts. We have developed the LICRED vector, which allows the fusion of P450 proteins to the N terminus of the reductase domain of P450RhF through a 16-amino-acid peptide linker, by using standard ligation-independent cloning protocols (Fig. 1). Such a system was successfully tested with two model P450 enzymes (P450cam and XplA) and its large-scale applicability was demonstrated by cloning 22 P450-encoding genes from *Nocardia farcinica*, followed by whole-cell activity assays and identification of new activity for two of the chimeric constructs (6, 7).

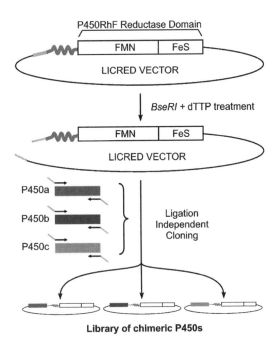

Fig. 1 Scheme of the general strategy for cloning P450 genes in the LICRED platform to generate libraries of self-sufficient P450s.

LICRED: A Versatile Drop-In Vector for Rapid Generation...   241

2   Materials

All solutions must be prepared using ultrapure water (18 MΩ/cm) and analytical grade reagents. Reagents can be prepared at room temperature, unless indicated otherwise. Restriction endonuclease and DNA polymerase enzymes should be kept on ice throughout the experiment.

2.1   Ligation-Independent Cloning Components

1. Purified LICRED plasmid (stock solution at 100 ng/μL in ultrapure water) (see Note 1).

2. Restriction enzymes: *Bse*RI endonuclease (NEB, 4 U/μL) and NEB buffer 2.

3. DNA manipulation equipment:

   Ultrapure agarose.

   Ethidium bromide.

   DNA electrophoresis apparatus.

   DNA ladder.

   Thermal cycler.

   DNA purification kit (such as Wizard® SV Gel and PCR Clean-Up System).

4. DNA polymerases: High-fidelity DNA polymerase (such as Phusion, KOD, Platinum) and T4 DNA polymerase (2.5 U/μL, LIC-qualified, Novagen-Merck) and relevant buffers.

5. Triphosphate nucleotide stocks:

   10 mM dNTP.

   25 mM dTTP.

   25 mM dATP (water stocks, can be stored at –20°C).

6. Chemicals:

   100 mM dithiothreitol (DTT).

   25 mM EDTA (water stocks, can be stored at –20°C).

7. Ligation-independent cloning primers for cloning, T7 primers for sequencing (20 μM stock in ultrapure water, stored at –20°C).

8. Competent *E. coli* cells for heat-shock transformation (such as DH5α or NovaBlue Singles, Novagen-Merck, stored at –80°C).

9. LB agar plates with kanamycin (30 μg/mL) (stored at 4°C for up to 6 months).

10. SOC medium (2% w/v tryptone, 0.5% yeast extract, 10 mM NaCl, 2.5 mM KCl, 10 mM $MgCl_2$, 20 mM glucose).

242   Federico Sabbadin et al.

2.2 Protein Expression Components

1. *E. coli* Rosetta 2 DE3 strain (Novagen-Merck) or analogous (to ensure rare codon expression) transformed with the LICRED vector carrying the desired P450 gene.

2. Purified LICRED plasmid (100 ng/μL) carrying the P450 gene of interest.

3. Chemicals:

   1 M isopropyl-β-D-thiogalactopyranoside (IPTG).

   1 M 5-aminolevulinic acid hydrochloride (ALA).

   0.5 M $FeCl_3$.

   Riboflavin stock (5 mg/mL) in dimethyl sulfoxide (DMSO).

4. Antibiotics:

   Kanamycin stock (100 mg/mL in water).

   Chloramphenicol stock (34 mg/mL in ethanol).

5. LB (Luria-Bertani) broth: Tryptone (10 g/L), NaCl (10 g/L), yeast extract (5 g/L).

6. M9 minimal medium: 90 mM $NaHPO_4$, 22 mM KH_2PO_4, 8.5 mM NaCl, 18.6 mM NH_4Cl, 2 mM $MgSO_4$, 22 mM d-glucose.

2.3 Protein Purification Components

1. Chemicals and solutions:

   0.1 M phenazine methosulfate (PMSF, protease inhibitor) in isopropanol.

   50 mM potassium-phosphate buffer, pH 7.

   Two different concentrations of imidazole (30 and 500 mM).

   DNAse I (5 mg/mL water stock); $MgCl_2$ (1 M stock in water).

2. Instrumentation: AKTA FPLC System (GE Healthcare).

3. Purification columns: 5 mL HiTrap HP nickel-agarose affinity column (GE Healthcare); Superdex™ 200 gel-filtration column (GE Healthcare) (see Note 2).

4. Protein visualization equipment: SDS-PAGE and Western blot apparatus, anti-His antibodies, Coomassie staining reagents.

5. Additional equipment: 50-mL syringes, 0.22-μm syringe filters, centrifugal protein-concentration filters.

2.4 In Vivo Biotransformations

1. *E. coli* Rosetta 2 DE3 strain (Novagen-Merck) or equivalent strain (to ensure rare codon expression) transformed with the LICRED vector carrying the desired P450 gene.

2. LB broth, M9 minimal medium, IPTG, ALA, $FeCl_3$, riboflavin stocks (as in Subheading 2.2), trichloroacetic acid (TCA) (240 mg/mL solution in water).

3. Substrate (generally from a 1 M stock in DMSO, DMF, or ethanol).

2.5 In Vitro Transformation

1. Purified P450-RhF-Red construct (5–30 mg/mL stock in 50 mM potassium-phosphate buffer pH 7, stored at −80°C).
2. 50 mM NADPH stock in 50 mM potassium-phosphate buffer pH 7 (to be prepared fresh every time).
3. Substrate (generally from a 1 M stock in DMSO, dimethylformamide (DMF), or ethanol).

3 Methods

3.1 Cloning P450 Genes in the LICRED Vector

1. Design ligation-independent cloning (LIC) (8) primers for the amplification of the desired P450 gene, bearing in mind that the stop codon must be removed (in order for the P450 to be successfully fused to the reductase domain of P450RhF) and that no start codon is required, since it is already present at the N terminus of the 6-his-tag sequence. The forward primer must be specific for the N-terminal sequence of the gene (possibly between 25 and 30 nucleotides, T_m ~65°C), preceded by the ligation-independent cloning extension 5′-CCAGGGACCAGCA-3′. The reverse primer must be specific for the C-terminal sequence of the gene (again, between 25 and 30 nucleotides, T_m ~65°C), preceded by the LIC extension 5′-GAGGAGAAGGCGCG-3′. Both primers should be at least 35 nucleotides in length, have a GC content around 50%, and end with one or two final G or C nucleotides, to increase annealing stability. Resuspend primers in ultrapure water to get the working primer concentration of 20 μM.
2. PCR: Set up 50-μL reaction mixes containing 25 ng of template DNA, 400 nM of each primer, 200 nM dNTPs, high-fidelity DNA polymerase, and the required buffer. Set PCR conditions as follows: initial denaturing step (depending on the polymerase used, generally between 94 and 98°C) for 2 min; 35 cycles of denaturation (94–98°C, 10–30 s), annealing (65°C, 30 s), and extension (72°C, time depending on the length of the gene); and final extension (72°C, 10 min) (see Note 3).
3. Verifying the PCR: Run 5 μL of the PCR products on an agarose gel, to verify a successful reaction, and clean up the remaining 45 μL with an appropriate commercially available kit. If the agarose gel analysis indicates unspecific or suboptimal amplification, it may be useful to repeat the PCR with a temperature gradient in the annealing step, ranging from 60 to 72°C. Calculate DNA concentration using a NanoDrop or with quantitative DNA markers.
4. Plasmid preparation: Transform a good pET host strain with the LICRED plasmid (derived from a pET-28a plasmid) and retrieve the plasmid via miniprep. Linearize the LICRED

plasmid in a 100-μL reaction volume containing 5 μg of vector and 20 U of *Bse*RI endonuclease plus the appropriate buffer. Incubate at 37°C for 110 min. Electrophorese the sample through a 1% (w/v) agarose gel, until the linearized vector is clearly separated from the uncut vector (this usually requires about 1.5 h if using a constant voltage of 110 V). It is very important to excise carefully the linearized vector, thereby minimizing the chance of undigested vector being carried through, as this could give false positives later on. Purify the linearized LICRED plasmid with an appropriate commercially available kit, and concentrate it to 50 ng/μL or more (see Note 4).

5. Plasmid-T4 DNA polymerase reaction: The plasmid outlined above usually results in enough digested vector to perform one 40-μL LIC reaction, which includes 2 μg of linear vector, 2 U of T4 DNA polymerase (LIC qualified), the appropriate buffer, 5 mM DTT (from water stock of 100 mM), and 2.5 mM dTTP (from water stock of 25 mM). Mix all components thoroughly and incubate at 22°C for 30 min, then stop the reaction by incubating at 75°C for 20 min. This process is best carried out on a programmed PCR machine. Purify the DNA using a commercial PCR clean-up kit (such as the Wizard® SV Gel and PCR Clean-Up Kit from Promega), elute in ultrapure water, and bring to a final concentration of 100 ng/μL.

6. Insert-T4 DNA polymerase reaction: Set up 20-μL reaction volumes containing 0.2 pmol of purified PCR product (use the equation: number of bp in the insert × 650 = pg/pmol), 1 U of T4 DNA polymerase, the appropriate buffer, 5 mM DTT, and 2.5 mM dATP. Mix and incubate at 22°C for 30 min, then stop the reaction with further incubation at 75°C for 20 min. This reaction mix does not need purification.

7. LIC-annealing reaction: Add 2 μL of T4-treated insert (100 ng/μL) to 1 μL of T4-treated vector (50 ng/μL) and incubate for 10 min at room temperature (20–22°C). Add 1 μL of 25 mM EDTA to give a final volume of 4 μL, mix with the pipette tip and leave at room temperature for a further 10 min. For the LIC control reaction, use water instead of the insert.

8. Transformation: Add 2 μL of LIC-annealing reaction to 50 μL of highly competent *E. coli* cells. Incubate on ice for 5 min then heat shock at 42°C for 2 min and immediately transfer to ice for another 5 min. Add 500 μL of SOC medium, incubate at 37°C for 1 h, then centrifuge the cells at low speed (1,500 × g for 5 min in a microfuge), resuspend them in 80 μL of SOC medium, spread on LB agar containing kanamycin (30 μg/mL), and incubate at 37°C overnight. A successful LIC reaction should result in ~100 colonies, while the negative control plate should remain blank.

LICRED: A Versatile Drop-In Vector for Rapid Generation...   245

9. Identification of positive clones: Colony PCR can be used to identify positive colonies, by using gene-specific primers, T7 primers (which anneal on both sides of the fusion construct), or even the LIC primers themselves. Colony PCR is recommended only when colonies grow on the negative control plate as well. The negative control plate should give an idea of the background contamination (plasmid with no insert); therefore, it would be ideal not to have any colonies on the negative control. However, if the same number of colonies grows on the reaction plate and on the negative control plate, proceed with colony PCR anyway because experimental evidence suggests that a significant number of positive clones might still be present. If the negative control has no colonies, colony PCR is not necessary and two colonies can be immediately picked and grown overnight in LB medium, for a miniprep purification. Restriction analysis of the purified plasmids can be performed, bearing in mind that BglII, XbaI, and NcoI endonucleases cut the plasmid at the 5' end of the construct, whereas NdeI, NheI, BamHI, EcoRI, SacI, SalI, $Hind$III, NotI, EagI, and XhoI cut at the 3' end. The expected size of the excised product can be calculated as the sum of the RhF-reductase (~1,000 bp) and the gene of interest. DNA sequencing confirmation of the insert is also recommended and can be carried out, at the 5' end, with T7 forward primer or with a gene-specific primer. If the insert is longer than 1,000 bp, it might be necessary to design special internal primers, in order to ensure a good coverage.

3.2 Protein Expression and Purification

1. Protein expression. Pick a single colony of *E. coli* Rosetta 2, transformed with the LICRED vector carrying the P450 gene of interest, and inoculate it into 5 mL of LB medium containing kanamycin (100 µg/mL) and chloramphenicol (34 µg/mL). After overnight incubation at 37°C, with shaking at 200 rpm, use 10 mL to inoculate 1 L of M9 minimal medium (see Note 5) containing the same antibiotics and incubate at 37°C with shaking. When the OD_{600} reaches 0.8, cool down the flask with tap water to about 20°C and induce expression with 1 mM IPTG, 1 mM ALA, 0.5 mM $FeCl_3$, and riboflavin (5 µg/mL). After incubation overnight at 20°C with shaking, harvest the cells by centrifugation (4,000×g for 5 min) and resuspend them in 70 mL of 50 mM potassium-phosphate buffer, pH 7 (storage buffer). Split the cell suspension into two 50-mL falcon tubes; add to each tube the required amount of a protease inhibitor (for example, 70 µL of PMSF from a stock of 0.1 M in isopropanol, giving a final concentration of 200 µM of PMSF). Perform cell lysis (by sonication or French press) and then clear the lysate of cell debris by centrifugation at 16,000×g for 15 min. Collect the supernatant (containing the

soluble fusion protein); add *DNAse I* to a final concentration of 5 µg/mL and MgCl₂ to a final concentration of 1 mM, to cleave the DNA and reduce viscosity before protein purification. Filter using a syringe and a 0.22-µm membrane.

2. Nickel-affinity chromatography and gel filtration. Load the filtered supernatant onto a 5-mL nickel-agarose affinity column (such as the HiTrap HP, GE Healthcare) with a syringe. The bound protein must then be washed with 100 mL of 50 mM potassium-phosphate buffer, pH 8, containing 30 mM imidazole plus 300 mM NaCl and eluted with a gradient (30–500 mM) of imidazole in the same buffer, by using an AKTA FPLC System (GE Healthcare). Protein concentration can be estimated from the A_{280} and the fractions corresponding to the absorbance peak are pooled and concentrated to a volume of 2 mL using centrifugal filters (10-kDa molecular weight cut-off). A second purification step of size-exclusion chromatography must then be carried out to remove imidazole, which is a strong inhibitor of P450. The concentrated LICRED fusion protein is therefore applied to a suitable gel-filtration column (such as a Superdex™ 200, GE Healthcare), pre-equilibrated with 50 mM potassium-phosphate buffer, pH 7. Collect the peak fractions and concentrate by using centrifugal filters. Stock solutions of LICRED fusion proteins (generally at 10 mg/mL) can be prepared by mixing with glycerol (in a protein solution:glycerol ratio of 2:1) and snap-freezing in liquid nitrogen. The glycerol stock can be stored at –80°C for several months (see Note 6).

3. Protein purity should be checked by SDS-PAGE and Western blotting, before activity assays.

3.3 In Vivo Biotransformation

1. Culture preparation. Use one colony of *E. coli* Rosetta 2 (Novagen-Merck), transformed with the LICRED plasmid containing the desired P450 gene, to inoculate 5 mL of LB medium containing kanamycin (100 µg/L) and chloramphenicol (34 µg/mL). Incubate overnight at 37°C with shaking at 200 rpm, then collect the cells by centrifugation at 4,000 × g and resuspend in 250 mL of fresh M9 medium containing kanamycin (100 µg/mL) and chloramphenicol (34 µg/mL). Incubate the cultures in 1-L flasks at 30°C with shaking until the OD_{600} is 0.8, then add IPTG and ALA to final concentrations of 0.4 and 0.5 mM, respectively, and transfer to 20°C for further incubation overnight with shaking. Harvest cells by centrifugation, transfer 300 mg of wet cell paste to a sterile 50-mL Falcon tube and resuspend cells in 3 mL of 50 mM potassium-phosphate buffer, pH7. After supplementation with substrate (generally 0.5 mM from a concentrated stock in DMSO, DMF or ethanol), place the tubes in a shaking

incubator at 4°C for several hours (depending on the speed of the reaction). After incubation, the supernatant can be analyzed by using HPLC-MS or GC-MS, depending on the substrate. As a generic protocol, samples for GC analysis can be prepared by mixing 100 μL of the reaction mixture with an equal volume of ethyl acetate (vortexing for 30 s) and harvesting the supernatant (containing the organic phase) by centrifugation in a microfuge at $18,400 \times g$ for 1 min. Samples for HPLC analysis can be prepared by stopping the reaction by addition of TCA to a final concentration of 24 mg/mL and collecting the supernatant after centrifugation at $18,400 \times g$ for 1 min.

3.4 In Vitro Biotransformation

1. Small reactions can be set up in Eppendorf tubes by using different enzyme concentrations (for the P450cam fusion protein, for example, concentrations between 0.2 and 2 μM were used), suitable substrate concentrations (e.g., 0–30 μM d-camphor), and 150 μM NADPH. An excess concentration of NADPH should be used and NADPH-recycling systems (such as alcohol dehydrogenase) can be added in order to regenerate the cofactor. Reactions can be performed at room temperature (20°C) with constant shaking at 1,000 rpm. For kinetic measurements, NADPH oxidation rate can be monitored at 340 nm using a spectrophotometer, and the following formula is applied to convert the difference in absorbance into the amount of cofactor oxidized per time unit:

$$\frac{\Delta C}{\Delta t} = \frac{K}{\varepsilon'}$$

where ΔC is the difference in [NADPH] after n min, Δt is the difference in time (min), K is the change of absorbance over time, and ε is the extinction coefficient of NADPH at 340 nm ($6.22 \times 10^{-3} \mu M^{-1} cm^{-1}$).

2. When using NADPH oxidation to determine the kinetic parameters of the enzyme, always use a negative control reaction (all reagents except the substrate) and subtract this as background in order to obtain the real activity values. This is important because all LICRED fusion proteins suffer from some level of uncoupling, a common issue among P450 enzymes.

4 Notes

1. The LICRED plasmid was engineered from the commercial pET28a and is, therefore, a low-copy-number plasmid. Standard miniprep kits allow purification of roughly 5 μg of

plasmidic DNA from a 5-mL overnight culture of *E. coli*, and such amount should be sufficient for the LIC procedure. However, since any linearized LIC vector must be prepared fresh each time, we suggest keeping a stock of at least 20 μg of circular plasmid in the freezer at –20°C. Also, the quality of the miniprepped plasmid is critical for efficient restriction with *Bse*RI; therefore, the DNA stock should always be checked for purity spectrophotometrically (a NanoDrop spectrophotometer is the best option, since it requires minimal volumes) and have an $A_{260/280} > 1.8$.

2. As an alternative to the 5-mL HiTrap HP nickel-agarose affinity column, nickel-NTA agarose slurry can be used in a batch system, i.e., mixing the resin with the lysate and performing several washing steps before elution. This procedure can be carried out in 50-mL Falcon tubes and can be useful when multiple proteins need to be purified at the same time and purity is not a priority. The eluate can then be dialyzed three times in a 5-L bucket containing buffer, in order to remove imidazole. When purity is critical, however, we suggest the use of a nickel-agarose column followed by a gel-filtration step.

3. The necessary addition of LIC-specific ends to the PCR oligonucleotides increases the chance of formation of primer dimers and hairpin structures, which can hinder the efficient and specific amplification of the insert. The use of special buffers (like the 5× GC Buffer of the Phusion® High-Fidelity DNA Polymerase Kit) can generally solve this problem. DMSO, in concentrations between 1 and 5%, can also improve the PCR. A touchdown PCR approach can also be used, in which the earlier steps have a high annealing temperature, which is then decreased for every subsequent set of cycles.

4. Purification of DNA from an agarose matrix often causes significant loss of material and low recovery (<30%). Decreasing the agarose concentration from the standard 1% (w/v) to 0.6% (w/v) allows quicker runs and easier melting of the excised band, which can significantly improve the DNA recovery afterwards (although band separation will be less accurate). When using commercial kits (such as the Wizard® SV Gel and PCR Clean-Up Kit from Promega), we also suggest performing the elution twice with 55 μL of pure water, pre-warmed at 65°C. This procedure usually gives about 100 μL of DNA at a concentration of 20 ng/μL, which can be easily concentrated to a smaller volume with a SpeedVac centrifugal evaporator, before the T4 DNA polymerase treatment.

5. In our experience, the use of M9 minimal medium for expression of artificial P450-RhF fusion proteins in *E. coli* is essential in order to obtain high yields of pure, active enzyme. When expression was carried out in LB medium, we observed significant

proteolysis of the chimeric construct and appearance of bands with lower molecular weight, when analyzed by SDS-PAGE and Western blotting.

6. Once defrosted, the protein-glycerol stock should be used within 24 h, after which spontaneous cleavage of the peptide linker between the P450 and reductase domains occurs. This can easily be verified by SDS-PAGE, by monitoring the appearance of several bands of lower molecular weight. It is therefore good practice to aliquot the glycerol stock into small, single-use batches before freezing.

Acknowledgement

This work was supported by the Centre of Excellence for Biocatalysis, Biotransformations and Biocatalytic Manufacture (CoEBio3).

References

1. Miura Y, Fulco AJ (1974) (Omega -2) hydroxylation of fatty acids by a soluble system from bacillus megaterium. J Biol Chem 249: 1880–1888

2. Noble MA, Miles CS, Chapman SK, Lysek DA, MacKay AC, Reid GA, Hanzlik RP, Munro AW (1999) Roles of key active-site residues in flavocytochrome P450 BM3. Biochem J 339:371–379

3. Roberts GA, Grogan G, Greter A, Flitsch SL, Turner NJ (2002) Identification of a new class of cytochrome P450 from a Rhodococcus sp. J Bacteriol 184:3898–3908

4. Roberts GA, Celik A, Hunter DJ, Ost TW, White JH, Chapman SK, Turner NJ, Flitsch SL (2003) A self-sufficient cytochrome p450 with a primary structural organization that includes a flavin domain and a [2Fe-2S] redox center. J Biol Chem 278:48914–48920

5. Hunter DJ, Roberts GA, Ost TW, White JH, Muller S, Turner NJ, Flitsch SL, Chapman SK

(2005) Analysis of the domain properties of the novel cytochrome P450 RhF. FEBS Lett 579:2215–2220

6. Robin A, Roberts GA, Kisch J, Sabbadin F, Grogan G, Bruce NC, Turner NJ, Flitsch SL (2009) Engineering and improvement of the efficiency of a chimeric [P450cam-RhFRed reductase domain] enzyme. Chem Commun 18:2478–2480

7. Sabbadin F, Hyde R, Robin A, Hilgarth EM, Delenne M, Flitsch S, Turner N, Grogan G, Bruce NC (2010) LICRED: a versatile drop-in vector for rapid generation of redox-self-sufficient cytochrome P450s. Chembiochem 11:897–994

8. Bonsor D, Butz SF, Solomons J, Grant S, Fairlamb IJ, Fogg MJ, Grogan G (2006) Ligation Independent Cloning (LIC) as a rapid route to families of recombinant biocatalysts from sequenced prokaryotic genomes. Org Biomol Chem 4:1252–1260

Chapter 21

Update on Allele Nomenclature for Human Cytochromes P450 and the Human Cytochrome P450 Allele (CYP-Allele) Nomenclature Database

Sarah C. Sim and Magnus Ingelman-Sundberg

Abstract

Interindividual variability in xenobiotic metabolism and drug response is extensive and genetic factors play an important role in this variation. A majority of clinically used drugs are substrates for the cytochrome P450 (CYP) enzyme system and interindividual variability in expression and function of these enzymes is a major factor for explaining individual susceptibility for adverse drug reactions and drug response. Because of the existence of many polymorphic *CYP* genes, for many of which the number of allelic variants is continually increasing, a universal and official nomenclature system is important. Since 1999, all functionally relevant polymorphic *CYP* alleles are named and published on the Human Cytochrome P450 Allele (CYP-allele) Nomenclature Web site (http://www.cypalleles.ki.se). Currently, the database covers nomenclature of more than 660 alleles in a total of 30 genes that includes 29 CYPs as well as the cytochrome P450 oxidoreductase (POR) gene. On the CYP-allele Web site, each gene has its own Webpage, which lists the alleles with their nucleotide changes, their functional consequences, and links to publications identifying or characterizing the alleles. CYP2D6, CYP2C9, CYP2C19, and CYP3A4 are the most important CYPs in terms of drug metabolism, which is also reflected in their corresponding highest number of Webpage hits at the CYP-allele Web site.

The main advantage of the CYP-allele database is that it offers a rapid online publication of CYP-alleles and their effects and provides an overview of peer-reviewed data to the scientific community. Here, we provide an update of the CYP-allele database and the associated nomenclature.

Key words Cytochrome P450, Pharmacogenetics, Adverse drug reactions, Drug response, NADPH cytochrome P450 reductase

1 Introduction

Interindividual variability in xenobiotic metabolism and drug response is extensive, and the causes can be of genetic, physiological, pathophysiological, or environmental origin. Genetic explanations for interindividual variability have been demonstrated for drug transport, drug metabolism, and drug–target interactions (1–4). Drug response and adverse drug reactions are often dependent on drug concentration and exposure. Ultrarapid drug metabolism (UM)

Ian R. Phillips et al. (eds.), *Cytochrome P450 Protocols*, Methods in Molecular Biology, vol. 987,
DOI 10.1007/978-1-62703-321-3_21, © Springer Science+Business Media New York 2013

can cause decreased drug efficacy, due to enhanced drug metabolism, or adverse reactions, due to increased bioactivation or formation of toxic metabolites. On the other hand, poor metabolism (PM) might cause adverse drug reactions (ADRs), due to inadequate metabolism and excessive levels of the parent drug, or, in the case of prodrugs, decreased bioactivation and reduced drug response (5). A recent important example illustrating this phenomenon is the CYP2C19-dependent bioactivation of the antiplatelet drug clopidogrel, which is impaired in patients with two defective *CYP2C19* alleles, thereby causing diminished drug response and in particular increased risk of stent thrombosis (cf. (1)). On the contrary, patients homozygous for the rapid *CYP2C19*17* allele appear to be at higher risk for bleeding complications (cf. (1)). In addition, the interindividual difference in susceptibility to carcinogens and environmental toxins is extensive and polymorphic CYP enzymes play a key role in their bioactivation and inactivation (6).

The genetics of drug-metabolizing enzymes, and of CYPs in particular, is a critical aspect in the design of improved drug therapies with higher response rates and reduced ADRs. In addition, the genetics of CYP enzymes is taken into great consideration at an early stage when new drug candidates are developed in the pharmaceutical industry. Both the US Food and Drug Administration (FDA) and the European Medicines Agency (EMA) have developed guidelines for using pharmacogenomics in drug development, where the polymorphic CYP enzymes play a major role (http://www.fda.gov/downloads/Drugs/GuidanceComplianceRegulatoryInformation/Guidances/UCM243702.pdf and http://www.ema.europa.eu/docs/en_GB/document_library/Scientific_guideline/2010/05/WC500090323.pdf). A fundamental basis for the research and applications regarding interindividual variability in xenobiotic metabolism by polymorphic CYPs is a common nomenclature for genetic variants and a system that allows researchers, clinicians, and regulators to be easily updated within the field.

2 Cytochromes P450

There are 57 active human *CYP* genes (http://drnelson.utmem.edu/CytochromeP450.html) and their corresponding protein products are involved in metabolism and activation of endogenous and exogenous substrates. The majority of CYPs involved in drug metabolism are polymorphic, and genetic variation in these genes can lead to abolished, reduced, altered, or increased enzyme activity. Abolished enzyme activity can be caused by gene deletion events, but can also have its origin in nucleotide variations causing, for example, altered splicing, introduction of stop codons, abolished transcription start sites, and deleterious amino-acid changes. Variations occurring in substrate-recognition sites can result in the synthesis of enzymes with altered substrate specificity, whereas increased enzyme activity can be the result of variations affecting the transcription rate, as is the case for *CYP2C19*17* (7). In addition, duplications of active *CYP* genes occur, leading to increased drug metabolism and the ultrarapid metabolizer phenotype, as observed, for example, in *CYP2D6*.

Cytochrome P450 is the main class of enzymes responsible for the metabolism of drugs and other xenobiotics. CYP enzymes are responsible for 75–80% of the phase I-dependent metabolism and 65–70% of the clearance of clinically used drugs (8, 9). CYP2C9, CYP2C19, and CYP2D6 are highly polymorphic and collectively account for about 40% of the metabolism of clinically used drugs. In addition, CYP1A2, CYP2A6, and CYP2B6 are polymorphic enzymes that significantly contribute to xenobiotic metabolism. The gene encoding the clinically most important drug-metabolizing enzyme, CYP3A4, has very few and rare allelic variants affecting enzyme function, despite a high level of interindividual variability in CYP3A4-catalyzed reactions. In general, genes encoding drug-metabolizing enzymes exist as many different allelic variants and the Human Cytochrome P450 Allele (CYP-allele) Nomenclature Database described in this chapter summarizes the general composition and properties of these allelic variants, which facilitates scientific, industrial, and regulatory work in the area of pharmacology and pharmacogenetics. For a summary of other databases in the pharmacogenetic area, see ref. 10.

3 The Human Cytochrome P450 Allele (*CYP*-Allele) Nomenclature Database Web Site

3.1 Introduction

Of 57 active human *CYP* genes, 29 are listed in the Human Cytochrome P450 Allele (CYP-allele) Nomenclature Database (http://www.cypalleles.ki.se). The CYP-allele Web site has the aim of unifying *CYP* allele nomenclature designation and to provide a database of genetic variation in CYPs as has been described in previous reviews (see ref. 11–13). Until today, a total number of more than 660 alleles have been designated (see Table 1). These alleles include 41 alleles of the cytochrome P450 oxidoreductase (POR) gene, which was incorporated into the Web site in 2008 (14). The inclusion of POR was motivated based on the link between POR and CYP activity, with POR engaging in electron donation to CYP enzymes for their catalytic activity.

The main aim for the CYP-allele Web site and database is to provide an official and unified nomenclature system as well as a summary of *CYP* alleles and their associated effects. The Webmaster, Sarah C. Sim, is responsible for the Web site and allele designation of new submissions. The Editorial Board of the Allele Nomenclature Committee consists of Magnus Ingelman-Sundberg, Ann K. Daly, and Daniel W. Nebert, and the Advisory Board is composed of 12 members listed at the Web site. References to nomenclature

Table 1
Gene list showing the number of unique alleles per gene in the CYP-allele database currently comprising 30 genes (ordered by the number of alleles)

Gene	Alleles	Gene	Alleles
CYP21A2	181	CYP2J2	10
CYP2D6	105	CYP5A1	9
POR	41	CYP2E1	7
CYP2A6	37	CYP4B1	7
CYP2C9	35	CYP2F1	6
CYP2B6	29	CYP2W1	6
CYP2C19	28	CYP19A1	5
CYP1B1	26	CYP2S1	5
CYP3A4	22	CYP8A1	4
CYP1A2	21	CYP26A1	4
CYP4A22	15	CYP3A7	3
CYP2C8	14	CYP4F2	3
CYP1A1	11	CYP3A43	3
CYP3A5	11	CYP2R1	2
CYP2A13	10	CYP4A11	1

The number of unique alleles represent those present on the CYP-allele database in January 2012

guidelines followed by the CYP-allele Nomenclature Committee can be found at the CYP-allele Web site (http://www.cypalleles.ki.se/criteria.htm).

Each gene on the CYP-allele Web site has its own Webpage with links to gene and protein reference sequences, a list of the alleles, the corresponding nucleotide variations, functional consequences, and *in vivo* and *in vitro* activity described together with relevant literature references (see Fig. 1). When the specific variant nucleotide and/or amino-acid variation causing a change in enzyme activity is known, the variation appears in bold. Links to single-nucleotide polymorphisms (SNPs) listed at the NCBI SNP database (http://www.ncbi.nlm.nih.gov/snp) are present when available and can be found both at the listed nucleotide and protein variations.

3.2 Inclusion Criteria for CYP-Alleles

The CYP-allele Web site describes sequence variations in the coding, intronic, and regulatory regions of the *CYP* genes. The designation of an allele ideally requires determination of all sequence variations in the complete gene, but at a minimum all exons and exon-intron borders must be covered.

P450 oxidoreductase (POR) allele nomenclature

Allele	Protein	Nucleotide changes		Effect	Verified clinical phenotype	In vitro activity	References
		cDNA	Gene				
POR*1	POR.1	None	None				
POR*2	POR.2	1370G>A	31187G>A	R457H	Decr.	None	Fluck et al., 2004 Arlt *et al.*, 2004 Adachi *et al.*, 2004 Sandee *et al.*, 2010
POR*3			27615G>A	Splicing defect	Decr.		Fluck et al., 2004
POR*4	POR.4	1475T>A	31663T>A	V492E	Decr.		Fluck et al., 2004
POR*5	POR.5	859G>C	29556G>C	A287P	Decr.	Decr.	Fluck et al., 2004 Arlt *et al.*, 2004 Sandee *et al.*, 2010

Fig. 1 A shortened but representative gene Webpage of the Human Cytochrome P450 Allele (CYP-allele) Nomenclature Web site. All gene Webpages present on the CYP-allele Web site contain links to gene and protein reference sequences, a list of the alleles, the corresponding nucleotide variations, functional consequences, and *in vivo* and *in vitro* activity, together with relevant literature references. When the specific variation causing a change in enzyme activity *in vitro* or *in vivo* is known, the variation appears in *bold*. Links to the NCBI SNP database (http://www.ncbi.nlm.nih.gov/snp) are present when available. Note that the POR allele Webpage example has been shortened, compared with that present on the CYP-allele Web site

The reference allele is designated *1. In order to receive a unique allelic name for a new variant sequence (e.g., *CYP2B6*4*), the nucleotide changes should affect transcription, splicing, translation, posttranscriptional, or posttranslational modifications, or result in at least one amino-acid change. When several effective polymorphisms are present on the same allele, the allelic number given is based on the polymorphism that causes the most severe functional consequence. For example, if a polymorphism that gives rise to a splicing defect (e.g., *CYP2C19*2*) also exists in combination with other less effective mutations, such as amino-acid changes, this allele will receive an extra letter (e.g., *CYP2C19*2B*), whereas the first allele described will receive an A (e.g., *CYP2C19*2A*). A combination of effective variations of equal size effects, such as amino-acid substitutions, will be given a new numeric allele name (e.g., *CYP2B6*6*), whether or not the variants have been shown to exist on their own. Importantly, linkage with other SNPs, especially those causing functional effects, must be experimentally addressed before allele designation. In the past, computer-predicted haplotypes have been designated and these are specified on the CYP-allele Web site; however, predicted alleles are no longer named.

Since the emphasis of the CYP-allele database is on variants causing functional effects, the CYP-allele Web site has recently introduced a more restricted designation system, in which novel alleles will only receive an allelic designation when carrying a novel sequence variation causing a functional alteration, or when new combinations of variations causing functional effects are described. Furthermore, if a new variation causing functional effects is shown to exist on different backgrounds of noneffective nucleotide changes, the sequence with the lowest number of additional variations will be chosen for allele designation. Thus, the previous policy of naming sub-alleles containing additional genetic variation not causing functional effects (e.g., *CYP2D6*10B*) is abandoned. The new restriction is the result of the highly increased activity in gene sequencing in the past years, whereby a very high number of new nonfunctional SNPs and combinations of these are detected.

There are some inconsistencies regarding the allelic names of early-characterized genes that do not follow today's guidelines; however, renaming of these alleles would result in confusion and has been avoided. Similarly, sub-alleles containing additional noneffective sequence variations that have been named in the past will be retained on the Web site. The updated inclusion criteria for new alleles can be found in Table 2 and at http://www.cypalleles.ki.se/criteria.htm.

3.3 Procedures for Submission of New Alleles

New alleles are submitted to the Webmaster (Sarah.Sim@ki.se), together with information sufficient to satisfy the criteria to be assigned an allele (see Table 2 and http://www.cypalleles.ki.se/criteria.htm). New variants are frequently submitted to the CYP-allele Web site, and most submissions concern reservation of allelic names for novel variants intended for publication in a journal. It is highly recommended to contact the Webmaster to reserve an allelic name to be included in a manuscript before submission to a journal, although journal publication is not a requisite for publication on the CYP-allele Web site. The use of allele names that have not been approved by the CYP-allele Nomenclature Committee is strongly discouraged, as is the use of invented nomenclature systems using the *-function. If an allelic name has not been reserved, authors should simply refer to novel alleles in their manuscripts as allele A or allele 1, etc. In case the authors do not want to release the information on the CYP-allele Web site before publication, the Webmaster will provide an allelic designation, but not release the information on the CYP-allele Web site or to any person until the manuscript has been accepted or published. Papers describing additional phenotype characterizations of previously identified alleles can continuously be submitted to the Webmaster for inclusion on the CYP-allele Web site.

3.4 Use of the CYP-Allele Database

The CYP-allele database is extensively used and referenced in the literature, and new alleles that have not been designated an allelic name by the nomenclature committee are rarely found in the literature. The total number of hits per week is close to 1,000. By far the most popular gene in the CYP-allele database is *CYP2D6*, which can receive more than 2,000 hits per month, followed by

Table 2
Criteria for inclusion of alleles in the Human Cytochrome P450 Allele (CYP-allele) Nomenclature Database

On this Web site only human CYP and cytochrome P450 oxidoreductase (POR) alleles are considered

The gene and the allele name is separated by an asterisk followed by Arabic numerals designating the specific allele (e.g., *CYP1A1*3*)

For allele designation, sequence information of at least all exons and exon-intron borders must be submitted

To be assigned a unique allele name, the sequence should contain at least one nucleotide change that has been shown to affect transcription, splicing, translation, posttranscriptional, or posttranslational modifications or result in at least one amino-acid change. So-called sub-alleles containing additional nonfunctional variations in addition to the functional ones described (e.g., *CYP2D6*10B*) will no longer be designated

If a new sequence variation causing functional consequences is shown to exist in different combinations with other sequence variations not causing functional effects, the sequence with the lowest number of additional variations will be chosen for allele designation

If a sequence variation causing detrimental effects (e.g., the 1846G > A variation in *CYP2D6*4* causing a splicing defect) is shown to exist together with other variations causing functional effects, the variation causing the detrimental effect will determine the name of the allele (e.g., *CYP2D6*4B*)

If sequence variations causing functional effects of similar magnitude (e.g., amino-acid substitutions) are found to exist together as well as on their own, the combination allele will be given a new allele name (see, e.g., *CYP2D6*102* and **103*)

For extra gene copies (n) placed in tandem on the same chromosome, the entire allelic arrangement should be referred to as, e.g., *CYP2D6*2Xn*

Nucleotides are numbered according to the base A in the initiation codon ATG as +1 and the base before A as –1

The names for the corresponding proteins have a period between the name of the gene product and the allele number (e.g., CYP2D6.2). If the allele is unable to produce full length protein, no protein name will be assigned

SNPs that are not easily assigned to a specific allele are listed at the bottom of the corresponding gene Webpage, in a separate SNP table

Novel sequence variations causing functional consequences that have an unclear linkage with other variations causing functional effects will not be assigned

All submissions are kept confidential and new allele names are, on request, kept confidential until publication. Authors are encouraged to submit their novel allele to the CYP-allele Web site before submission to a journal, especially since many editors request naming of new alleles by the CYP-allele Web site before publication. Journal publication is, however, not a requirement for publication on the CYP-allele Web site

Usage of star (*) allele designations resembling those officially designated by the CYP-allele Nomenclature Committee is highly discouraged because of the apparent risk of confusion

Further information on *in vitro* or *in vivo* activity analyses for known *CYP* alleles can at any time be submitted to the CYP-allele Web site for inclusion with the respective allele

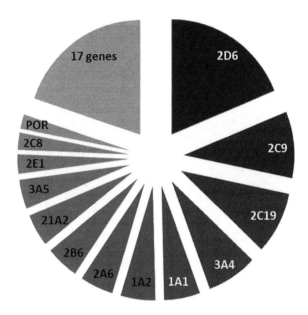

Fig. 2 The relative number of gene-specific CYP-allele Webpage hits. Seventeen genes with fewer than 170 hits per month are shown as a group. As a whole, the CYP-allele Web site receives close to 1,000 hits per day. The relative number of Webpage hits was measured over a 6-month period in 2011

CYP2C9, *CYP2C19*, and *CYP3A4* (see Fig. 2). The *CYP2D6* gene is the second highest in terms of number of alleles assigned (see Table 1). The gene with the largest number of alleles assigned is, however, *CYP21A2*, with 181 unique alleles, many of which affect the rate of biosynthesis of steroid hormones and can cause congenital adrenal hyperplasia.

4 Conclusions

The CYP-allele Web site serves as an important platform for scientists in the cytochrome P450 field, by the designation of new alleles, providing a means of rapid online publication and a database of CYP-alleles and their impact on enzyme function. The Web site is highly established in the scientific community and extensively referred to in the literature.

Acknowledgement

The Human Cytochrome P450 Allele (CYP-allele) Nomenclature Database is financed by the Section of Pharmacogenetics, Karolinska Institutet.

References

1. Sim SC, Ingelman-Sundberg M (2011) Pharmacogenomic biomarkers: new tools in current and future drug therapy. Trends Pharmacol Sci 32:72–81
2. Paugh SW, Stocco G, McCorkle JR, Diouf B, Crews KR, Evans WE (2011) Cancer pharmacogenomics. Clin Pharmacol Ther 90:461–466
3. Ma Q, Lu AY (2011) Pharmacogenetics, pharmacogenomics, and individualized medicine. Pharmacol Rev 63:437–459
4. Wang L, McLeod HL, Weinshilboum RM (2011) Genomics and drug response. N Engl J Med 364:1144–1153
5. Ingelman-Sundberg M, Sim SC, Gomez A, Rodriguez-Antona C (2007) Influence of cytochrome P450 polymorphisms on drug therapies: pharmacogenetic, pharmacoepigenetic and clinical aspects. Pharmacol Ther 116:496–526
6. Rodriguez-Antona C, Gomez A, Karlgren M, Sim SC, Ingelman-Sundberg M (2010) Molecular genetics and epigenetics of the cytochrome P450 gene family and its relevance for cancer risk and treatment. Hum Genet 127:1–17
7. Sim SC, Risinger C, Dahl ML, Aklillu E, Christensen M, Bertilsson L, Ingelman-Sundberg M (2006) A common novel CYP2C19 gene variant causes ultrarapid drug metabolism relevant for the drug response to proton pump inhibitors and antidepressants. Clin Pharmacol Ther 79:103–113
8. Bertz RJ, Granneman GR (1997) Use of in vitro and in vivo data to estimate the likelihood of metabolic pharmacokinetic interactions. Clin Pharmacokinet 32:210–258
9. Evans WE, Relling MV (1999) Pharmacogenomics: translating functional genomics into rational therapeutics. Science 286:487–491
10. Sim SC, Altman RB, Ingelman-Sundberg M (2011) Databases in the area of pharmacogenetics. Hum Mutat 32:526–531
11. Ingelman-Sundberg M, Daly AK, Oscarson M, Nebert DW (2000) Human cytochrome P450 (CYP) genes: recommendations for the nomenclature of alleles. Pharmacogenetics 10:91–93
12. Sim SC, Ingelman-Sundberg M (2010) The Human Cytochrome P450 (CYP) Allele Nomenclature website: a peer-reviewed database of CYP variants and their associated effects. Hum Genomics 4:278–281
13. Oscarson M, Ingelman-Sundberg M (2002) CYPalleles: a web page for nomenclature of human cytochrome P450 alleles. Drug Metab Pharmacokinet 17:491–495
14. Sim SC, Miller WL, Zhong XB, Arlt W, Ogata T, Ding X, Wolf CR, Fluck CE, Pandey AV, Henderson CJ, Porter TD, Daly AK, Nebert DW, Ingelman-Sundberg M (2009) Nomenclature for alleles of the cytochrome P450 oxidoreductase gene. Pharmacogenet Genomics 19:565–566

Chapter 22

Simultaneous In Vivo Phenotyping of CYP Enzymes

Sussan Ghassabian and Michael Murray

Abstract

As major determinants of the duration of drug action the CYP enzymes strongly influence drug efficacy and toxicity. In vivo phenotyping for CYP activities using cocktails of well-tolerated CYP-specific substrates may be valuable in the development of personalized medicine protocols, particularly for drugs that have significant toxicity profiles. However, the use of the cocktail approach in the clinic is dependent on the rapid provision of patient-specific information to the clinician. Here we describe the application of liquid chromatography–tandem mass spectrometry (LC–MS–MS) for the simultaneous phenotyping of five major drug-metabolizing CYPs in patients within a 5-min assay.

Key words CYP phenotyping, Drug cocktail, Liquid chromatography–tandem mass spectrometry, Solid-phase extraction

1 Introduction

Pharmacogenomic approaches involving the analysis of *CYP* gene polymorphisms are emerging for the optimization of drug therapy. However, because many *CYP* genes are strongly inducible or susceptible to inhibition by a range of therapeutic agents and other xenobiotics, the value of genotyping may be relatively restricted. Instead, in vivo phenotyping, in which small doses of CYP-specific drug substrates are administered to patients, offers a convenient and relatively noninvasive approach to obtain information on CYP function in individuals.

A number of phenotyping "cocktails" that consist of different combinations of CYP-specific drug substrates have been developed. Essential features of suitable cocktails include (a) the use of well-tolerated probe substrate drugs whose elimination is mediated by individual CYPs, (b) minimal interaction between component drugs, and (c) drugs that are free from concerns over their availability and ethical use (1–3). The Inje cocktail meets these requirements (4). In this cocktail, CYP1A2 activity is assessed by the conversion of caffeine to paraxanthine, CYP2C9 by the oxidation of losartan to its carboxylic acid metabolite Exp-3174, CYP2C19 by

Ian R. Phillips et al. (eds.), *Cytochrome P450 Protocols*, Methods in Molecular Biology, vol. 987, DOI 10.1007/978-1-62703-321-3_22, © Springer Science+Business Media New York 2013

the conversion of omeprazole to its 5-hydroxy metabolite, CYP2D6 from dextromethorphan O-demethylation to dextrorphan, and CYP3A with its substrate midazolam.

Despite the apparent potential of the cocktail approach in streamlining therapy its clinical uptake has been limited. There may be several reasons for this. For many drugs the relationship between pharmacokinetics and pharmacodynamics is not always clear. There is also skepticism among some clinicians that dose selection prior to therapy represents an advance over the more customary approach of retrospective dose titration. However, one significant obstacle is the speed with which phenotyping information can be supplied to clinicians, who seek to initiate therapy as rapidly as possible. Many of the cocktails that were initially developed require multiple patient plasma or urine samples and multiple analytical approaches that significantly delay the provision of patient-specific information.

More recently developed cocktails offer advantages over the earlier approaches. High-throughput solid-phase extraction and LC–MS–MS or HPLC with fluorometric detection have been combined to streamline the analysis of multiple CYP substrates and metabolites (5–7). The simultaneous analysis of five or six phenotyping substrates enables patient-specific information to be derived efficiently, thus preventing delays to therapy. Recently, we have developed a high-throughput LC–MS–MS approach based on the Inje cocktail because of the advantages offered by this drug combination and which provides information on five CYPs within a 5-min assay run time (8, 9).

2 Materials

All solutions used in analytical methods are prepared with Milli-Q water or similar. Glassware used in the analysis of drugs and metabolites in plasma should be silanized before use.

2.1 Drugs Used in the In Vivo Phenotyping Cocktail

Details of drugs used in the phenotyping cocktail are:

1. Caffeine (No Doz®, Key Pharmaceuticals, Macquarie Park, NSW, Australia): Dose 100 mg.

2. Dextromethorphan (Bisolvon Dry Oral Liquid® 10 mg/5 mL, Boehringer Ingelheim, North Ryde, NSW, Australia): Dose 30 mg.

3. Midazolam (Midazolam Injection®, 1 mg/mL, Pfizer, Bentley, WA, Australia): 5 mL was diluted to 75 mL with 5% dextrose. Dose: 30 mL of this solution is administered to each patient.

4. Losartan (Cozaar®, Merck Sharp & Dohme, South Granville, NSW, Australia): Dose 25 mg.

5. Omeprazole (Meprazole®, Sandoz, Pyrmont, NSW, Australia): Dose 20 mg.

Simultaneous In Vivo Phenotyping of CYP Enzymes   263

2.2 Instrumentation

1. Solid-phase extraction is performed on a Gilson ASPEC XL4 instrument (Villiers-le-Bel, France) controlled by 735 Sampler software (Version 6.0).

2. Chromatography is undertaken on an Altima C_{18}, 5 μm (150 × 2.1 mm I.D.) narrow-bore column (Alltech Associates, Castle Hill, Australia) coupled to an Agilent HP 1090 liquid chromatograph (Agilent Technologies, Sydney).

3. A Thermo Finnigan TSQ 7000 MS system (San Jose, CA, USA) was used for the quantification of all analytes.

2.3 Analysis of Plasma Samples

1. Caffeine, paraxanthine, dextromethorphan, dextrorphan, and the internal standard phenacetin, all from Sigma-Aldrich (Castle Hill, NSW, Australia).

2. Midazolam was a generous gift from Dr John Vine of Racing Analytical Services (Flemington, Victoria, Australia). 1'-Hydroxymidazolam (Toronto Research Chemicals, Pickering, Ontario, Canada).

3. Losartan was donated by Merck Sharp & Dohme (Guildford, NSW, Australia). Exp-3174 (losartan carboxy acid) (SynFine Research, Inc., Richmond Hills, Ontario, Canada).

4. Omeprazole (Trapeze Associates Pty, Rhenochem AG, Basel, Switzerland), and 5-hydroxyomeprazole (Ramidus AB, Lund, Sweden).

5. HPLC-grade acetonitrile and methanol. Formic acid. Analytical reagent-grade chloroform and propan-2-ol.

6. Oasis HLB cartridges (Waters Corporation, Milford, MA, USA).

7. Drug-free human plasma (Australian Red Cross Blood Service, Melbourne, Victoria, Australia).

3 Methods

3.1 Patient Studies

1. Pre-phenotyping: From at least 12 h before day 1 of the study until its conclusion patients should avoid foods and beverages containing caffeine, such as tea, coffee, and cola. Patients should also avoid grapefruit juice because of the potential for intestinal CYP3A inhibition (10) and cruciferous vegetables, such as broccoli and Brussels sprouts, because of the potential for CYP1A2 induction (11). Exposure to medications that are inhibitors or inducers of CYPs must also be minimized.

2. On day 3 of the study withdraw a 10-mL blood sample, for establishment of baseline concentrations of all drug substrates and metabolites in individuals (see Note 1).

3. Subjects then received the drugs in the phenotyping cocktail simultaneously (Subheading 2.1 above). Monitor patients closely

264 Sussan Ghassabian and Michael Murray

for adverse effects (see Note 2). Withdraw blood samples at 1, 2, 4, and 6 h after administration (see Note 3). After immediate centrifugation at $4,000 \times g$ for 15 min, isolate plasma and store at 4°C. When all samples have been collected, transfer them to –80°C before analysis.

3.2 Solid-Phase Extraction of Phenotyping Drugs and Their Metabolites

1. Add the internal standard phenacetin (1 μg) to each plasma sample (1 mL).

2. Add acetonitrile (3 mL) to each sample, then vigorously vortex and centrifuge for 15 min at $4,000 \times g$.

3. Remove the acetonitrile layer and evaporate under a stream of nitrogen at 40°C.

4. Add double-distilled water (2 mL) to the dried samples and load 1.5-mL aliquots of the resultant aqueous phase onto Oasis HLB solid-phase extraction cartridges (3 mL, 60 mg). Solid-phase extraction is performed on a Gilson ASPEC XL4 instrument (Villiers-le-Bel, France) controlled by 735 Sampler software (Version 6.0).

5. Condition cartridges with methanol (1 mL) and then water (1 mL).

6. Load samples (1.5 mL) onto cartridges (0.5 mL/min), wash twice with double-distilled water (2×1 mL) and then with 10% methanol in water (2×1 mL).

7. Elute samples from cartridges with methanol (3×1 mL) at a flow rate of 0.5 mL/min.

3.3 Liquid–Liquid Extraction of Caffeine and Paraxanthine

1. Solid-phase extraction cartridges proved unsuitable for analysis of paraxanthine, and a separate liquid–liquid extraction procedure for caffeine and paraxanthine was established (see Note 4).

2. To plasma samples (200 μL) that had been spiked with phenacetin (250 ng) add acetonitrile (600 μL) and vortex vigorously to precipitate plasma proteins.

3. After centrifugation at $4,000 \times g$, remove the acetonitrile layer and evaporate under a stream of nitrogen at 40°C.

4. Then add double-distilled water (200 μL), followed by chloroform:isopropanol (85:15; 3 mL).

5. Shake the mixture for 15 min and centrifuge for 15 min at $4,000 \times g$. Remove the organic phase and evaporate it to dryness under nitrogen at 40°C.

6. Reconstitute extracts from both solid-phase extraction (100 μL; Subheading 3.2) and liquid–liquid (100 μL; Subheading 3.3) extraction in a mixture of water and acetonitrile (50:50; 100 μL) from which 20 μL is subjected to LC–MS–MS.

Table 1
Chromatographic resolution and fragmentation parameters for analytes

CYP	Analyte	Retention time (min)	Precursor ion (*m/z*)	Product ion (*m/z*)
CYP1A2	Caffeine	1.44	195	138.0
	Paraxanthine	1.28	181	123.9
CYP2C9	Losartan	2.95	423	207.0
	Exp-3174	4.14	437	207.0
CYP2C19	Omeprazole	1.46	346	198.2
	5'-Hydroxyomeprazole	1.30	362	214.0
CYP2D6	Dextromethorphan	1.60	272	171.0
	Dextrorphan	1.29	258	157.0
CYP3A4	Midazolam	1.59	326	291.0
	1'-Hydroxymidazolam	1.59	342	324.0
Internal standard	Phenacetin	2.39	180	109.9

3.4 Analysis of Plasma Extracts by LC–MS–MS

1. LC–MS–MS with electrospray ionization–tandem mass spectrometry is used to analyze the phenotyping drugs and their metabolites.

2. Isocratic elution of analytes is with acetonitrile:water (1:1), containing 0.1% formic acid, at a flow rate of 0.3 mL/min (Table 1).

3. The LC/MS system is operated in positive ionization mode with argon as the collisional gas at 2.0 mTorr and a corona current of 5 μA. The temperature of the heated capillary is set to 275°C.

4. Selected reaction monitoring of product ions achieves quantification of high sensitivity (Table 1). This may be enhanced with newer instruments.

3.5 Analysis of Phenotyping Data

1. The ratio of the areas under the plasma drug concentration versus time curve between 0 and 6 h (AUC_{0-6h}) for the phenotyping drug and its metabolite is the preferred phenotyping index for CYP2D6, CYP2C9, and CYP3A4.

2. The ratio of the plasma concentrations of paraxanthine and caffeine at 4 h post-dose is the preferred phenotyping index for CYP1A2.

3. The ratio of the plasma concentrations of 5-hydroxyomeprazole and omeprazole at 4 or 6 h post-dose is the preferred phenotyping index for CYP2C19.

4 Notes

1. Although participants abstain from caffeine, traces may be detectable in plasma. After quantification this amount should be subtracted from subsequent measurements.

2. Although uncommon, some patients experience adverse reactions to drugs used in phenotyping cocktails.

3. Omeprazole is available as an enteric-coated tablet so that variable absorption rates may occur due to differential gastric emptying (4). Omeprazole may be absent from early blood samples obtained from some subjects.

4. Caffeine is regularly used for CYP1A2 phenotyping. Other investigators have used solid-phase extraction in quantification of the caffeine/paraxanthine pair. In our hands, recovery from Oasis HLB cartridges was reproducible but very low (~30%); other cartridges were tested but performed similarly.

Acknowledgement

This work was supported in part by an International Postgraduate Research Scholarship to S.G. and by the Australian National Health and Medical Research Council.

References

1. Frye RF, Matzke GR, Adedoyin A, Porter JA, Branch RA (1997) Validation of the five-drug "Pittsburgh cocktail" approach for assessment of selective regulation of drug-metabolizing enzymes. Clin Pharmacol Ther 62:365–376

2. Streetman DS, Bertino JS, Nafziger AN (2000) Phenotyping of drug-metabolizing enzymes in adults: a review of in vivo cytochrome P450 phenotyping probes. Pharmacogenetics 10:187–216

3. Christensen M, Andersson K, Dalen P, Mirghani RA, Muirhead GJ, Nordmark A, Tybring G, Wahlberg A, Yaşar U, Bertilsson L (2003) The Karolinska cocktail for phenotyping of five human cytochrome P450 enzymes. Clin Pharmacol Ther 73:517–528

4. Ryu JY, Song IS, Sunwoo YE, Shon JH, Liu KH, Cha IJ, Shin JG (2007) Development of the "Inje cocktail" for high-throughput evaluation of five human cytochrome P450 isoforms in vivo. Clin Pharmacol Ther 82:531–540

5. Scott RJ, Palmer J, Lewis IA, Pleasance S (1999) Determination of a "GW cocktail" of cytochrome P450 probe substrates and their metabolites in plasma and urine using automated solid phase extraction and fast gradient liquid chromatography tandem mass spectrometry. Rapid Commun Mass Spectrom 13:2305–2319

6. Yin OQ, Lam SS, Lo CM, Chow MS (2004) Rapid determination of five probe drugs and their metabolites in human plasma and urine by liquid chromatography/tandem mass spectrometry: application to cytochrome P450 phenotyping studies. Rapid Commun Mass Spectrom 18:2921–2933

7. Jerdi MC, Daali Y, Oestreicher MK, Cherkaoui S, Dayer P (2004) A simplified analytical method for a phenotyping cocktail of major CYP450 biotransformation routes. J Pharm Biomed Anal 35:1203–1212

8. Ghassabian S, Chetty M, Tattam BN, Glen J, Rahme J, Stankovic Z, Ramzan I, Murray M, McLachlan AJ (2009) A high-throughput assay using liquid chromatography-tandem mass spectrometry for simultaneous in vivo phenotyping of 5 major cytochrome P450 enzymes in patients. Ther Drug Monit 31:239–246

9. Ghassabian S, Chetty M, Tattam BN, Glen J, Rahme J, Stankovic Z, Ramzan I, Murray M, McLachlan AJ (2010) The participation of

cytochrome P450 3A4 in clozapine biotransformation is detected in people with schizophrenia by high-throughput in vivo phenotyping. J Clin Psychopharmacol 30:629–631

10. Lown KS, Bailey DG, Fontana RJ, Janardan SK, Adair CH, Fortlage LA, Brown MB, Guo W, Watkins PB (1997) Grapefruit juice increases felodipine oral availability in humans by decreasing intestinal CYP3A protein expression. J Clin Invest 99:2545–2553

11. Murray S, Lake BG, Gray S, Edwards AJ, Springall C, Bowey EA, Williamson G, Boobis AR, Gooderham NJ (2001) Effect of cruciferous vegetable consumption on heterocyclic aromatic amine metabolism in man. Carcinogenesis 22:1413–1420

Chapter 23

Detection of Regulatory Polymorphisms: High-Throughput Capillary DNase I Footprinting

Matthew Hancock and Elizabeth A. Shephard

Abstract

We describe a method for high-throughput analysis of protein-binding sites in DNA using 96-well plates and capillary electrophoresis. The genomic DNA or plasmid DNA to be analyzed is amplified using fluorescent primers, incubated with an appropriate nuclear extract and treated with DNase I. Separation of the DNase I-generated fragments and co-analysis of their base sequences identify the position of protein-binding sites in a DNA fragment. The method is applicable to the identification of base changes, e.g., single-nucleotide polymorphisms (SNPs), that eliminate protein binding to DNA.

Key words DNase I footprinting, DNA–protein binding, Regulatory polymorphism, High-throughput

1 Introduction

The detection of the base sequence of a protein-binding site using DNase I footprinting and traditional gel electrophoresis is time-consuming and laborious. Recent advances in the methodology of DNase I footprinting, using capillaries to analyze DNA fragments, permit a high-throughput approach to the detection of protein-binding sites in DNA (1, 2). The DNA fragments to be analyzed are labeled fluorescently, which makes the technique safer and more reproducible when compared with traditional DNase I footprinting using radiolabeled DNA. The adaptation of capillary technology to DNase I footprinting permits high-throughput analysis of multiple samples. The DNA fragments are separated using the ABI prism genetic analyzer 3730, allowing 96 samples to be run simultaneously, compared with just a few samples traditionally analyzed on a single polyacrylamide gel. The machine can analyze a maximum of 12 plates concurrently.

We used the SV40 promoter as a control to compare the results obtained by the high-throughput and traditional DNase I footprinting techniques (Figs. 1 and 2). We chose to incorporate this new methodology for a novel application in which human *FMO1*

Ian R. Phillips et al. (eds.), *Cytochrome P450 Protocols*, Methods in Molecular Biology, vol. 987,
DOI 10.1007/978-1-62703-321-3_23, © Springer Science+Business Media New York 2013

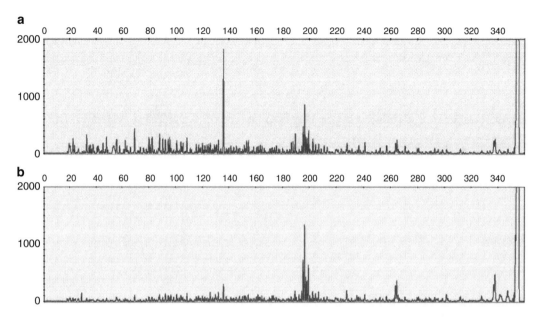

Fig. 1 DNase I digestion of the FAM-labeled SV40 DNA. The optimal range of DNase I concentration was shown to be large, with a digestion time of 5 min at 25°C. 0.05 U (**a**) and 0.005 U (**b**) both give an acceptable size range of DNA fragments at this temperature and time. Footprints are observed in the correct location throughout the 350-bp sequence (see Note 12)

promoter polymorphisms could be screened for their effect on DNA–protein binding (Fig. 3). As *FMO1* promoters are tissue specific it is desirable to analyze the influence of a polymorphism through the use of extracts from different tissue types. This is possible as the method allows a large number of samples to be analyzed. Any change in DNA–protein binding observed is likely to change the amount of transcript produced and, subsequently, the amount of protein translated. Genetic polymorphisms shown to change DNA–protein binding by this method could then be used in association studies to identify polymorphisms associated with variable drug response.

The technique outlined below is the first to detect, in a high-throughput manner, the effect of genetic variation on DNA–protein interaction. By allowing the screening of genetic variation in a large number of samples, polymorphisms within regulatory regions of groups of genes could, in future, be efficiently screened. An example would be the screening of all known regulatory regions for a family of genes. These SNPs could then be further analyzed through association studies or further characterized by additional functional studies.

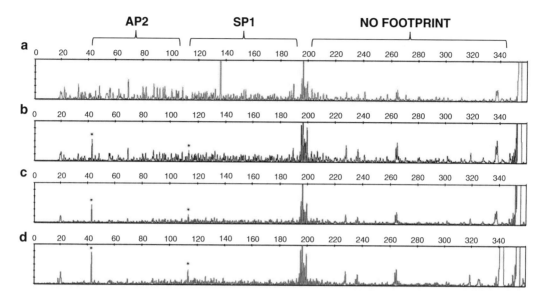

Fig. 2 Chromatograms showing the footprints observed with DNase I-digested fragments from the 5′ FAM-labeled SV40 promoter after incubation with COS-7 nuclear-protein extract. The fragments were separated through capillaries and each peak represents cleavage of a phosphodiester linkage. The *x*-axis gives the number of base pairs from the 5′ label and the *y*-axis is the intensity of the peak. From (**a**) to (**d**) the total COS-7 nuclear protein is increased: (**a**) 0 μg, (**b**) 10 μg, (**c**) 20 μg, and (**d**) 40 μg. The traces were obtained using optimized conditions, which were used for subsequent experiments. The promoter shows intense DNase I-hypersensitive regions either side of the AP1, AP2, and SP1 consensus sites. Further downstream the trace shows no footprinted regions. This area of sequence is known to contain no protein-binding sites. Therefore the results of the high-throughput method confirm the DNA–protein-binding sites detected within the SV40 promoter by the traditional DNase I footprinting method

2 Materials

2.1 Plasmid Construction, Ligation, and Cloning

1. TOPO TA cloning kit for sequencing (Invitrogen): The kit includes PCR4-TOPO vector, 10 ng/mL in 50% (v/v) glycerol, 50 mM Tris–HCl, (pH 7.4 at 25 °C), 1 mM EDTA, 2 mM dithiothreitol (DTT), 0.1% (v/v) Triton X-100, bovine serum albumin (BSA) (100 mg/mL), and 30 μM phenol red. Topoisomerase is contained within the solution, enabling ligation (see Note 1).

2. Salt solution: 1.2 M NaCl, 0.06 M $MgCl_2$.

3. SV40 promoter sequence: This sequence can be purchased as part of the core footprinting kit (Promega, E3730).

4. SOC Medium (Sigma-Aldrich, Poole, UK).

5. Luria Bertani (LB) medium and LB-agar (Anachem, Luton, UK): prepare by dissolving the appropriate number of capsules in the desired volume of distilled water and autoclaving.

6. Ampicillin, 50 mg/mL stock solution.

7. LB-agar plates containing ampicillin (100 μg/mL).

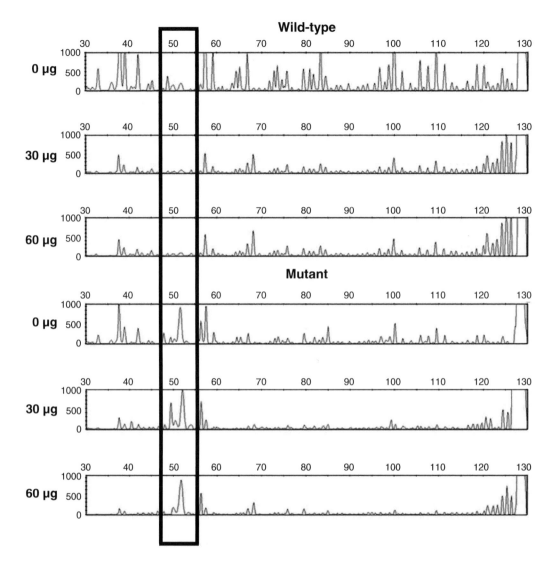

Fig. 3 Detection of the loss of DNA–protein binding within the *FMO1* P2 promoter. The chromatograms show the DNase I-digested fragments of the wild-type and mutant 5′ FAM-labeled *FMO1* P2 promoter region (−255 to −128). The fragments were separated through capillaries, and each peak represents a single base pair. The *x*-axis shows the number of base pairs from the 5′ label and the *y*-axis is the intensity of the peak. The amount of protein added to each reaction is given adjacent to the trace. The *black box* highlights the region that shows loss of protein binding in the mutant sequence. Protein concentrations used were 0, 30, and 60 μg. The chromatogram traces were aligned using GeneMapper software. The termination-sequencing reactions for the P2 promoter were carried out for each of the four DNA nucleotides. Each termination-reaction sequencing trace was aligned and the base pair coordinates were noted. The coordinates for the *FMO1* sequences were then located within the footprinted regions of the wild-type and mutant P2 promoter sequence

8. XL1-Blue competent cells (Stratagene, La Jolla, CA). Genotype: *recA1 endA1 gyrA96 thi-1 hsdR17 supE44 relA1 lac* (F′ *proAB lac*IqZΔM15 Tn*10* (Tetr)).

9. Plasmid isolation kit, e.g., mini-prep kit (QIAGEN).

Detection of Regulatory Polymorphisms...   273

2.2 PCR Labeling of DNA and Gel Electrophoresis

1. 5′AP2 primer FAM-CTT CTG AGG CGG AAA GAA CC (see Note 2).
2. T7 primer 5′-TAATACGACTCACTATAGGG-3′ (see Note 3).
3. *FMO1* P2 F primer 5′FAM-GCTACAGCCTTGACAATTG TACG 3′.
4. *FMO1* P2 R primer 5′AGCGGGTCTCTGGTGTGTGAATG 3′.
5. 10× Tris–borate–EDTA buffer (TBE): 0.89 M Tris-base, 0.09 M boric acid, 20 mM EDTA (pH 8).
6. Agarose (Bioline, London, UK).
7. 6× loading buffer: 0.25% (w/v) bromophenol blue, 0.25% (w/v) xylene cyanol, 30% (v/v) glycerol, store at room temperature.
8. DNA ladder, e.g., hyperladder I and V (Bioline, London, UK).
9. SYBR Safe® DNA gel stain in 0.5× TBE (Invitrogen, Paisley, UK).
10. Fujifilm PhosphorImager FLA-200 system.
11. Nanodrop machine.
12. Agarose gel electrophoresis equipment.

2.3 Nuclear-Protein Extraction

1. Phosphate-buffered saline (PBS): 137 mM $NaCl_2$, 2.7 mM KCl, 10 mM Na_2HPO_4, 1.8 mM KH_2PO_4.
2. Complete Protease Inhibitor Cocktail Tablets (Roche, West Sussex, UK). A mixture of several protease inhibitors with broad inhibitory specificity.
3. 1 M DTT, store at –20°C.
4. Hypotonic buffer: 10 mM HEPES (pH 7.9), 10 mM KCl, 0.2 mM EDTA, 0.1 mM EGTA. Immediately before use, supplement by adding one protease inhibitor cocktail tablet per 10 mL of buffer and add DTT to a final concentration of 1 mM.
5. NP-40 (Sigma).
6. Nuclear-lysis buffer: 20 mM HEPES (pH 7.9), 0.4 M NaCl, 1 mM EDTA. Immediately before use supplement by adding one protease inhibitor cocktail tablet per 10 mL of buffer and add DTT and NP-40 to final concentrations of 1 mM and 0.02% (v/v), respectively.
7. Dialysis buffer: 20 mM HEPES (pH 7.6), 1 mM EDTA (pH 8). Autoclave and store at 4°C. Immediately before use add DTT to a final concentration of 1 mM.
8. Slide-a-Lyzer® Dialysis cassette (Pierce, Rockford, IL).

2.4 DNA–Protein-Binding Reaction

1. FAM-labeled DNA (50 ng) (see Subheading 2.2, item 1).
2. Binding buffer: 50 mM Hepes (pH 7.6), 2.5 mM DTT, 60% (v/v) glycerol, 250 mM NaCl, 0.25% (v/v) NP-40. Store in aliquots at –20°C.
3. Poly dI/dC (Sigma): Dissolve in 5 mM NaCl to a final concentration of 4 μg/μL and store in aliquots at –20°C.

274 Matthew Hancock and Elizabeth A. Shephard

2.5 DNase I Digestion

1. 1× DNase I reaction buffer: 10 mM Tris–HCl (pH 7.6), 2.5 mM $MgCl_2$, 0.5 mM $CaCl_2$ (New England Biolabs, Inc (NEB), Hitchin, UK).

2. DNase I, 2,000 U/μL (NEB).

3. 0.5 M EDTA, pH 7.6.

2.6 DNA Purification

1. Phenol:chloroform:isoamyl alcohol (25:24:1, v/v/v) mixture (pH 6.6/7.9), liquid (Fischer Scientific, Loughborough, UK).

2. 3 M sodium acetate, pH 5.2.

3. Ethanol: 100 and 70%.

4. Phase Lock Gel Light 1.5-mL tubes (5 PRIME, Nottingham, UK).

5. QIAquick PCR Purification Kit (QIAGEN, Crawley, UK).

6. Buffer PB (QIAGEN): This DNA-binding buffer is provided in the Qiagen PCR purification kit.

7. Buffer PE (QIAGEN): This DNA-elution buffer is provided in the Qiagen PCR purification kit.

2.7 Dye-Terminator Sequencing Reactions

1. Plasmid DNA (template).

2. Thermo Sequenase Dye Primer Manual Cycle Sequencing Kit (USB, High Wycombe, UK) containing reaction buffer (concentrate): 150 mM Tris–HCl (pH 9.5), 35 mM $MgCl_2$.

3. ddA Term mix: 300 μM each of dATP, dCTP, dTTP and 7-deaza-dGTP, 3 μM ddATP.

4. ddC Term mix: 300 μM each of dATP, dCTP, dTTP and 7-deaza-dGTP, 3 μM ddCTP.

5. ddG Term mix: 300 μM each of dATP, dCTP, dTTP and 7-deaza-dGTP, 3 μM ddGTP.

6. ddT Term mix: 300 μM each of dATP, dCTP, dTTP and 7-deaza-dGTP, 3 μM ddTTP.

7. Thermo Sequenase DNA polymerase (20 U/μL) with Thermoplasma acidophilum inorganic pyrophosphatase (0.03 U/μL) in 50 mM Tris–HCl, pH 8.0, 1 mM EDTA.

8. 1 mM DTT, 0.5% (v/v) Tween-20, 0.5% (v/v) Nonidet™ P-40, 50% (v/v) glycerol.

9. FAM-labeled primer.

2.8 Capillary Electrophoresis

1. Genetic analyzer 3730 (Applied Biosystems, Cheshire, UK).

2. Hi-Di formamide (Applied Biosystems).

3. LIZ standard (Applied Biosystems).

2.9 Data Analysis

1. GeneMapper software (Applied Biosystems).

2. Microsoft Excel.

3 Methods

3.1 Plasmid Construction and Cloning

PCR products produced by the Bio-Taq enzyme have adenosines at the 3′ end of each DNA strand. The TOPO vector is a linearized vector and possesses single thymidine overhangs. The adenosine overhangs of the PCR product can be annealed to the thymidine overhangs of the TOPO vector and the DNAs are ligated with the DNA ligase activity of topoisomerase, which is covalently bound to the linearized vector.

The SV40 promoter has previously been analyzed using traditional DNase I footprinting. The promoter is well characterized and the DNA sequences to which various transcription factors bind are known (3). The SV40 promoter can therefore be used as a positive control for the optimization of DNase I capillary footprinting using a nuclear-protein extract (Fig. 2).

1. Mix 4 μL of the PCR product mixture with 1 μL of salt solution and 1 μL of PCR 4-TOPO vector (which includes topoisomerase) and incubate at room temperature for 5 min.

2. Use 2 μL of the reaction mixture from step 1 to transform 50 μL of XL1-blue competent cells. Mix DNA and cells gently and leave on ice for 30 min. Heat shock cells at 42°C for 30 s and return to ice. Add 250 μL of SOC medium and incubate at 37°C for 1 h.

3. Spread 50 μL onto LB-agar plates containing ampicillin. Incubate plates at 37°C overnight.

4. Use a pipette tip to pick individual colonies from the plate and transfer the tip to a 5-mL culture of LB broth. Cultures are grown overnight at 37°C, with shaking at 225 rpm.

5. Isolate plasmid DNA using a suitable kit according to manufacturer's instructions.

3.2 PCR Labeling of DNA and Gel Electrophoresis

The ABI prism 3730 genetic analyzer can detect up to five separate dyes when using the G5 dye set. To label the DNA, one needs to use the appropriate dye. It is recommended that the fluorescent dye FAM be used. If analysis of the reverse strand is required, the 3′ reverse oligo can be labeled with a separate dye. We have tested the use of HEX, a second dye, and have found it to be suitable.

A FAM-labeled fragment can be produced from plasmid DNA or genomic DNA. PCR is carried out with a 5′ FAM-labeled forward primer located, ideally, 60 bp upstream of the DNA sequence of interest, and a 3′ reverse primer located 60 bp downstream of the sequence of interest. The PCR is carried out using conditions specific to the primer sets being used.

1. Load the PCR product on to a TBE agarose gel (agarose percentage suitable for size of product) without the addition of a dye for DNA visualization. An appropriate DNA ladder is

276   Matthew Hancock and Elizabeth A. Shephard

loaded into an adjacent well. The ladder is visualized by the addition of SYBR® Safe (added at a 1:100 dilution). The samples are electrophoresed through the gel at 70 V (6×7.5 cm agarose gel).

2. Image the DNA using a fluoresence imaging machine (Fujifilm PhosphorImager FLA-200 System) at an excitation of 473 nm. This will determine whether the PCR product is present, is the correct size, and has incorporated sufficient dye.

3. Quantify the amount of DNA spectrophotometrically using a Nanodrop machine. $A_{260/280}$ ratio of between 1.8 and 2 is considered as pure DNA.

3.3 Nuclear-Protein Extraction from Cultured Cells

1. Pellet approximately 1×10^7 cells by centrifugation at 2,500 rpm for 5 min in an Eppendorf microcentrifuge, wash with $1 \times$ cold PBS, resuspend in 1 mL of cold hypotonic buffer, and incubate on ice for 15 min.

2. Add 0.4% (v/v) NP-40 to the cell suspension and pass through a 25-G needle 5–6 times.

3. Centrifuge the cell lysate at 13,000 rpm for 2 min at 4°C. Discard the supernatant and add 300 μL of nuclear-lysis buffer to the nuclear pellet.

4. Agitate the cells gently by rolling on a Denley Spiramix for 30 min at 4°C.

5. Transfer the supernatant (nuclear extract) into a fresh tube. Dialyze the sample to remove excess salt. To dialyze take 500 μL of sample (or less) and place in a Slide-a-Lyzer® Dialysis cassette. Attach the cassette to a buoy and immerse in 1 L of cold dialysis buffer in a glass beaker. Place a magnetic stir bar within the buffer and place on a magnetic stirrer. Carry out the dialysis procedure in a cold room or cold cabinet. After 2 h replace with fresh buffer. After an additional 2 h of dialysis remove the sample from the cassette and transfer into a 1.5-mL Eppendorf tube. Aliquot the sample and store at –70°C.

3.4 DNA–Protein-Binding Reaction

The labeled DNA is incubated with a suitable nuclear-protein extract to allow the proteins to bind to the DNA. Nuclear extracts should be chosen that represent the tissue in which the promoter of the gene is active. COS-7 (a cell line derived from African green monkey kidney fibroblasts) nuclear extract was chosen when analyzing the SV40 promoter. COS-7 contains endogenous levels of the SP1 protein. The SV40 promoter contains several SP1 consensus binding sites.

1. A total reaction volume of 50 μL is used for DNA–protein binding. 10 μL of 5× binding buffer, primers (1 μL of each), 1 μL of poly dI-dC, dH_2O (see Note 4), and protein are added to a chilled tube, which is then left on ice for 5 min. The amount

Detection of Regulatory Polymorphisms... 277

of water and protein will vary according to the amount of protein added (see Subheading 3.4, step 3).

2. Set up a range of nuclear-protein concentrations to show that protein binding varies with concentration. It is recommended that three or four different protein concentrations are used, between 0 and 60 μg (see Note 5).

3. Add template DNA (50 ng) and mix gently using a pipette tip. Leave the binding reaction for 30 min to allow for DNA–protein interactions to take place and the reaction mixture to equilibrate on ice.

4. As a control, set up a duplicate set of samples and add 100 μg of proteinase K (see Note 6). Heat at 37°C for 10 min.

3.5 DNase I Digestion (Fig. 1, See Notes 5 and 7)

The DNA is digested using DNase I. The concentration of the enzyme needs to be optimized so that each phosphodiester linkage along the length of the fragment is cleaved. However, the conditions should be such that each fragment in the population of DNAs is ideally cleaved just once. As these cleavages occur randomly, the population of DNA strands produced by enzymatic digestion should represent all possible DNA lengths. If too much enzyme is used then a higher percentage of DNA fragments will be too short; likewise, if too little enzyme is used a higher percentage of fragments will be too long. To be able to analyze a length of promoter sequence the enzyme needs to digest randomly and evenly along the length of the fragment.

1. Add 50 μL of DNase I reaction buffer to the DNA–protein reaction mixture. Mix gently.

2. Add DNase I enzyme (0.005 U) and incubate for 5 min at room temperature. Terminate the reaction by the addition of 100 μL of 0.5 M EDTA (pH 8.0). The DNA is now ready to purify.

3.6 DNA Purification (See Note 8)

DNA can be purified by either of the methods described below.

3.6.1 Phenol/Chloroform Method

1. Place 150 μL of DNA mix (from Subheading 3.5) in a 1.5-mL Eppendorf tube. Add 150 μL of Tris-buffered phenol: chloroform:isoamyl alcohol (25:24:1, v/v/v).

2. Make sure the lid of the tube is firmly closed. Hold the tube firmly and with a piece of paper towel covering the lid. Vortex vigorously. Transfer to a phase lock gel tube and centrifuge at 13,000 rpm for 2 min in a microcentrifuge to separate the aqueous and organic phases.

3. The aqueous phase is the upper phase. Denatured protein will be at the interphase and the organic phase is the lower phase. Remove as much aqueous phase as possible and transfer to a fresh tube.

278   Matthew Hancock and Elizabeth A. Shephard

4. Precipitate DNA by the addition of 1/10 volume of 3 M sodium acetate, pH 5.2, and 3 volumes of 100% ethanol.

5. Place sample at –20°C for 1 h.

6. Centrifuge at 4°C for 10 min in a microfuge at 13,000 rpm.

7. Discard the supernatant and add 1 mL of 70% (v/v) ethanol to the tube. Centrifuge at 4°C for 5 min at 13,000 rpm.

8. Discard the supernatant.

9. Air-dry the pellet at room temperature and resuspend in 50 µL of H_2O.

3.6.2 Silica-Column Method

1. Add 5× volume of buffer PB to the sample (from Subheading 3.5).

2. Mix, load on to the silica column, and centrifuge in a microfuge for 1 min at 13,000 rpm.

3. Discard the eluate and add 750 µL of buffer PE to the column. Centrifuge at 13,000 rpm for 1 min.

4. Discard the eluate and recentrifuge for 1 min at 13,000 rpm to remove any residual buffer PE.

5. Add 50 µL of dH_2O to the column and centrifuge for 1 min at 13,000 rpm to elute the DNA.

3.7 Dye-Terminator Sequencing Reactions (See Note 9)

Set up the dye-termination sequencing reaction mixes as follows:

1. Make up a master mix containing 200 ng of plasmid DNA (template), 2.2 µL of reaction buffer (Subheading 2.7, item 2), and 1 µL of labeled primer (FAM). Add nuclease-free water to give a final volume of 19 µL. Add 1 µL of the Thermo Sequenase enzyme and mix.

2. Aliquot 4 µL of the master mix into each of four tubes (keep on ice). One tube for each of the terminating dideoxynucleotides.

3. Add the appropriate dideoxynucleotide termination mix to each tube.

4. Take 1 µL of a reaction mix and dilute to a total volume of 5 µL.

5. The dye-terminator sequencing reactions are now carried out using a specific amplification procedure: 40 cycles comprising denaturation, at 95°C for 30 s; annealing, at 55°C for 30 s; and elongation, at 72°C for 60 s. Annealing temperature can be optimized for different primers.

3.8 Capillary Electrophoresis

The DNase I footprinting samples are analyzed on the same 96-well plate as the DNA sequencing reactions.

1. Load 5 µL of the DNA fragments (prepared as in Subheading 3.6 and in step 5 of Subheading 3.7) into a well of a 96-well plate. In addition, a tenfold dilution of each sample should also be

Detection of Regulatory Polymorphisms...   279

analyzed. This is to ensure that any differences in protein concentration or buffer conditions will be minimal and so that signal intensities will not be affected (see Note 10).

2. Add 4.9 µL of HiDi formamide and 0.1 µL of LIZ standard to each sample. The LIZ standard contains DNAs of known size and is used to align the sequence traces at the end of the DNA separation. The standard contains a LIZ dye, which can be detected using the G5 dye set (see Note 11).

3. Analyze samples using an Applied Biosystems genetic analyzer 3730 and the G5 dye set genotyping module. Use a voltage of 10 kV.

3.9 Data Analysis

1. The footprinting traces can be analyzed using the GeneMapper software.

2. Start a new project from the "File" tab and insert the relevant data.

3. Select microsatellite analysis and select the correct size standard. This will be the standard you used during the fragment separation. In this instance LIZ 500 is used.

4. Press "analyze" and the computer will align all traces.

5. Highlight the traces you wish to visualize for analysis (a specific DNA fragment with different concentrations of protein is recommended) and press the "display plots" tab.

6. Go to the drop-down "pane" tab and choose the appropriate number of traces to be visualized (this will depend on the number of different protein concentrations used).

7. Once visualized the axis can be set appropriately to visualize the peaks. Press the magnifying icon over the axis to change.

8. The data are now ready to be normalized (see Note 12). Ensure that only the marker dye is selected. Press the sizing table to reveal the size of the markers.

9. Highlight size-marker values for five markers for each trace and export to Excel.

10. Average these values in Excel for each trace. Identify the trace that gives the smallest average value for the five size markers. Take this smallest average value and divide it into each size-marker average value from all traces you wish to compare. This will make the size-marker average values relative to each other and gives a unique relative value for each trace.

Return to GeneMapper and ensure the legend for each trace is present. Turn off all dyes except the dye that was used to label the DNA, in this instance blue (FAM).

11. Use the legend to insert the relative value for each trace. This will automatically change the size of each peak and they will

now be relative to each other (normalized) based on the size of the markers within each trace.

12. Once normalized the footprints can be manually identified.

13. To identify the position of the footprint align the dye-termination sequencing to the footprinting trace and read off the DNA sequence for the footprint (see Note 9).

14. To identify whether a polymorphism affects the binding of a protein locate the position of the polymorphism by aligning the sequencing reaction. Once identified, compare the different sequences for the effect of the addition of increasingly more nuclear extract on the binding of the protein. If the signal is no longer reduced by the addition of protein to the mutant template, compared with that obtained from the non-mutated template, then the polymorphism is identified as a potential functional polymorphism. Figure 3 shows the effect of a mutation on the binding of a GC-box-binding protein within the *FMO1* P2 promoter. When the polymorphism is present, the signal is no longer reduced, validating the technique.

15. The traces can be exported by printing as a PDF.

4 Notes

1. Topoisomerase is contained within the TOPO vector solution. When the DNA is present, the 3′ A overhangs present at the end of the PCR product anneal to the overhanging 3′ deoxy-thymidine (T) residues of the vector.

2. Fluorescein (FAM) is a fluorescent dye used as a label to detect the DNA within the genetic analyzer 3730. The FAM label has an optimal absorbance of 495 nm and an emission of 520. The labeled primers were purchased from MWG Eurofins (Germany). This label gave the strongest signal; however, if the researcher wishes to study both strands of DNA, a second dye, HEX, was found to work efficiently. HEX has an optimal absorbance at 535 nm and the excitation wavelength is 556 nm.

3. When labeling a DNA template for DNase I footprinting from the TOPO vector the T7 primer can be used as a primer for any template contained within the vector.

4. The addition of competitor poly dI-dC DNA is required to prevent nonspecific DNA–protein interactions. Different amounts were tested. The minimum amount used is 4 μg. Increases in this amount did not improve the specificity of DNA–protein binding. The total amount of DNA to be used in a reaction is recommended as between 50 and 200 ng.

5. It was observed that when different nuclear protein concentrations were used, DNase I digestion would occur with different efficiencies. This appears to be due to the change in buffering conditions as nuclear protein was added. Adding BSA to the DNA–protein-binding reaction mix to keep total protein concentration the same in each reaction mix can help to overcome this problem. In addition, increasing the volume of the digestion reaction mix to 100 µL reduces the differences in digestion significantly. The increase in volume helps to diminish any differences in buffer conditions between samples.

6. Proteinase K is added to the binding reaction mix to control for nonspecific protein binding to the DNA. All protein binding should be eliminated by this treatment.

7. The optimal range of DNase I concentration was shown to be large (Fig. 1). 0.05 U (A) and 0.005 U (B) both give an acceptable size range of DNA fragments when incubated for 5 min at 25°C. Footprints are observed throughout the 350-bp SV40 promoter sequence. Because the higher DNase I concentration (A) shows a slight skew towards smaller DNA fragments, we chose to use 0.005 U in our experiments. The result of this optimization shows that the DNase I capillary technique is flexible in the amount of DNase I enzyme that can be used and would not, in most circumstances, require optimization. This makes the technique applicable for high-throughput screening as a wide range of enzyme concentration removes the need to optimize the concentration when analyzing a number of different promoter fragments of different lengths at the same time.

8. Before analyzing the digested DNA, it is necessary to remove the bound protein. Method described is suitable but more DNA is recovered using the silica columns.

9. To identify the position of DNA–protein interactions along a DNA fragment it is necessary to also sequence the DNA fragment. Therefore dye-terminator sequencing reactions are carried out using a 5′ FAM-labeled primer. The dye-terminators are not labeled and therefore the dye signal will come from the 5′ FAM label. The sequencing reactions were carried out with the same 5′ FAM primer used to amplify the DNA sequence to be footprinted. This is essential because this allows the sequence fragment traces to be aligned (based on size) correctly with the DNase I footprint traces. The inclusion of the LIZ standard ensures that the sequence and footprinting reactions can be aligned. The software GeneMapper was used to align the individual dye-terminator sequencing reactions. A different color was applied for each dye-terminator reaction for ease of comparison. The dye-terminator sequence trace was subsequently aligned with the DNase I footprinting traces to identify the sequence location of the DNase I footprints (Fig. 2). This results

in the identification of the base pairs to which protein has bound. A similar method was used previously (2).

10. When adding dye-terminator sequencing and DNase I footprinting reaction mixes to the 96-well plate for analysis a tenfold dilution should also be used for each sample. This is to ensure that any differences in protein concentration or buffer conditions are minimal and so that signal intensities are not affected.

11. The experiment can be further controlled by the addition of the LIZ marker before extracting the DNA. This would then serve as an internal control for the extraction of the DNA.

12. Because of differences in the signal intensity between samples when analyzed in the genetic analyzer 3730, the samples are normalized to the LIZ standard intensity. In a traditional footprint experiment, using radioactivity and gel electrophoresis, problems with different signal intensities between samples are overcome by normalizing to a region of the DNA that is predicted not to be a footprint. The capillary method allows us to improve the normalization method. This is because markers, which are independent of the FAM-labeled DNA fragments, are run within the same sample.

References

1. Wilson DO, Johnson P, McCord BR (2001) Nonradiochemical DNase I footprinting by capillary electrophoresis. Electrophoresis 22: 1979–1986

2. Zianni M, Tessanne K, Merighi M, Laguna R, Tabita FR (2006) Identification of the DNA bases of a DNase I footprint by the use of dye primer sequencing on an automated capillary DNA analysis instrument. J Biomol Tech 17:103–113

3. Dynan WS, Tjian R (1983) The promoter-specific transcription factor Sp1 binds to upstream sequences in the SV40 early promoter. Cell 35:79–87

Chapter 24

Isolation of Mouse Hepatocytes

Mina Edwards, Lyndsey Houseman, Ian R. Phillips, and Elizabeth A. Shephard

Abstract

Primary hepatocyte cultures better reflect the properties of the liver in vivo than do cell lines derived from the liver. Here we describe a method for the isolation and culture of mouse primary hepatocytes. The cells are viable, can be transfected by DNA, and retain key properties of liver cells such as the induction of cytochrome P450 gene expression by drugs such as phenobarbital.

Key words Collagenase, Matrigel, Mouse hepatocyte isolation, Mouse hepatocyte viability, Hepatocyte plating density, Perfusion of liver

1 Introduction

Primary cultures of hepatocytes are a useful tool in which to study cellular processes, e.g., changes in gene expression and endogenous and xenobiotic metabolism. Primary hepatocytes are not wholly representative of a normal liver given that the cells are removed from their in vivo situation and forced to grow in vitro. Nevertheless, the functional capabilities of freshly isolated hepatocytes more closely mirror the capacities of a normal liver in vivo than do liver-derived cell lines (1). For instance, HepG2 cells are more representative of cancerous cells, a dedifferentiated system, than of a normal healthy liver. The gene expression profiles of primary human hepatocytes (cultured in a sandwich system to increase the similarity of the cellular environment to that of a liver) were 77% similar to that of liver biopsies, whereas HepG2 cells displayed less than 48% similarity (2). Primary hepatocytes are also much better than HepG2 cells at biotransformation reactions, with increased levels of oxidation (phase I metabolism) as well as sulfation and glucuronidation (phase II metabolism) (3). Rat hepatocytes cultured on substratum also retain the ability of cytochrome P450 (CYP) genes to be induced by phenobarbital, a feature that is lost in continually dividing cells (4).

Ian R. Phillips et al. (eds.), *Cytochrome P450 Protocols*, Methods in Molecular Biology, vol. 987, DOI 10.1007/978-1-62703-321-3_24, © Springer Science+Business Media New York 2013

There has been a great increase in the number of knockout mouse lines developed to investigate various metabolic processes and gene regulation. Cultures of primary mouse hepatocytes isolated from knockout mouse lines provide an adjunct to experimentation on the knockout mice themselves and, importantly, will reduce the number of animals needed. There are many papers published on the isolation and culture of primary rat hepatocytes. Isolation of mouse hepatocytes is more difficult than that of rat hepatocytes, e.g., the small size of the mouse makes inserting the cannula to perfuse the liver with collagenase more difficult. In addition, during the isolation and plating stages of the cells, mouse hepatocytes behave in several ways quite differently from rat hepatocytes. In this chapter we describe a method for isolation of mouse hepatocytes that we have found to give consistent results. The method is based on that of Klaunig et al. (5) and Bahjat et al. (6). The hepatocytes are cultured on Matrigel, which has been shown to retain liver-specific features in primary cultures (7). The mouse hepatocytes can be transfected with DNA constructs, used for quantification of RNA and protein expression and for metabolic studies. The cells are healthy and retain the ability to induce the expression of several *Cyp* genes in response to phenobarbital.

2 Materials

1. Ketamine: 5.7 mg/mL in deionized water (stable for 6 months at −80°C).

2. Xylamine hydrochloride: 2.6 mg/mL in deionized water, prepare fresh.

3. Anesthetizing combined solution (ACS): Mix 1.5 mL of ketamine and 1.5 mL of xylamine hydrochloride solutions. Mix well, and sterilize by passing through a 0.22-μm filter.

4. Sodium heparin: (50 IU/mL). Sterilize by filtration through a 0.22-μm filter and store at 4°C. Stable for months at 4°C.

5. Trypan blue: 0.4% (w/v) in 0.85% (w/v) NaCl.

6. Alcohol, 70% (v/v).

7. Sodium pyruvate: 100 mM in distilled water. Filter sterilize. Stable up to 24 months at 4°C.

8. EGTA (ethylene glycol tetraacetic acid), pH 7.4: 50 mM (prepare 100 mL). Filter sterilize. Stable at room temperature.

9. D-Glucose (dextrose): 25% (w/v) (prepare 50 mL). Sterilize by autoclaving and allow to cool to room temperature. Store at 4°C.

10. Perfusion buffer 1, pH 7.4: Ca^{2+}-and Mg^{2+}-free Hanks balanced salt solution (HBSS): 33 mM KCl, 0.441 mM KH_2PO_4,

4.17 mM NaHCO$_3$, 137.93 mM NaCl, 0.338 mM Na$_2$HPO$_4$, D-glucose (dextrose) (4.75 mg/mL), 26.6 μM phenol red, 0.5 mM EGTA. Sterilize by passing through a 0.22-μm filter. Stable for 6 months at 4°C (see Note 1).

11. Bovine serum albumin (BSA) fraction IV (96% w/v). Store at 4°C.

12. CaCl$_2$: Use anhydrous powder to prepare a 100 mM solution. Filter sterilize. Can be stored at 4°C for 1–2 weeks.

13. Perfusion buffer 2, pH 7.4: Ca^{2+}- and Mg^{2+}-containing HBSS: 1.26 mM CaCl$_2$, 0.493 mM MgCl$_2$, 0.407 mM MgSO$_4$, 5.33 mM KCl, 0.441 mM KH$_2$PO$_4$, 4.17 mM NaHCO$_3$, 137.93 mM NaCl, 0.338 mM Na$_2$HPO$_4$, D-glucose (dextrose) (4.75 mg/mL), 26.6 μM phenol red, 0.72% (w/v) BSA (see Note 2).

14. Perfusion buffer 2, pH 7.4 supplemented with 3 mM CaCl$_2$. Prepare fresh (see Note 2).

15. Collagenase, type H (*Clostridium histolyticum*) (Roche). Warm bottle to room temperature before use and weigh required amount directly into perfusion buffer 2 (see Item 16).

16. Perfusion buffer 2, pH 7.4 supplemented with 3 mM CaCl$_2$ and collagenase (see Note 2). Add collagenase H (to a final concentration of 0.08 U/mL) to perfusion buffer 2 supplemented with 3 mM CaCl$_2$ immediately before use and mix well.

17. Dialyzed fetal bovine serum (D-FBS). Store at –20°C.

18. Antibiotic, antimycotic solution (100×): Penicillin (10,000 U/mL), streptomycin (10 mg/mL), amphotericin B (25 μg/mL), cell-culture tested. Store at –20°C.

19. Insulin-transferrin-selenium liquid supplement (ITS) (100×): Insulin (1.0 g/L), sodium selenite (0.67 mg/L), transferrin (0.55 g/L), sodium pyruvate (11.0 g/L). Store at 4°C.

20. Standard WE medium: William's Medium E (WE), containing D-glucose (dextrose) (2 mg/mL) and L-glutamic acid (50 mg/mL). Store at 4°C.

21. Dexamethasone powder suitable for cell culture: Prepare a 20 μg/mL solution (see Note 3).

22. Matrigel basement membrane matrix: Matrigel (BD Biosciences) (8.4 mg/mL) in Dulbecco's Modified Eagle's Medium (phenol-red free), gentamycin (10 μg/mL). Store Matrigel at –20°C. Stable for 3 months (see Note 4).

23. Matrigel-coated culture plates (see Note 5).

24. Hepatocyte-culture medium (HCM): Standard WE medium supplemented with D-FBS (7% v/v), ITS supplements (10 mL/L), antibiotic, antimycotic solution (10 mL/L), and to final concentrations of 30 mM sodium pyruvate and 5 nM dexamethasone.

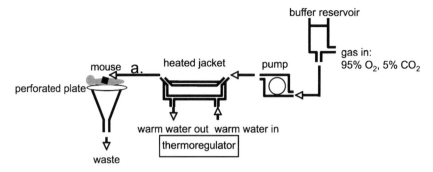

Fig. 1 Layout of components of the perfusion apparatus. *Horizontal arrows* indicate connections made by tubing to allow gassed and warmed perfusion buffer to be pumped through liver. a. Connection to cannula

25. Hepatocyte-washing medium (HWM): HCM medium without dexamethasone.
26. Sterile Nylon 70-μm cell strainers.
27. Sterile volumetric pipettes.
28. Sterile Cannula: Autoclave a 24 G gavage needle (Popper & Sons).
29. 60-mm and 100-mm sterile cell-culture plates. One of each.
30. Two 50-mL sterile centrifuge tubes.
31. Hemocytometer.
32. Perfusion apparatus (Fig. 1).
33. Laminar-flow cabinets (category II safety cabinets). One for perfusion apparatus and one for washing and plating hepatocytes.
34. Inverse-phase light microscope.
35. Perfusion pump (505S/RL, Watson Marlow) and appropriate plastic tubing (e.g., TYGON S-50-HL flexible plastic tubing, surgical grade (NORTON)).
36. Thermocirculator (Harvard Apparatus, Ltd).
37. One cylinder of medical gas mixture: 95% oxygen (O_2) and 5% carbon dioxide (CO_2).
38. CO_2 incubator. Preferably one that has a disinfection cycle.
39. Refrigerated, low-speed centrifuge.
40. Daylight magnifying lens illuminator.
41. Thermostated water bath.
42. Dissection mat.
43. Surgical instruments: Deschamps ligature needle, Mayo surgical scissors with blunt tips and curved blades, medium and small sharp scissors with flat blades, clamps (06-0010, Lawton), homemade vein-lifter (see Note 6).

Isolation of Mouse Hepatocytes   287

44. Syringes and needle Micro-Fine 0.33 mm (29 G) × 12.7 mm 0.5 mL U-100 insulin syringe (Becton Dickinson).

45. Thread: Braided silk, sterile, nonabsorbable suture, size 3 Mersilk (Ethicon).

3 Methods

3.1 Sterilization of the Perfusion Apparatus

1. Install the perfusion apparatus in the laminar-flow hood.

2. The day before the hepatocyte isolation fill the apparatus with 70% ethanol and leave it completely full for 24 h.

3. The following day circulate the ethanol for 15 min.

4. Turn on the connected thermocirculator to 37°C to heat the water of the thermocirculator's heating jacket.

5. Rinse the apparatus six times with sterile, double-distilled water. Each rinse should circulate through the apparatus for 10 min.

6. The perfusion apparatus is now ready for use.

3.2 Coating the Culture Plates with Matrigel

These steps are carried out before isolation procedure (see Notes 4 and 5).

1. Precool 12-well culture plates and keep on ice.

2. Dilute ice-cold Matrigel 1:9 with ice-cold William's E medium. Keep the solution on ice at all times.

3. Use a cold pipette tip to dispense 200 µL of diluted Matrigel into a well. Use a cold, standard sterile cell scraper that fits the well, to help spread the Matrigel evenly.

4. Place the coated plate in an incubator set at 37°C, 5% CO_2 for 1 h.

5. When the hepatocytes are almost ready to be plated, rinse the plates once with William's E medium (no supplements) to remove any traces of non-gelled Matrigel.

3.3 Cannulation and Perfusion of Liver (See Note 7)

1. Fill the perfusion apparatus reservoir with 50 mL of perfusion buffer 1. Circulate for 5–10 min, and ensure that there are no air bubbles. The reservoir for buffer 1 must be full before connection of the cannula (Subheading 3.3, step 14).

2. Set the temperature on the thermocirculator to 42°C.

3. Spray and clean the dissection mat with 70% ethanol. Place surgical box containing sterile instruments next to the mat.

4. Aseptically, place the gavage needle (which is the cannula) into a sterile 60-mm cell-culture plate containing 8 mL of heparin solution. Fill the gavage needle with heparin using an insulin syringe and needle (see Note 8).

5. Weigh the mouse by placing it in a beaker on a bench balance.

6. Anesthetize the animal by injecting intraperitoneally 0.1 mL/20 g body weight of the anesthetizing, combined solution (ACS). When the animal fails to respond to stimuli, swab the abdomen thoroughly with 70% ethanol. Place the animal on its back on the dissection mat and secure in place.

7. Maintain sterile conditions and open the abdominal cavity by making a U-shaped incision. Move the intestines to the left of the animal's torso, to reveal and expose the inferior vena cava (IVC).

8. Thread a ligature needle with thread. Loosely tie a ligature around the IVC, with the help of a vein-lifter (see Note 6), just below the junction to the kidney (*vena subrenalis*).

9. Make a slight incision with a small sharp pair of flat scissors, parallel to the wall of the IVC, below the ligature and large enough to insert the cannula (gavage needle). The cannula has previously been filled with heparin (Subheading 3.3, step 4) (see Note 8).

10. Check to ensure that there are no air bubbles in the cannula (see Note 8).

11. Insert the cannula at the lower end of the IVC. Do not push too far into the vessel.

12. After the insertion of the cannula into the IVC, tie the loose ligature (Subheading 3.3, step 8) with a double knot around the cannula.

13. Place a second ligature around the IVC just below the heart and tie with a double knot. This prevents the flow of perfusion buffers throughout the body.

14. Check the perfusion apparatus and make sure there are no air bubbles anywhere in the system. Check also that the thermo-circulator has reached 42°C; this is to obtain a 37°C temperature in the liver.

15. Carefully transfer the mouse from the dissecting mat and place gently on the platform of the perfusion apparatus. Take care to ensure that the cannula remains inserted in the correct position.

16. Connect the cannula to the perfusion apparatus tubing, which must be free of air bubbles. From here on one has to work very quickly. Steps 17–20 must take no longer than 1–2 min in total.

17. Start the perfusion immediately. Perfuse buffer 1 through the liver at a rate of 2 mL/min. If the perfusion is successful, the liver will bleach to a light beige color within a few seconds and have a shiny and soft surface (see Note 9).

18. Immediately make a small incision in the portal vein, to allow the perfusion buffers to flow freely out of the liver.

19. Increase the flow rate to 7 mL/min.

Isolation of Mouse Hepatocytes 289

20. Cut the diaphragm and clamp the IVC just below the heart.

21. Massage the liver gently between the thumb and the forefinger, to assist the digestion of the organ and prevent clot formation.

22. Rinse the abdominal cavity of the animal with distilled, sterile water, to clear out the blood.

23. Continue perfusion with buffer 1 for about 7 min.

24. Change to buffer 2 (see Note 2) and continue the perfusion for about 8–10 min.

25. The liver is now very soft and ready to remove for the isolation of hepatocytes (see Subheading 3.4).

3.4 Isolation of Hepatocytes

1. Carefully remove the liver. Do not cut through the intestines as this could cause contamination. Transfer the liver to a 100-mm cell-culture plate containing 10 mL of ice-cold hepatocyte-washing medium (HWM).

2. Using a pair of round-ended forceps, carefully peel away the Glisson's capsule enclosing the liver, in order to disperse the hepatocytes.

3. Disperse the cells further through a large-bore pipette, e.g., 25-mL pipette.

4. Place a 70-µm nylon cell-strainer on the top of a 50-mL sterile tube. Filter the cell suspension through the strainer and leave the cells to settle for 10 min.

5. Carefully remove the supernatant, which contains mostly dead cells.

6. Dilute the settled cells in 20 mL of ice-cold HMW medium.

7. Centrifuge at 50 g for 2 min at 4°C.

8. Decant the supernatant and resuspend the cells very gently in 20 mL of ice-cold HWM.

9. Repeat steps 6–8 twice.

10. Remove the supernatant carefully with a sterile pipette and resuspend the cell pellet very gently in 20 mL of ice-cold HCM.

11. Keep cell suspension on ice.

12. Count the cells with a hemocytometer, to determine the number and percentage of viable cells, using the trypan blue exclusion test (see Note 10).

13. Dispense 1 mL (5×10^6 cells/mL) of viable hepatocytes into each well of a Matrigel-coated 12-well culture plate (see Note 11).

14. Incubate the cells at 37°C with 5% CO_2 for 2–3 h.

15. Check cell attachment under the microscope. Most of the cells should be attached.

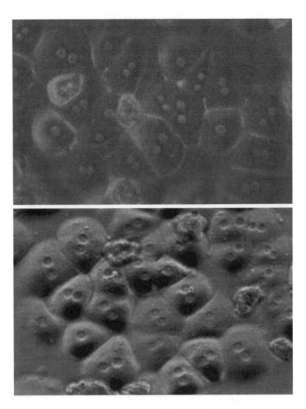

Fig. 2 Hepatocytes cultured on Matrigel. *Top panel*, light source from below to demonstrate defined gap junctions and pavement morphology. *Bottom panel*, light source from above to show rounded and healthy morphology

16. Using a pipette, remove the medium and dead or unattached cells. Add fresh HCM with or without D-FCS depending on how the hepatocytes are to be used in further experimental work. Figure 2 shows the morphology of hepatocytes isolated as above and plated on Matrigel. Figure 3 shows that the hepatocytes maintained healthy responses to stimuli and maintained transcriptional activity, when dosed with phenobarbital (see Note 12).

4 Notes

1. Before perfusion, prepare 100 mL of buffer 1 and saturate for 15 min with 95% O_2 and 5% CO_2. Check the pH and readjust to pH 7.4 with either 1 M HCl or 0.5 M NaOH. Sterilize by filtration through a 0.22-μM filter unit. Prewarm buffer 1 and keep at 37°C until ready to use. Before starting the dissection of the mouse, circulate 50 mL of the buffer through the perfusion apparatus to ensure an air-bubble-free unit.

2. Before perfusion, prepare 100 mL of buffer 2 and supplement with $CaCl_2$ to a final concentration of 3 mM. Check the pH and

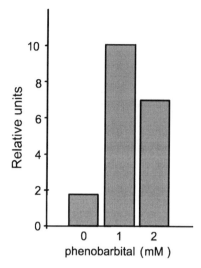

Fig. 3 Induction of CYP2B protein in phenobarbital-treated primary mouse hepatocytes. Hepatocytes were treated with phenobarbital for 24 h. Protein was detected by western blotting of cell lysates using an anti-rat CYP2B1/2 antibody

readjust to pH 7.4 with either 1 M HCl or 0.5 M NaOH. Sterilize by filtration through a 0.22-μM filter unit. Prewarm buffer and keep at 37°C until ready to use. Collagenase H is added to buffer 2 just before use, to a final concentration of 0.08 U/mL. The solution is mixed well and used immediately.

3. Store the powder at 2–8°C. Prepare a 20 μg/mL stock solution by first dissolving 1 mg of dexamethasone powder in 1 mL of absolute ethanol; swirl gently to dissolve and then add 49 mL of sterile HCM. Prepare working aliquots and store at −20°C. Avoid repeated freezing and thawing.

4. Matrigel is supplied in Dulbecco's Modified Eagle's Medium (phenol-red free). We dilute in standard William's E medium. Matrigel solidifies rapidly if warmed. Thaw the frozen matrix at 4°C overnight on ice. Keep the thawed Matrigel on ice all the time, mix it well (in an ice-water bath) to homogeneity. Dispense appropriate volumes aseptically into precooled tubes, using precooled pipettes and tissue-culture plasticware. We find the most reliable way to precool plasticware is to store the plasticware needed in a −20°C freezer and remove as required.

5. Before and during the coating with Matrigel, the plates and other plasticware must be kept ice-cold.

6. We use an L-shaped metal instrument, 3-cm long, for ease of handling.

7. A successful liver perfusion requires speed and coordination. It is essential to practice inserting the cannula, tying the ligatures, and transferring the mouse to the perfusion platform, before

embarking on a full-scale isolation procedure. On the day of the perfusion check your list and make sure that everything one needs is at hand.

8. Any air bubbles in the cannula (gavage needle) will seriously impair the quality of the perfusion. Heparin is used in the cannula to help prevent blood coagulation in the liver lobes. Coagulation can hinder complete perfusion of the organ.

9. If, on starting the perfusion with buffer 1, the liver does not bleach to a light beige color then stop the perfusion and discard the liver.

10. Nonviable cells will take up the trypan blue stain into their nuclei. A very low viability generally gives poor results and suggests a possible problem with the procedure and/or the reagents used. The viability obtained using this test with mouse hepatocytes is a lot lower than that observed for isolated rat hepatocytes. We found that a viability of 65% produces good cultures of mouse hepatocytes. The key criteria for viability are speed and how quickly the liver bleaches.

11. Mouse hepatocytes do not respond well when plated at too high or too low a density. Under our conditions we found 5×10^6 cells per well of a 12-well plate to be optimum.

12. To confirm that the hepatocytes are healthy and retain liver-specific properties cells can be dosed with phenobarbital. In healthy cells CYP2B should be induced. As shown in Fig. 3, CYP2B protein is inducible in the mouse hepatocytes and, therefore, they have maintained crucial factors required for induction of expression of the mouse *Cyp2B* gene. A concentration of 1 mM phenobarbital was found to give the highest induction. Previous studies also show that induction of CYP2B by phenobarbital is greatest at similar concentrations (0.75–1 mM) (8).

Acknowledgements

We thank Dr. Rick Moore (NIEHS, North Carolina, USA) for advice. L.H. is a recipient of a Drummond Scholarship.

References

1. Wilkening S, Stahl F, Bader A (2003) Comparison of primary human hepatocytes and hepatoma cell line Hepg2 with regard to their biotransformation properties. Drug Metab Dispos 31:1035–1042

2. Olsavsky KM, Page JL, Johnson MC, Zarbl H, Strom SC, Omiecinski CJ (2007) Gene expression profiling and differentiation assessment in primary human hepatocyte cultures, established hepatoma cell lines, and human liver tissues. Toxicol Appl Pharmacol 222:42–56

3. Nyberg SL, Remmel RP, Mann HJ, Peshwa MV, Hu WS, Cerra FB (1994) Primary hepatocytes outperform Hep G2 cells as the source of biotransformation functions in a bioartificial liver. Ann Surg 220:59–67

4. Smirlis D, Muangmoonchai R, Edwards M, Phillips IR, Shephard EA (2001) Orphan receptor promiscuity in the induction of cytochromes P450 by xenobiotics. J Biol Chem 276:12822–12826

5. Klaunig JE, Goldblatt PJ, Hinton DE, Lipsky MM, Chacko J, Trump BF (1981) Mouse liver cell culture. I. Hepatocyte isolation. In Vitro 17:913–925

6. Bahjat FR, Dharnidharka VR, Fukuzuka K, Morel L, Crawford JM, Clare-Salzler MJ, Moldawer LL (2000) Reduced susceptibility of nonobese diabetic mice to TNF-alpha and D-galactosamine-mediated hepatocellular apoptosis and lethality. J Immunol 165:6559–6567

7. Page JL, Johnson MC, Olsavsky KM, Strom SC, Zarbl H, Omiecinski CJ (2007) Gene expression profiling of extracellular matrix as an effector of human hepatocyte phenotype in primary cell culture. Toxicol Sci 97:384–397

8. Kocarek TA, Schuetz EG, Guzelian PS (1990) Differentiated induction of cytochrome P450b/e and P450p mRNAs by dose of phenobarbital in primary cultures of adult rat hepatocytes. Mol Pharmacol 38:440–444

Chapter 25

Highly Efficient SiRNA and Gene Transfer into Hepatocyte-Like HepaRG Cells and Primary Human Hepatocytes: New Means for Drug Metabolism and Toxicity Studies

Véronique Laurent, Denise Glaise, Tobias Nübel, David Gilot, Anne Corlu, and Pascal Loyer

Abstract

The metabolically competent hepatocyte-like human HepaRG cells represent a suitable alternative in vitro cell model to human primary hepatocytes. Here, we describe the culture procedure required to expand progenitor HepaRG cells and to differentiate them into hepatocyte-like cells. Transient transfection of gene and siRNA into cultured cells, using nonviral strategies, is an invaluable technique to decipher gene functions. In this chapter, we detail transfection protocols for efficient transfer of plasmid DNA or siRNAs into proliferating progenitor or quiescent differentiated HepaRG cells as well as into primary hepatocytes.

Key words Transfection, Nucleofection™, Human hepatocytes, HepaRG cells, Drug metabolism, In vitro toxicity

1 Introduction

Normal hepatocytes perform many biochemical reactions including the detoxification of endogenous toxins and xenobiotics, via the catalytic activities of phase I, II, and III drug-metabolism enzymes (DMEs). To study the signaling pathways regulating hepatic expression of DMEs and to characterize their enzymatic activities, two main types of in vitro models are available: primary cultures of normal hepatocytes and hepatoma cell lines (1). Primary cultures are probably the most relevant in vitro systems for studying various aspects of the liver biology, including the differentiation/proliferation status (2, 3), the catalytic activities of DMEs, and the regulation of their expression (4). However, large variations in functional activities, especially for cytochrome P450 enzymes (CYP), have been observed in human hepatocytes isolated from different donors (1). In addition, the relative shortage and unpredictable availability of human biopsies, along with the limited growth and lifespan of

Ian R. Phillips et al. (eds.), *Cytochrome P450 Protocols*, Methods in Molecular Biology, vol. 987,
DOI 10.1007/978-1-62703-321-3_25, © Springer Science+Business Media New York 2013

296 Véronique Laurent et al.

hepatocytes in primary cultures, limit considerably the use of these models. Most hepatoma cell lines express low amounts of DMEs, compared with primary hepatocytes (5). Thus, highly selected subclones of hepatoma cells, such as the HepG2C3A, expressing higher amounts of CYP have been established (6). Alternatively, recombinant-derived hepatoma cell lines have been generated through stable transfection of expression vectors encoding DMEs or liver-specific transcription factors that drive transcription of DME genes (7). These cell lines, however, show limited levels of expression of DMEs and development of novel differentiated hepatocyte-like cell lines remains a relevant task.

Undifferentiated human HepaRG cells are bipotent hepatic progenitors with high proliferation potential and capable of differentiating into biliary and hepatocyte-like cells (8). This recently established cell model represents a unique "in vitro system" to investigate the molecular pathways of hepatocyte differentiation (9). Moreover, HepaRG hepatocyte-like cells that express most of the major CYPs, such as CYP3A4, 2E1, and 1A2 (10), provide a valuable in vitro alternative model to primary culture of hepatocytes for studying metabolism, toxicity, and genotoxicity of xenobiotics (11–15). Here, we detail a procedure for expanding the progenitor population of HepaRG cells, the protocols routinely used in our laboratory for the optimal differentiation of hepatocyte-like HepaRG cells, and a simple method for purifying these hepatocyte-like cells in order to prepare a pure culture of highly differentiated hepatocyte-like HepaRG cells.

Using nonviral strategies, transfer of nucleic acids (plasmids and siRNAs) into primary hepatocytes and differentiated HepaRG hepatocyte-like cells has been a difficult task for many years and was a limiting step for studying the regulation of DMEs. In this chapter, we review the most efficient transfection protocols, involving the use of newly available transfection reagents (lipids and polymer formulations) and cutting-edge electroporation devices, for delivering siRNAs and DNA plasmids into both progenitor and metabolically competent differentiated HepaRG cells. We also indicate the procedures that can be extended to primary human and rodent hepatocytes. The efficient procedures of transfection are illustrated with recent data from our laboratory related to the regulation of CYP450 expression in HepaRG cells.

2 Materials

2.1 Culture Reagents: Solutions and Media

Phosphate-buffered saline (PBS) without Ca^{2+} and Mg^{2+}. Trypsin (0.05%)–EDTA (5.3 mM) 1× solution. OptiMEM® medium. William's E medium supplemented with (final concentrations) L-glutamine 2 mM, 1% penicillin–streptomycin (5,000 UI/mL to 5,000 μg/mL), recombinant human insulin (5 μg/mL), 50 μM

RNA and DNA Transfer in Hepatocytes   297

hemisuccinate hydrocortisone (see Note 1 for references and preparation of stock solutions of these additives), and 10% fetal calf serum (see Note 2 for selection of fetal calf serum). For full hepatocyte differentiation, the same medium is supplemented with 2% dimethyl sulfoxide (DMSO). Culture media are kept at 4°C protected from excessive light exposure.

2.2 Transfection Reagents

1. SynNanoVect cationic lipids KLN47 and BSV10. The lipophosphoramidate compounds KLN47 and BSV10 were purchased from the SynNanoVect platform dedicated to the synthesis of transfection reagents (www.synnanovect.ueb.eu). Whereas KLN47 liposomes are formulated in water in the absence of co-lipids, the BSV10 liposomes are formed in water in the presence of 50% of the neutral co-lipid dioleylphosphatidylethanolamine (DOPE), for optimal transfection efficiencies (16). Liposomal concentrations of KLN47 and BSV10-DOPE are 1 mg/mL.

2. SiPORT™ Amine (Applied Biosystems, Life Technologies™) is a polyamine-based transfection reagent specifically designed for transfer of small RNAs (catalog numbers AM4502 and AM4503 for 0.4 mL and 1 mL, respectively).

3. DharmaFECT (Thermo Fisher Scientific). DharmaFECT reagent is available in four distinct formulations (1–4) for best transfection efficiencies in a wide range of cell types. For primary hepatocytes and HepaRG cells, formulations 1 (catalog number T-2001; three packaging sizes of 0.2, 0.75, and 1.5 mL) and 4 (T-2004) give the best compromise between efficiency and toxicity.

2.3 Electroporation Devices

1. Neon™ transfection system (Invitrogen, Life Technologies, catalog number, MPK5000) is a benchtop electroporation device that delivers nucleic acids (genes and RNAs) and proteins into mammalian cells. This system relies on the use of a pipette tip inserted into an electroporation chamber generating a homogeneous electric field characterized by a large gap size between the two electrodes. Reagents for electroporation provided by the manufacturer include the electrode tips (10 or 100 μL), electrolyte, and cell-suspension buffers. The touch screen allows modulation of three crucial parameters of the electric pulse (voltage, width, and number of pulses) to optimize transfection efficiency. Cell-specific Neon transfection protocols are available on the company Web site (www.invitrogen.com/neon).

2. Amaxa™ 4D-Nucleofector™ technology (Lonza) is a modular electroporation platform comprising a core unit and the X unit as the first functional benchtop system (www.lonza.com/4d-nucleofector). The X unit allows transfection of

298 Véronique Laurent et al.

cells in suspension in either 100-μL Nucleocuvettes™ or 16-well 20-μL Nucleocuvette™ strips. This core system can be completed by connecting additional devices to the core unit: the 96-well Shuttle™ system and the Y unit enabling electroporation of adherent cells in 24-well plates. Reagents for electroporation provided by the manufacturer include the Nucleocuvettes or the Nucleocuvette strips, cell-suspension buffers, and plasmid DNA encoding the green fluorescent protein (GFP), for protocol optimization. Cell-specific Nucleofection™ transfection protocols are available on the company Web site: http://www.lonzabio.com/resources/product-instructions/protocols/.

2.4 Bacterial Amplification of Plasmids

The purifications of expression vectors for mammalian cell transfection must be performed using endotoxin-free kits such as the NucleoBond® Xtra Midi or Maxi EF (MACHEREY-NAGEL) or EndoFree™ plasmid kits (QIAGEN).

2.5 Chemically Synthesized Double-Stranded siRNAs

ON-TARGETplus SMART pool siRNAs targeting mRNAs of Aryl hydrocarbon Receptor (AhR) and cy-3-labeled siRNAs (siGLO) (Dharmacon) (see Note 3).

3 Methods

Carry out all procedures using culture-grade reagents and disposable tissue-culture dishes, flasks, and pipets. The use of solvents, fixation reagents (formaldehyde-based fixators), and cell-lysis buffers in the tissue-culture rooms must be prohibited. Mycoplasma screening must be performed on each batch of thawed HepaRG cells one passage after thawing.

3.1 Culture Procedures for Long-Term Maintenance of the Progenitor Status of the HepaRG Cells and Their Potential for Differentiation into Hepatocyte-Like Cells

The optimal process of differentiation of the HepaRG progenitors towards hepatocytes largely depends upon proper culture conditions detailed in this section. When this protocol is not strictly adopted, the ability of progenitors to differentiate rapidly decreases and amplification of a population of "small hepatocytic cells," with poor expression of hepatocyte-specific functions, is observed. The optimal protocol of culture is organized in 2-week periods between two successive passages (Fig. 1).

1. Two weeks after the previous passage, detach confluent HepaRG cells. For a 75-cm² flask, discard the culture medium, rinse the cell monolayer with 10 mL of PBS, add 5 mL of trypsin solution, allow the trypsin solution to cover the entire surface, and discard 3 mL. Incubate the culture flask in an incubator at 37°C for 5 min. Suspend the cells in 8 mL of medium without DMSO, to give a final volume of 10 mL. Disrupt clumps by pipetting the cell suspension up and down.

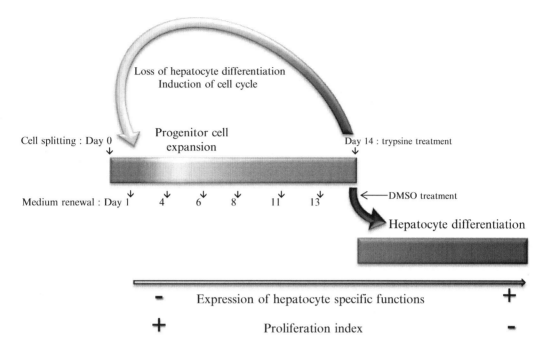

Fig. 1 Schematic representation of the optimal HepaRG cell culture protocol for long-term stability of the differentiation potential into hepatocyte-like cells. The proposed protocol includes trypsinizing the cells on Thursday (day 0), renewing the medium on Friday (days 1 and 8), and on Monday (days 4 and 11) and Wednesday (days 6 and 13). Two weeks after splitting cells, when cultures are confluent, two options are possible: (1) to further expand the cell population following the same procedure and (2) to differentiate the cells by culturing confluent cells in the same culture medium supplemented with 2% DMSO (8, 10)

2. Count the cells. Usually, for a 75-cm² flask, the cell density is between 1 and 1.5×10^7 cells/10 mL of suspension. Plate the cells at a density of 2.6×10^4 cells/cm² (2×10^6 cells/75-cm² flask) in new dishes.

3. Renew the medium the day after seeding and then every 2 or 3 days (Fig. 1).

4. Two weeks later, either expand the progenitor cells, following the procedure described above, or culture the cells for another 2 weeks in the same medium supplemented with 2% DMSO (full differentiation medium). Keep renewing the differentiation medium every 2 days (see Note 4 for optimal differentiation process).

5. During the 2-week cell expansion protocol and the procedure of hepatocyte differentiation in presence of DMSO, carefully examine cell morphology by phase-contrast microscopy, to monitor proper expansion of progenitor cells, hepatocyte-like cell differentiation, and emergence of "small hepatocyte-like" cells and fibroblasts (Fig. 2, see Note 5).

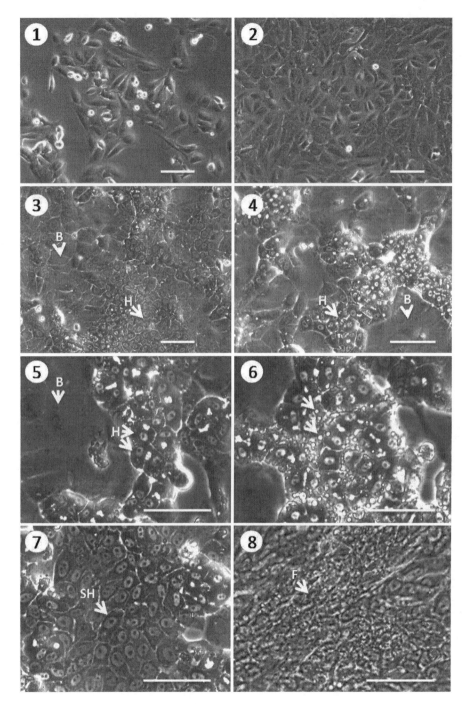

Fig. 2 Morphology in phase contrast of HepaRG cells at different stages of differentiation. (1) Bipotent progenitors at low density, day 1 post-trypsination, (2) HepaRG cells at day 7 posttrypsination, (3) committed HepaRG hepatocyte-like (*H*) and biliary (*B*) cells at day 14 posttrypsination, (4) coculture hepatocyte-like HepaRG and biliary cells at day 30 posttrypsination, including the last 2 weeks with 2% DMSO treatment (enhancement of liver-specific function levels), (5) high magnification of hepatocyte-like (*H*) and biliary cells (*B*), (6) hepatocyte-like cells containing lipid droplets (*arrows*) at 40 days posttrypsination, (7) small hepatocyte-like cells (*SH*) that fail to fully differentiate into hepatocytes, and (8) emergence of fibroblast-like cells (*F*) in inappropriate culture conditions and late passages. Bar: 100 μm

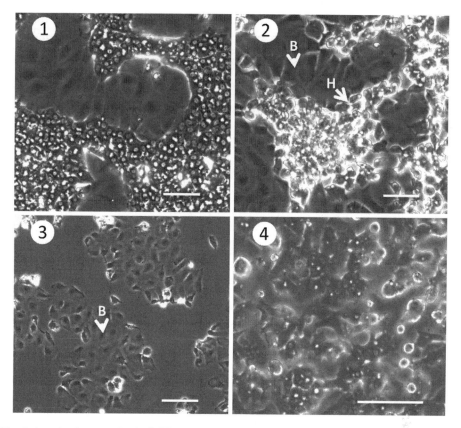

Fig. 3 Morphology in phase contrast of differentiated HepaRG cells (1) and selective detachment by trypsin (2). Hepatocytes: H, biliary cells: B. Biliary cells (3) after selective detachment of hepatocyte HepaRG-like cells. Photograph of pure hepatocyte-like HepaRG cells 24 h after selective detachment and seeding (4). Bar: 150 μm

3.2 Selective Detachment and Seeding of Differentiated Hepatocyte-Like HepaRG Cells

From DMSO-treated HepaRG cultures (30 days of culture, Figs. 2-4 and 3-1), hepatocyte-like cells can be purified, with purity ≥80%, by a simple trypsin treatment. Use flasks, not plates, for this procedure.

1. For 75-cm² flasks, rinse cultures twice with 10 mL of PBS. Add 2.5 mL of PBS and 2.5 mL of trypsin solution. Close the flask tightly and mix the contents gently. Incubate at room temperature for 5–10 min and monitor, by phase-contrast microscopy, the morphological change of hepatocyte-like cells (Fig. 3-2).

2. When hepatocyte-like cells have rounded up (Fig. 3-2), shake the flask containing the PBS–trypsin solution laterally, to allow complete detachment of hepatocyte-like cells (see Note 6).

3. Collect and combine the 5 mL of diluted trypsin solution containing the cells with 5 mL of William's E medium. Pellet the cells by centrifugation (50×g, 2 min), resuspend the pellet in 5 mL of William's E medium, and count the hepatocyte-like cells. Typically, ~3–5 × 10⁶ hepatocyte-like cells can be purified

302   Véronique Laurent et al.

from a 75-cm² flask. Plate the cells at a density compatible with subsequent experiments: 5–7×10^4 cells/cm² in the William's E medium without DMSO, for studying hepatocyte proliferation, or $\geq 2 \times 10^5$ cells/cm² in culture medium supplemented with 2% DMSO, for maintaining the expression of hepatocyte-specific functions and to keep cells quiescent.

3.3 Transfection of Proliferating Progenitor and Quiescent Differentiated HepaRG Cells Through the Use of Chemical Transfection Reagents

Here, we describe some reagents that can be used to chemically transfect proliferating progenitor or quiescent differentiated HepaRG cells with expression vectors or siRNA (see Note 7). All protocols use 24-well culture plates unless otherwise specified. Procedures can be scaled up by keeping constant the ratio between cell density and amounts of transfection reagents and DNA.

1. Cationic lipids KLN47 and BSV10 (SynNanoVect lipophosphoramidate compounds (16), see Note 8). The day before transfection, plate proliferating progenitor HepaRG cells at 10^5 cells/well (24-well plates, 2 cm²/well). Sonicate liposome solutions (1 mg/mL) twice for 5 min in a sonication water bath. In two separate Eppendorf tubes, add 4 μL of KLN47 or 1.8 μL of BSV10/DOPE liposomes in 100 μL of Opti-MEM medium. Prepare two tubes of DNA by diluting 0.8 μg of plasmid DNA into 100 μL of Opti-MEM medium (see Note 9). Vortex briefly. Combine liposome-Opti-MEM solutions with the DNA-Opti-MEM mix. Vortex for 20 s and incubate at room temperature for 30 min. Discard culture medium from the culture plates and renew with 300 μL of fresh William's E medium without antibiotics. Add the 200 μL of liposome-DNA mix into wells dropwise. Incubate liposome–DNA complexes with cells for 4–6 h in the incubator (see Note 10). Discard transfection medium and renew with medium containing antibiotics. Analyze expression of the protein of interest between 24 and 48 h post-transfection (see Note 11).

2. SiPORT™ Amine (Applied Biosystems, Life Technologies™, see Note 12). The day before transfection, plate proliferating progenitor HepaRG cells at 10^5 cells/well (24-well plates, 2 cm²/well) or quiescent differentiated hepatocyte-like HepaRG cells (density $\geq 2 \times 10^5$ cells/cm²) selectively detached (see Subheading 3.2). Dilute 1.5 μL of siPORT Amine transfection agent in 25 μL of Opti-MEM medium and incubate for 10 min at room temperature (see Note 13). Dilute siRNA (1–10 pmol, see Note 14) in 25 μL of Opti-MEM medium. Combine the diluted SiPORT™ Amine agent and diluted siRNA, mix gently, and incubate for 10 min at room temperature. Add the mix onto the cells that should be kept in their regular culture medium without or with DMSO, for progenitor and differentiated HepaRG cells, respectively (final volume 500 μL). Alternatively, cells and siRNA-SiPORT™ Amine

RNA and DNA Transfer in Hepatocytes 303

transfection mix can be combined and plated simultaneously (reverse transfection). This last procedure can also be used for primary human and rodent hepatocytes (see Note 15). Renew the medium 24 h after transfection. Assay for target-gene repression 24–72 h after transfection (see Note 16).

3. DharmaFECT (Thermo Fisher Scientific, see Note 17). Progenitor or differentiated HepaRG cells or rodent or human primary hepatocytes (see Note 15) are trypsinized and replated in 24-well plates at appropriate densities (5×10^4 cells/cm^2 in the William's E medium without DMSO, for progenitors, and $\geq 2 \times 10^5$ cells/cm^2 in culture medium supplemented with 2% DMSO, for differentiated HepaRG cells and primary hepatocytes). Dilute 3 µL of DharmaFECT agent and 1.25–10 µL of siRNA (5 µM stock solution) in two separate tubes containing 60 µL of Opti-MEM medium. Incubate for 5 min at room temperature. Combine the diluted DharmaFECT reagent and diluted siRNA, mix gently, and incubate for 20 min at room temperature. Add 380 µL of the William's E culture medium without or with DMSO (see Note 18), for progenitors and differentiated HepaRG cells, respectively. Remove culture medium from the 24-well plate and add the transfection mix diluted in culture medium. The final volume is ~500 µL and siRNA concentration is 25 nM (see Note 19). Discard the transfection culture medium after 6 h and renew the medium. Assay for target gene repression 24 to 72 h after transfection (see Note 16).

3.4 Electroporation of Progenitor and Differentiated HepaRG Cells Using Neon™ System

1. Rinse cultures of progenitor or differentiated hepatocyte-like cells with PBS and detach by trypsin treatment (see Subheading 3.2 for differentiated cells). Collect cells in William's E medium (final volume 10 mL for a 75-cm^2 flask) and count the cells.

2. Pellet the number of cells needed for the electroporation experiment by centrifugation ($50 \times g$, 2 min), resuspend the pellet with 10 mL of cold PBS. Pellet the cells by centrifugation ($50 \times g$, 2 min), resuspend the cells with 1 mL of cold PBS, and transfer the cell suspension to a microfuge tube (see Note 20). Pellet the cells by centrifugation using a benchtop centrifuge ($50 \times g$, 30 s) and resuspend cell pellet in the transfection buffer provided by the manufacturer (R or T, see Note 21) at a cell density of 10^6 cells/mL.

3. Aliquot the cell suspension according to the different transfection conditions (different DNA plasmids and siRNAs) used. Transfection is performed using electrode tips of either 10 µL (electroporation of 10^5 cells) or 100 µL (electroporation of 10^6 cells). Add to cell suspension (in buffer R or T) plasmid DNA (see Note 22) or siRNA (see Note 23). The total volume

of cells and nucleic acid mix (plasmid DNA or siRNA) must include at least the equivalent of 1 electrode tip volume more than the volume required for the number of electroporations calculated for the experiment (see Note 24).

4. Set up the Neon electroporation station by inserting a disposable Neon™ tube. Fill the tube with 3 mL of electrolyte buffer (Buffer E for the 10-µL tip and E2 for 100-µL tip).

5. Quickly proceed to the electroporation (see Note 25) using the following parameters for both progenitor and differentiated HepaRG cells: voltage 1,500 V, width 20 ms, and 1 pulse (16). Carefully avoid bubbles in the electrode tip. Electrode tips can be used twice for the same cells and nucleic acid master mix.

6. Transfer cells to culture dishes at cell densities compatible with subsequent experiments: 5×10^4 cells/cm^2 in the William's E medium without DMSO, for proliferating HepaRG progenitors, and $\geq 2 \times 10^5$ cells/cm^2 in culture medium supplemented with 2% DMSO, for quiescent differentiated hepatocyte-like cells. Remove antibiotics from the culture medium during the first 24 h following the electroporation (Fig. 4).

7. Renew culture medium 24 h after transfection. Assay for target gene repression 24–48 h after transfection. Using green fluorescent protein (GFP) expression vectors (pmaxGFP™ or pEGFP™), transfection efficiencies reach 80–85% of GFP-positive cells (Fig. 5-1–2). Using plasmid pCDNA3.1 (Invitrogen) encoding CYP450 2E1, we obtained enforced expression of CYP2E1 in both progenitor and differentiated cells (Fig. 5, right panel) (16) and confirmed the regulation of the CYP2E1 expression by insulin at a posttranscriptional level.

3.5 Transfection of Progenitor and Differentiated HepaRG Cells Through the Use of 4D-Nucleofector™ X-Unit

1. Rinse cultures of progenitor or differentiated hepatocyte-like cells with PBS and detach by trypsin treatment (see Subheading 3.2 for differentiated cells). Collect cells in William's E medium (final volume 10 mL for a 75-cm^2 flask) and count the cells.

2. Wash cells once with PBS. Resuspend cell pellet in the transfection buffer (see Note 26) at a density of 1–2.5×10^7 cells/mL.

3. Prepare culture dishes to collect cells after transfection. Add culture medium and pre-warm the dishes while proceeding with the transfection procedure.

4. Aliquot the cell suspension according to the different transfection conditions (different DNA plasmids and siRNAs) used. Add plasmid DNA or siRNA to cell suspension (see Note 27). Transfection is performed using either 20-µL Nucleocuvette™ strip (electroporation of 2–5×10^5 cells) or 100-µL single Nucleocuvette™ (electroporation of 1–2.5×10^6 cells) with the 4D-Nucleofector™ X-unit. Alternatively, Nucleofection™ can be performed using the 96-well Shuttle™ device (see Note 28).

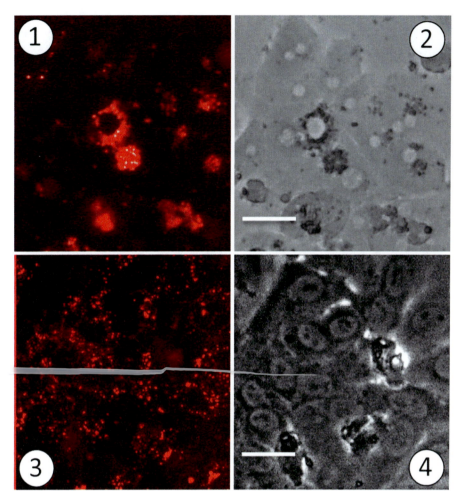

Fig. 4 Detection of cy-3-labeled siRNA (siGLO, Dharmacon) (1, 3) in primary human hepatocytes (2, phase contrast) and in progenitor HepaRG cells (4, phase contrast) transfected using SiPORT™ Amine. Le Vee et al. (17) have been obtained similar results using DharmaFECT to deliver siRNA in HepaRG differentiated cells and primary human hepatocytes. Bar: 20 μm. Protocols described in steps 2 and 3 in Subheading 3.3 have been used to repress expression of AhR in primary human hepatocytes and HepaRG cells to investigate the involvement of AhR in the regulation of DMEs following exposure to xenobiotics (14, 17)

5. Quickly proceed to the electroporation, using the following programs: CM-150, CM-137, with the solution from the SE 4D-Nucleofector™ X kit, or the EN-138, EN-150, with the solution from the SF 4D-Nucleofector™ X kit. Specific protocols for electroporation of human primary hepatocytes have been established by the manufacturer (see Note 29).

6. After transfection, incubate Nucleocuvette™ strips for 10 min at room temperature. Resuspend cells with pre-warmed culture medium, by gently pipetting up and down, and plate the cells in culture dishes.

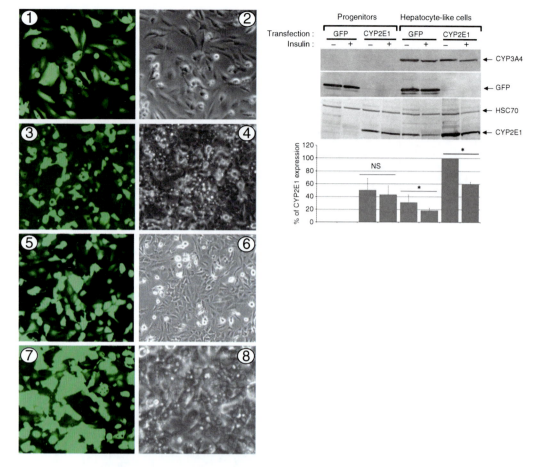

Fig. 5 Detection of GFP-positive HepaRG cells (*left panel*, 1–8). Progenitor (1–2 and 5–6) and differentiated (3–4 and 7–8) HepaRG cells were electroporated by the pmaxGFP-expression vector, using the Neon™ system (1–2) and 4D-Nucleofector™ X (3–4) or Y (5–8) units. GFP-positive cells (1, 3, 5, 7) were detected using inverted fluorescence microscopy and cell morphology was examined by phase-contrast microscopy (2, 4, 6, 8). Enforced expression of CYP2E1 in progenitor and differentiated HepaRG cells (*right panel*). HepaRG cells were transfected either with GFP or CYP2E1 expression vectors, using the Neon™ system, and expression of endogenous CYP2E1 and CYP3A4, heat-shock protein (HSC) 70, and ectopic GFP and CYP2E1 were studied by immunoblotting. Endogenous CYP2E1 and CYP3A4 expression was found in differentiated HepaRG cells only, as previously shown (16). GFP and CYP2E1 were strongly expressed in transfected cells (16). Removal of insulin further enhanced CYP2E1 expression in differentiated HepaRG cells, confirming the hypothesis that CYP2E1 expression is negatively regulated by insulin (18). However, CYP2E1 expression in progenitor cells does not seem affected by the presence of insulin. *NS* nonsignificant, Mann–Whitney test significant at *$p < 0.05$

7. Twenty-four hours after transfection, renew the medium. Assay for target gene repression 24–48 h after transfection. Using green fluorescent protein (GFP) expression vectors (pmaxGFP™ or pEGFP™), transfection efficiencies reach 80–90% of GFP-positive cells (Fig. 5-3–4).

RNA and DNA Transfer in Hepatocytes   307

3.6 Nucleofection™ of Progenitor and Differentiated HepaRG Cells Using 4D-Nucleofector™ Y-Unit

To the best of our knowledge, the 4D-Nucleofector™ Y unit is the first electroporation device that allows transfection of cells in adherence (in a 24-well plate culture format, see Note 30).

1. Plate cells in 24-well plates at least 24 h before transfection (see Note 31). Optimal transfection efficiency is obtained for 70–80% confluent cells (2–4×10^5 cells/well).

2. Prepare the Nucleofection™ substrate solution: For 1 well, combine 350 μL of Nucleofection™ solution with 16 μg of plasmid DNA (see Note 32).

3. Remove medium from the 24-well plate and add slowly the Nucleofection™ substrate solution. Insert the dipping electrode array into the 24-well plate. Avoid air bubbles underneath the electrode array by holding the 24-well plate at a 45–75° angle while inserting the electrode array, to push out bubbles. Quickly proceed to the electroporation using the following programs: for progenitor cells use program ED-142 or CA-215, and for differentiated cells use CM-215, CD-215, CG-215, or CA-228.

4. After transfection, carefully remove the dipping electrode array from the culture plate, eliminate the Nucleofection™ substrate solution, and replace with pre-warmed culture medium.

5. Twenty-four hours after transfection renew the medium. Assay for target gene repression 24–48 h after transfection. Using green fluorescent protein (GFP) expression vectors (pmaxGFP™ or pEGFP™), transfection efficiencies reach 75–90% of GFP-positive cells (Fig. 5-5–8).

4 Notes

1. Stock solutions for the HepaRG cell culture medium. Commercial L-glutamine and penicillin–streptomycin solutions ($\times 100$) are thawed and aliquoted into 5-mL sterile tubes and stored frozen. One aliquot of each is thawed and added to each 500-mL bottle of William's E medium. Recombinant human insulin (Sigma) is solubilized in sterile ultrapure water, to which 37% HCl is added dropwise until complete dissolution, to obtain a stock solution at 5 mg/mL. Insulin stock solution can also be aliquoted (500 μL/tube) and frozen. Hemisuccinate hydrocortisone (Upjohn Pharmacia), packaged as 100 mg of lyophilized product/vial, is dissolved with 4 mL of PBS (stock solution at 5×10^{-2} M) and aliquots of 500 μL are frozen. For each 500-mL bottle of William's E medium, one aliquot of insulin and one of hemisuccinate hydrocortisone are thawed and added to the medium.

2. Fetal calf serum (FCS). We observed that FCS was crucial to properly maintain HepaRG cell proliferation and the capacity to differentiate. We screen new batches of sera when the lot in use becomes low. We obtain small batches from companies specialized in tissue-culture reagents and we test these batches over at least 4–5 passages (1 passage meaning 2 weeks of culture, 4 passages will require at least 2 months; this time-consuming screen needs to be planned long in advance). During the screen, we monitor the cell toxicity, the proliferation rate of progenitor HepaRG cells, and their ability to differentiate in the presence of 2% DMSO (see Subheading 3.1). We also pay particular attention to selecting sera that limit the expansion of non-fully differentiated small hepatocyte-like cells and fibroblasts (see pictures in Fig. 2-7–8). We found great differences between test batches, although all sera are United States Department of Agriculture certified. When an optimal batch is identified, we purchase a large volume of this serum for long-term storage at –20°C.

3. Stock solutions of siRNAs. We recommend preparing stock solution of siRNAs using RNAse-free water at a final concentration of 100 μM, for long-term storage. However, for preparation of transfection reagent mix, it is essential to prepare diluted solutions of RNAs in the range of 20–1 μM, to avoid pipetting too large or too small volumes of siRNAs that will dilute the transfection buffers (see Note 23).

4. Differentiation of HepaRG cells towards hepatocyte-like cells. At day 14 after cell splitting, cells form a confluent monolayer. These cells ($1.2–1.5 \times 10^7$ cells/75-cm^2 flask) are already committed to hepatocyte or biliary cell differentiation. Addition of 2% DMSO enhances expression of hepatocyte-specific functions, especially cytochromes P450 (10). Ten days to 2 weeks of DMSO treatment are required to obtain the highest expression of these markers. The cell number is often slightly reduced from 1.2 to 1.5×10^7 cells, before addition of DMSO, to $\sim 1–1.2 \times 10^7$ cells/75-cm^2 flask, after addition of DMSO. During the first week following DMSO addition, if too many cells die and detach from the plastic dish it means that DMSO was added too early or that combination of serum and DMSO induces toxicity. During DMSO treatment, remaining hepatocyte-like cells complete their differentiation and undergo drastic morphological changes, to give rise to well-defined colonies of hepatocytes characterized by a dark cytoplasm, a large nucleus with a single nucleolus, and functional neo-canaliculi (9).

5. Emergence of cells with poor hepatocyte differentiation level and fibroblasts. We systematically observe the decrease in the capability of HepaRG progenitor cells to generate hepatocyte-like cells after 20–30 passages, depending on the batches of

serum and the protocol used for expanding the progenitor population. Conversely, "small-like" hepatocytes (Fig. 2-7) and fibroblasts (Fig. 2-8) increase with higher passage numbers. It is crucial to expand HepaRG cells, to store a frozen master and working banks in liquid nitrogen. The master bank should be prepared from cells with low passage number (<20 passages) and the working bank with early passages. The appearance of these "derived cells" is limited by selecting appropriate sera (see Note 2) and with strict culture conditions (avoid keeping cells out of the incubator for too long, prewarm culture medium at room temperature before renewal, and avoid clumps while splitting cells). When these derived-cell phenotypes appear, the cell batch must be discarded and a new vial of cells from the working bank must be thawed.

6. Selective detachment of differentiated hepatocyte-like cells. This procedure requires some practice. It is essential to monitor the cell detachment under the microscope to ensure the selectiveness. Hepatocytes detach first. As soon as they appear round, shake the flask from left to right side with the trypsin solution covering the cell monolayer. Collect the hepatocytes. The biliary cells can also be detached by incubating with pure trypsin at 37°C for 2–3 min after a PBS wash. Note that hepatocyte-like cells that are treated for more than 2 weeks with 2% DMSO are much more difficult to selectively detach.

7. Two distinct procedures must be distinguished: transfection of expression vectors (plasmids) and short RNA or DNA sequences to knockdown gene expression. Expression vectors require nuclear translocation for transcription by the endogenous RNA polymerase(s), whereas siRNAs inhibit gene expression by targeting mRNA degradation in the cytoplasm. Using chemical reagents, transfection of expression vectors in proliferating HepaRG progenitors is achievable, whereas gene transfer into quiescent differentiated hepatocytes remains poorly efficient. Conversely, siRNA-mediated gene targeting in nondividing hepatic cells can be achieved. The rationale for these differences in transfection efficiency is that chemical reagents allow the uptake of large expression vectors by cells (the plasma membrane is no longer a limiting barrier), but only a small fraction translocate to the nucleus, most likely during mitosis (19). Thus, transfection of DNA plasmids in differentiated quiescent hepatocytes leads to low expression levels of the gene of interest. Because siRNA-mediated gene knockdown does not require nuclear translocation, the rate of proliferation does not impact significantly the gene repression.

8. KLN47 and BSV10/DOPE formulations are suitable for transfection of expression vectors (plasmids) in proliferating progenitor HepaRG cells (16) and should not be used with

quiescent differentiated hepatocyte-like cells. Preliminary experiments suggest that these lipids can also be used for delivering siRNAs both to proliferating and quiescent HepaRG cells. Other compounds, such as glycine–betaine lipids, allow efficient gene transfer in primary rodent hepatocytes (20).

9. Amounts of transfection reagents versus DNA (charge ratio). The charge ratio is given by $R = $ (mass of reagent/molecular weight of reagent)/(mass of DNA/molecular weight of DNA). We recommend $R = 2$ (4 µL/0.8 µg DNA) or $R = 4$ (8 µL/0.8 µg DNA) for the KLN47 and $R = 1$ (1.8 µL/0.8 µg DNA) for the BSV10/DOPE formulations (16).

10. Liposomes combined with DNA (lipoplexes) are incubated with cells for at least 5 h. We recommend incubating 6–8 h, but longer exposures tend to increase cell toxicity.

11. Expression of genes of interest appears maximal between 24 and 48 h and decreases significantly at day 3 regardless of the proliferation rate.

12. SiPORT™ Amine (Applied Biosystems, Life Technologies™) is suitable for transfection of siRNA in both proliferating progenitors and quiescent differentiated HepaRG cells (either coculture of hepatocyte-like and biliary cells or the purified hepatocyte-like cells, see Subheading 3.2).

13. SiPORT™ Amine reagent is provided in a volatile solvent. Pipet the transfection agent quickly and tightly close the tube of SiPORT™ Amine, to prevent excessive evaporation. The reagent should be brought to room temperature before pipetting the volume required for the experiment.

14. SiRNAs can be used in a wide range of concentrations from 1 nM up to a final concentration of 8 nM (~10 pmol) in the culture dish (1 well of a 24-well plate) without significant increase in toxicity. We recommend titering the amount of siRNA required for optimal knocking-down of your gene of interest. One microliter of a siRNA stock solution at 20 µM must be used to reach 8 nM (10 pmol). For the low final concentration of 1 nM, the siRNA stock solution must be diluted (down to 1 or 2 µM), to avoid pipetting too small a volume. Importantly, the ratio between siRNA and SiPORT™ Amine reagent must be kept relatively constant: 1.5 µL SiPORT™ Amine reagent for 1.5 µL of siRNA (10 µM). Higher amounts of SiPORT™ Amine reagent (2–3 µL, with constant amounts of siRNAs) can be tested to improve gene targeting.

15. This last procedure also applies to primary human and rodent hepatocytes. There are two options for transfecting primary hepatocytes. They can be transfected either immediately following purification (liver dissociation by collagenase perfusion) (1, 2) or after culturing them for several days in culture conditions

that keep them well differentiated. We found that for rat and mouse hepatocytes, reverse transfection using freshly isolated hepatocytes works better. For human hepatocytes, viability following perfusion of small liver biopsies is always lower. Transfection of freshly isolated human hepatocytes usually results in severe cell death. We recommend plating the human hepatocytes in 35- or 75-cm^2 flasks at high density (density $\geq 2 \times 10^5$ cells/cm^2) in the William's E medium (without DMSO) for 1 day, before supplementation with 2% DMSO on the second day. Hepatocytes recover from the liver dissociation over 3–5 days and can subsequently be detached by mild trypsin treatment (trypsin is diluted 1/5 in PBS) before reverse transfection by SiPORT™ Amine or DharmaFECT reagents. They can also be electroporated using the Neon™ system (see Subheading 3.4) or the 4D-Nucleofector™ X unit (see Subheading 3.5). Protocols proposed in the Methods section indicate the amounts/volumes of siRNAs and transfection reagents for cells plated in 24-well plates (density $\geq 2 \times 10^5$ cells/cm^2, well surface 2 cm^2).

16. In most cases, the strongest gene/protein repression is observed at 48 h after transfection, but we found that proteins with very short half-life (<12 h) can be strongly down-regulated within 24 h post-transfection. In contrast, AhR half-life seems much longer (>48 h) and siRNA-mediated knockdown of AhR protein is usually optimal at 72 h. We recommend performing a time-course experiment (24, 48, and 72 h) to evaluate gene/protein expression.

17. DharmaFECT (Thermo Fisher Scientific) is suitable for transfection of siRNA in both proliferating progenitors and quiescent differentiated HepaRG cells (either the coculture hepatocyte-like and biliary cells or the purified hepatocyte-like cells, see Subheading 3.2) as well as rodent and human primary hepatocytes (see Note 15).

18. DMSO (2% added to the culture medium) strongly improves transfection efficiency of siRNAs in differentiated HepaRG cells and primary human hepatocytes (17). Although the molecular mechanism is not elucidated, variation in plasma membrane fluidity is suspected. Do not add DMSO to culture medium of progenitor HepaRG cells.

19. We recommend titering the amounts of siRNAs required for optimal knocking-down of your gene of interest. A range of concentrations from 5 to 50 nM can be tested. Similarly, volumes of DharmaFECT between 3 and 10 μL can be tested, to improve gene targeting.

20. For cell number <5 × 10^6, cells are transferred to an Eppendorf tube after the first wash. Make sure you do not lose too many cells during this step.

21. The manufacturer provides two resuspension buffers: R and T. The most appropriate buffer can be selected from the Neon™ database (Invitrogen). At this step, it is crucial to reduce as much as possible the dilution of the buffers R or T by traces of PBS.

22. Manufacturer recommends using 0.5–2 μg of plasmid DNA/ electroporation with 10-μL electrode tip. For HepaRG cells, we found that 0.5 μg of plasmid DNA allows maximal transfection efficiencies (16). Amounts higher than 0.8 μg increase cell toxicity. The plasmid DNA stock solution must be at least 1 μg/μL, to avoid dilution of the cell suspension. Volumes of DNA should not be more than 10% of the buffers R or T.

23. For siRNA, we used between 10 and 100 pmol for 10^5 cells in the 10-μL electrode tips. We recommend titering the amounts of siRNA for each gene targeted. siRNA stock solution must be concentrated enough to avoid adding more than 10% of siRNA (usually dissolved in RNAse-free water) into the buffers R or T containing the cells. Alternatively, lyophilized siRNA can be dissolved in buffer R or T.

24. For both 10- and 100-μL electrode tips, you must prepare additional mix (cells-plasmid DNA/siRNA mix). The tip is used to pipet the cells. When the volume of mix is too small, it becomes impossible to pipet cells without forming bubbles in the electrode tip. We recommend preparing at least 1 extra tip volume for each electroporation condition.

25. Cells should not remain for more than 20 min in buffers R or T. Longer incubation in these buffers reduces cell viability. Cells resuspended in buffer R or T must be kept on ice before electroporation.

26. There are three cell line electroporation kits available for the Amaxa™ 4D-Nucleofector™: the SE, SF, and SG Amaxa™ 4D-Nucleofector™ X kits containing distinct solutions (SE, SF, and SG). However, an optimization cell line kit, containing the three solutions, is also available. We found that buffers SE and SF were suitable for transfecting progenitor and differentiated HepaRG cells, but we highly recommend buffer SE, which generates less fusion between cells than does SF. Buffers need to be completed with the supplement reagent: for one 20-μL electroporation, mix 16.4 μL of buffer and 3.6 μL of supplement reagent.

27. For one Nucleofection™ in 20-μL Nucleocuvette™ strip (electroporation of $2–5 \times 10^5$ cells), we use 0.5–1 μg of plasmid DNA and between 5 and 100 pmol of siRNAs. We recommend titering the amounts of siRNA for each gene targeted.

28. The Amaxa™ 96-well Shuttle™ device allows transfections of cells in 96-well plates using a 20-μL reaction/well, similar to

the Nucleocuvette™ strip. This device is particularly suitable for high-throughput screenings of plasmids or siRNA libraries. For more information on specific applications of these devices, consult the Lonza Web site (see Note 29).

29. Specific protocols for primary cells and cell lines are available on the database at the following Web site: http://www.lonzabio.com/resources/product-instructions/protocols/. For human primary hepatocytes, the P3 kit must be used.

30. The Y unit is a flexible system, which allows transfection of cells in adherence. This device is particularly suitable for primary cultures that cannot be trypsinized following plating after isolation from organs. In addition, the 24-well format of culture is compatible with DNA, RNA, or protein analysis, as well as in situ experiments.

31. Cells must be plated in a 24-well plate compatible with the dipping electrode array. Order plates from CELLSTAR™, Greiner Bio-One, or Nunc Multidishes Nucleon™. Nucleofection™ of cells attached to glass coverslips inserted in 24-well plates is possible, allowing confocal microscopy experiments. The 24-well dipping electrode array provided in the kit cannot be split. However, only part of the electrodes can be used if all manipulations are kept sterile using a culture hood. It is essential to plate cells only for the wells that will be used for electroporation. The electrodes used must be indicated, to make sure that they will not be used again in another experiment.

32. The AD1 and AD2 4D-Nucleofector™ kits, containing two distinct solutions (AD1 and AD2), are provided by the manufacturer. For progenitor HepaRG cells, the AD1 kit must be used, whereas the AD2 kit gives better results for differentiated cells. The Nucleofection™ substrate solution (for 1 well of a 24-well plate) is composed of 287 μL of AD1 or AD2, 63 μL of the supplement solution, and 16 μg of plasmid DNA (use stock solution of DNA at 1–2 μg/μL).

Acknowledgements

We would like to thank all the members of the SynNanoVect platform (www.synnanovect.ueb.ue) and Catherine Ribault for excellent technical assistance. We are very grateful to Denis ToVan and Sandrine Allamane (LONZA) for helping with the setting up of the 4D-Nucleofector™ technology. This work was supported by Inserm, the GIS-IBiSA, Biogenouest®, and the European Commission FP7 program "LIV-ES" (HEALTH-F5-2008-223317).

References

1. Guillouzo A, Guguen-Guillouzo C (2008) Evolving concepts in liver tissue modeling and implications for in vitro toxicology. Expert Opin Drug Metab Toxicol 4:1279–1294
2. Corlu A, Ilyin G, Cariou S, Loyer P, Guguen-Guillouzo C (1996) The coculture: a system for studying the regulation of liver differentiation/proliferation activity and its control. Cell Biol Toxicol 13:235–242
3. Loyer P, Ilyin G, Cariou S, Corlu A, Guguen-Guillouzo C (1996) Progression through G1 and S phase of adult rat hepatocytes. Prog Cell Cycle Res 2:37–47
4. Abdel-Razzak Z, Loyer P, Fautrel A, Gautier JC, Corcos L, Turlin B, Beaune P, Guillouzo A (1993) Cytokines down-regulate major cytochrome P450 in primary culture of human hepatocytes. Mol Pharmacol 44:707–715
5. Wilkening S, Stahl F, Bader A (2003) Comparison of primary human hepatocytes and hepatoma cell line HepG2 with regard to their biotransformation properties. Drug Metab Dispos 31:1035–1042
6. Kelly JH (1994) Permanent human hepatocyte cell line and its use in a liver assist device (LAD). US patent 5,290,684
7. Jover R, Bort R, Gomez-Lechon MJ, Castell JV (1998) Re-expression of C/EBP alpha induces CYP2B6, CYP2C9 and CYP2D6 genes in HepG2 cells. FEBS Lett 431:227–230
8. Gripon P, Rumin S, Urban S, Le Seyec J, Glaise D, Cannie I, Guyomard C, Lucas J, Trepo C, Guguen-Guillouzo C (2002) Infection of a human hepatoma cell line by hepatitis B virus. Proc Natl Acad Sci U S A 99:15655–15660
9. Cerec V, Glaise D, Garnier D, Morosan S, Turlin B, Drenou B, Gripon P, Kremsdorf D, Guguen-Guillouzo C, Corlu A (2007) Transdifferentiation of hepatocyte-like cells from the human hepatoma HepaRG cell line through bipotent progenitor. Hepatology 45:957–967
10. Aninat C, Piton A, Glaise D, Le Charpentier T, Langouet S, Morel F, Guguen-Guillouzo C, Guillouzo A (2006) Expression of cytochrome P450, conjugating enzymes and nuclear receptors in human hepatoma HepaRG cells. Drug Metab Dispos 34:75–83
11. Legendre C, Hori T, Loyer P, Aninat C, Ishida S, Glaise D, Lucas-Clerc C, Boudjema K, Guguen-Guillouzo C, Corlu A, Morel F (2009) Drug-metabolising enzymes are down-regulated by hypoxia in differentiated human hepatoma HepaRG cells: HIF1-a involvement in CYP3A4 repression. Eur J Cancer 12: 2882–2892
12. Kanebratt KP, Andersson TB (2008) Evaluation of HepaRG cells as an in vitro model for human drug metabolism studies. Drug Metab Dispos 36:1444–1452
13. Jossé R, Aninat C, Glaise D, Dumont J, Fessard V, Morel F, Poul JM, Guguen-Guillouzo C, Guillouzo A (2008) Long-term functional stability of HepaRG hepatocytes and use for chronic toxicity and genotoxicity studies. Drug Metab Dispos 36:1111–1118
14. Dumont J, Jossé R, Lambert C, Anthérieu S, Laurent V, Loyer P, Robin MA, Guillouzo A (2010) Preferential induction of the AhR gene battery in HepaRG cells after a single or repeated exposure to heterocyclic aromatic amines. Toxicol Appl Pharmacol 249:91–100
15. Pernelle K, Le Guevel R, Glaise D, Stasio CG, Le Charpentier T, Bouaita B, Corlu A, Guguen-Guillouzo C (2011) Automated detection of hepatotoxic compounds in human hepatocytes using HepaRG cells and image-based analysis of mitochondrial dysfunction using JC-1 dye. Toxicol Appl Pharmacol 254:256–266
16. Laurent V, Fraix A, Montier T, Cammas-Marion S, Ribault C, Benvengu T, Jaffres P-A, Loyer P (2010) Highly efficient gene transfer into hepatocyte-like cells: new means for drug metabolism and toxicity studies. Biotechnol J 5:314–320
17. Le Vee M, Jouan E, Fardel O (2010) Involvement of aryl hydrocarbon receptor in basal and 2,3,7,8-tetrachlorodibenzo-p-dioxin-induced expression of target genes in primary human hepatocytes. Toxicol In Vitro 24:1775–1781
18. Moncion A, Truong NT, Garrone A, Beaune P, Barouki R, De Waziers I (2002) Identification of a 16-nucleotide sequence that mediates post-transcriptional regulation of rat CYP2E1 by insulin. J Biol Chem 277:45904–45910
19. Brunner S, Sauer T, Carotta S, Cotten M, Saltik M, Wagner E (2000) Cell cycle dependence of gene transfer by lipoplex, polyplex and recombinant adenovirus. Gene Ther 7:401–407
20. Gilot D, Miramon M-L, Benvegnu T, Ferrieres V, Loreal O, Guguen-Guillouzo C, Plusquellec D, Loyer P (2002) Cationic lipids derived from glycine betaine promote efficient and non-toxic gene transfection in cultured hepatocytes. J Gene Med 4:415–427

INDEX

A

Allelic replacement 81
Amaxa™ 4D-Nucleofector™ technology 297, 312
δ-Aminolevulinic acid (δ-ALA) 109, 110, 192, 193, 208, 209, 215, 218, 222
Amodiaquine 28–31, 35
Amplex Red 149, 150, 152–155
   stock solution preparation 151
Aromatase 69, 125
Automation 33, 43, 130, 133

B

7-Benzyloxy-4-trifluoromethylcoumarin (BFC),
   as substrate 137, 140, 144, 158, 159
Bergamottin 63, 64
Bicistronic membranes, with P450
   and its reductase 158
Bilayer 115–117, 123
   nanodisc 115–124
Biocatalysis 239, 240
Bioconjugation 133
Bioluminescent enzyme assays 1–8
Biotransformation, *in vivo* 242, 246–247
BSL-3 laboratory 87, 89, 90
BSV10, lipophosphoramidate 297, 302, 309, 310
Bufuralol 28–31, 35

C

Caffeine 131, 132, 261–266
Cassette of substrates 12
Catalase 64, 149, 150, 152–155
   stock solution preparation 151
Chimeric constructs 240, 249
Cholate 118–121, 123, 228, 229, 232
Cholesterol degradation 81
Cholesterol Derived Metabolites 92
Chromatography, thin-layer (TLC) 229, 235
ChuA, heme transporter 108
Cocktail of substrates 13, 14, 17, 18
Complementation 89–90
Coumarins, use of in P450
   assays 62, 137, 144, 146
CPR. *See* Cytochrome P450 reductase (CPR)

Cumene hydroperoxide (CHP) 149
CYP. *See* Cytochrome P450 (CYP)
Cytochrome c 209, 217, 231, 233–234
Cytochrome P450 (CYP)
   alleles 251–258
   assay of 1–8, 61–68
   CYP17, incorporation into Nanodisc 124
   CYP19, incorporation into Nanodisc 116, 124
   CYP119 .... 108, 110, 111
   CYP125 .... 81, 84
   CYP1A2 .... 5–7, 11, 12, 16, 25, 67, 137, 144, 146, 185, 199, 203, 253, 254, 261, 263, 265, 266
   CYP3A 7, 12, 177, 262, 263
   CYP3A4
      assay of 157–161
      biophysical analysis 116
      functional properties 116, 117, 120
   CYP2B4, probing heme with 13C
      methylisocyanide 51–59
   CYP2B6 .... 63, 67, 68, 144, 146, 253–255
   CYP116B2 240
   CYP2C9 .... 5–7, 11, 12, 16, 25, 137, 144, 197–199, 203, 206, 207, 211–213, 216, 253, 254, 258, 261, 265
   CYP2C11 .... 206, 211–213, 216
   CYP2C19 .... 5–7, 11, 12, 16, 25, 137, 144, 199, 206, 211–213, 216, 252–254, 258, 261, 265
   CYP2D6 .... 5–7, 11, 12, 16, 25, 137, 144, 253, 254, 256, 258, 262, 265
   CYP2E1, gene transfer 137, 304, 306
   CYP107U1 .... 74, 75
   inhibition screening 25–49
   nomenclature 230, 251–258
   orphans 71–76, 240
Cytochrome P450 reductase (CPR)
   effect of membrane electrostatics 117
   incorporation into Nanodisc 120
Cytosolic fraction 133

D

Dehydroepiandrosterone (DHEA) 234, 235
Detergent 87, 93, 115–117, 120, 122

Cytochrome P450 Protocols
Index

Dextromethorphan 12–14, 262, 263, 265
Diclofenac 12–16, 28–31, 35, 130–133, 197, 198, 203
Dimyristoylphosphatidylcholine (DMPC) 121, 123
Dipalmitoylphosphatidylcholine (DPPC) 121, 123
Directed evolution 205, 206
DNA damage
    DNA adduct... 133, 134
    metabolite-DNA adduct 133, 134
DNA family shuffling.............................. 177, 178, 180, 183, 206, 207
DNA, genomic 86–89, 93, 220, 226, 229, 275
DNase I footprinting, high throughput
    capillary electrophoresis 274, 278–279
    DNase I treatment, optimization 275, 281
    DNA sequencing...278
    dye terminator sequencing reactions 274, 278, 281
    fluorescent tagging of DNA................ 269, 275, 280
    materials.. 271–274
    nuclear protein (extraction) 271, 273, 275–277, 281
    polymerase chain reaction 273–276, 280
DNA shuffling 177–187, 205–223
Drug–drug interaction (DDI)....................... 11–20, 25, 62, 141, 143, 145, 157
Drug metabolism.............................. 25, 135, 138, 141, 225, 251–253, 295–313
Dye production 209, 211, 212, 215–217, 221

E

E. coli Rosetta 2 DE3..242
E. coli RP523... 100, 101, 104
Electrophiles.. 163–175
Electrospray, ionization in mass spectrometry (ESI-MS)............... 33, 40, 73, 103, 131, 265
Enzyme kinetics........... 33, 68, 117, 124, 200, 239, 247
7-Ethoxy trifluoromethyl coumarin (7-EFC)........ 62–66
Eukaryotic P450
    CYP17 124
    CYP3A4..................................... 4–8, 11, 16, 19, 26, 68, 116, 117, 120, 124, 137, 144, 194, 199, 203, 253, 254, 258, 265, 296, 306
    CYP19A1..254
    depth of insertion into membrane............... 115, 117
Evolution *in vitro* 205, 216

F

Ferredoxin (and reductase)....................................72, 73
Ferrochelatase...108
Flavoprotein.. 117
Fluorescence assay........... 2, 61–68, 136, 143, 157–161
Fluorescence, assays with P450 .... 61–68, 136, 143, 157–161

Furafylline 5, 19, 28–32, 138, 144
Fusion protein 246–248

G

Gene 17, 81–90, 96, 105, 137, 178, 184, 206, 207, 213–214, 225, 240, 242–246, 252–258, 261, 270, 276, 283, 284, 292, 295–313
Genotyping 225–236, 261, 279
GFP. *See* Green fluorescent protein (GFP)
Glycolipids.. 80
G protein coupled receptors (GPCRs)...................... 116
    oligomerization state 115–117
Green fluorescent protein (GFP)...... 298, 304, 306, 307

H

Haplotypes 230, 255
Heat-shock protein 90 (Hsp90) 164–175
Heme nitric oxide/oxygen-binding (H-NOX) proteins................................... 97
Heme probes... 51–59, 95
    See also Methylisocyanide
Heme synthesis
    delta-aminolevulinic acid 193
    ferrochelatase 108
    iron-depleted medium 108
Heme transporter ChuA 108
Hemoprotein.................... 96, 107, 196, 199, 205, 215
HepaRG cells... 295–313
    culture conditions...................... 298, 300, 309, 310
Hepatocyte gene transfer 295–313
Hepatocyte, isolation from mouse
    centrifugation of cell suspensions 289
    liver
        dissection 287, 288
        perfusion... 287–289
    materials.. 284–287
    perfusion apparatus, sterilization 287
    plating
        Matrigel-coated plates 287
    viability assay 292
High-performance liquid chromatography (HPLC), in assays 72–76, 84, 92, 98, 100, 102, 105, 131, 157, 171, 247, 262, 263
High-resolution mass spectrometry (HRMS)......................................73, 74, 81
High throughput
    assay 1, 62
    screening.................................. 5, 141, 180, 281, 313
    screening assay...149
His-tag. *See* Histidine affinity tag (His-tag)
Histidine affinity tag (His-tag) 118, 121, 122, 243

HRMS. *See* High-resolution mass spectrometry (HRMS)

Hsp90. *See* Heat-shock protein 90 (Hsp90)

17α-Hydroxylase 234

Hydroxynonenal (HNE)................. 164–167, 172–175

7-Hydroxy trifluoromethyl coumarin (7-HFC)...62, 63, 67

7-Hydroxy-4-trifluoromethylcoumarin, production in assay (HFC)...................67, 137, 139, 140

I

IC50...45, 144, 146

Indigo 205–223

Indigoid pigment production........................... 205–223

Indirubin.. 215, 222

Indole.. 205, 209, 210, 216–217, 220

  metabolism..215

  5-aminoindole..209

  5-bromoindole..209

  6-cyanoindole..209

  5-fluoroindole..209

  5-hydroxyindole..209

  5-hydroxyindole-3-acetic acid...............................209

  5-methoxyindole..209

  2-methylaindole..209

  5-methylindole..209

  5-nitroindole..209

Iron-depleted medium...108

K

Ketoconazole.............. 5, 19, 28–30, 32, 138, 144, 159

  as inhibitor of P450 3A4 .... 158, 161

Ki 146, 147

K_{inact}...62, 63, 65, 67

KLN47, lipophosphoramidate..........297, 302, 309, 310

L

LICRED 239–249

Ligation independent cloning 240, 241, 243

Lipid

  analysis..83–84, 91

  DMPC..121, 123

  DPPC..121, 123

  POPC..118–123

  profile 91

  solubilization

    cholate 120

    Triton X-100 ....118, 232

Liquid chromatography-tandem mass spectrometry (LC-MS/MS)................. 13, 15, 16, 129, 134, 164, 166–168, 170, 173, 174

Liquid handling 26, 27, 32

Losartan 261–263, 265

17,20 Lyase 234

M

Magnetic bioreactor particles 130

Magnetic particle 130–132

Mass spectrometer (MS) 15, 20, 26, 33, 40–42, 47, 48, 73, 84, 91, 92, 131, 133, 170, 171

Mass spectrometry.. ....2, 12, 26, 33, 81, 84, 103, 149, 157, 163, 164, 166, 265

  in assays...33, 73, 166, 265

Mechanism-based inactivation (MBI).........................62

Membrane mimetics

  ATP synthase..117

  epidermal growth factor receptor (EGFR)............117

  G protein coupled receptors (GPCRs)...............116

  ion channels 116

  Nanodisc 116–124

Membrane scaffold protein (MSP) 118, 120, 122, 123

  MSP1D1, MSPE3D1 121

Metabolite profiling 71, 129–134

Metabolomics, use with P450 71–76

Methylisocyanide

  ^{13}C labeled, NMR spectrum................................55

  ^{13}C labeled, synthesis...................................51–59

  NMR spectrum when bound to CYP2B4........51–59

Methyl-malonyl-coenzyme A (MMCoA)....................81

Microsomes 3, 7, 12–15, 19, 20, 26, 130, 132, 133, 143, 193

Microtiterplate layout.................................65, 152, 153

Midazolam................................. 12–14, 16, 18, 28–31, 35, 203, 262, 263, 265

M9 minimal medium 242, 245, 248

Molecular breeding................................204, 205, 223

Montelukast 28–32

MRM. *See* Multiple-reaction monitoring (MRM)

MSP. *See* Membrane scaffold protein (MSP)

Multi-channel arm 32, 33

Multiple-reaction monitoring (MRM)................15, 16, 20, 27, 28, 39–42, 48

Mycobacterium tuberculosis 79–93

  allelic replacement 81

  BSL-3 laboratory.....................................87, 89, 90

  complementation...89, 90

  genomic DNA.............................86, 88, 89, 93

  transposon mutatgenesis.......................................81

N

Nanodiscs

  cryospectroscopy 124

  functional reconstitution of membrane proteins complexes 116

  lipid microenvironment 116

  protein-lipid interaction.....................115–117, 123

  stabilization of membrane proteins 117

  stopped-flow 122, 123

Cytochrome P450 Protocols
Index

Nanodiscs (*Cont.*)
surface immobilization of membrane
proteins 116
surface plasmon resonance spectroscopy 116
Neon™ transfection system 297, 306, 312
Nicotinamide adenine dinucleotide
phosphate (NADPH) 3, 6, 15,
18, 19, 29–31, 33–38, 46, 47, 62, 63, 65–68,
71–73, 120, 131, 132, 158, 159, 161, 178,
179, 185, 187, 197, 205, 207, 209, 216,
217, 219, 225, 229, 233, 235, 243
as cofactor 13, 136, 149, 239, 247
Nitric oxide synthase
Fe(III) mesoporphyrin IX 96, 97, 100
Mn(III) protoporphyrin IX 96, 97, 100
Nocardia farcinica 240
Non viral gene transfer
electroporation 296–298,
303–305, 307, 312, 313
transfection with cationic lipids 297, 302

O

O-deethylation/*O*-deethylase 62, 63
Oligomerization state
chemotactic receptors 116
G protein coupled receptors (GPCRs) 116
Omeprazole 262, 263, 265, 266
On-line solid phase extraction 25
Organic peroxide 149–153, 155
stock solution preparation 151
Orphan P450 .... 71–73, 240

P

P450
decoupling 154
eukaryotic 115
human 62, 63, 67, 141, 158,
185, 231–232, 234–235
oxidoreductase 225–236, 253–255, 257
P450 3A4, assay of 158
p450cam 57, 107, 240, 247
P450RhF 239, 240, 243
P450 107U1 .... 74, 75
Palmitoyloleoylphosphatidylcholine
(POPC) 118–123
Peak area 17, 44, 48, 49, 132, 174
Peptide linker 239, 240, 249
Pharmacogenetics 225, 253
Phenotyping cocktail, *in vivo* 262, 263, 266
Pigment production 205–223
Plate readers, in P450 assays 64, 66, 67, 142,
143, 151, 153, 155, 157–159, 191, 192,
195–200, 203, 217
Polymorphism 225, 230, 254, 255, 261, 269–282

Porphyrins
incorporation 96, 98, 102, 105
substitution 95–97
unnatural 95–97
Post-assay pooling 25
Pregnenolone (Preg) 234
Probe substrate 1, 2, 8, 11, 12, 26,
28–31, 33, 44, 46, 47, 62, 67, 68, 135, 141,
157, 178, 261
Progesterone 229, 234
Pro-luciferins 2–6
Propionyl CoA 81
Protein adducts 163–175
Protein purification 102, 109–110, 242, 246
Proteomics 163
Protoporphyrin IX 96, 97, 100, 104
cobalt 107–113
iron 108, 112

Q

Quinidine 5, 19, 28–30, 32, 138, 144

R

RapidFire 33, 40–42, 48
Recombinant expression 196
Reference inhibitor 26, 28–32, 34, 43, 44, 46
Reporter enzyme-based P450 assay 149–155
Resorufin
calibration curve 151–152
stock solution preparation 151
Restriction enzymes 88, 178–181,
184, 186, 206, 207, 209, 214, 241

S

Self-assembly 118, 120
assembly temperature 119
Self-sufficient P450 240
Single-nucleotide polymorphisms (SNPs) 225,
230, 254–257, 270
detection 256, 270
S-mephenytoin 12–14, 16, 28–31, 35, 44
Solid-phase extraction/tandem mass spectrometry
(SPE/MS/MS) 33
Solubilization 115, 118, 121, 232
Steroid hormone 225, 234, 258
Steroid metabolite analysis 84
Substrate-independent P450 screen
classification of results 153
data analysis 153–154
reaction profile 154
Substrate-recognition sites 206, 253
Sulfaphenazol 5, 19, 28–32, 138, 144
Supersomes 63, 68, 133

Index 319

T

Tacrine 12–16, 28–31, 35
Tandem mass spectrometry 164, 265
T4 DNA polymerase 241, 244, 248
Tert-butylhydroperoxide 149
Test compounds 1, 2, 7, 8, 14, 26–28, 33, 34, 36, 37, 44, 46, 47, 49
Time-dependent inhibition 62
Transformation, *in vitro* 243
Transposon mutagenesis 81
Tranylcypromine 28–30, 32, 138, 144

Triton X-100 .... 118–120, 122, 123, 228, 229, 232, 271
Troglitazone 130–133
Tyloxapol 82, 83, 90, 91, 93

U

UPLC, use in chromatography systems 72, 76

X

XCMS software, use of................................... 71, 73–75
XLFit.. 33, 43–45
XplA 240

Printed by Publishers' Graphics LLC